THE FAR-FUTURE UNIVERSE

THE FAR-FUTURE UNIVERSE

Eschatology from a Cosmic Perspective

EDITED BY GEORGE F. R. ELLIS

Published in association with the Pontifical Academy of Sciences and the Vatican Observatory

Vatican City

Templeton Foundation Press
PHILADELPHIA AND LONDON

Templeton Foundation Press
Five Radnor Corporate Center, Suite 120
100 Matsonford Road
Radnor, Pennsylvania 19087
www.templetonpress.org

Designed and typeset by Kachergis Book Design
Printed by Sheridan Books

LIBRARY OF CONGRESS CATALOGING-IN-PUBLICATION DATA
The far-future universe : eschatology from a cosmic perspective / edited by George F.R. Ellis.
p. cm.
"The present book about events all the way at the end of time grew out of a symposium sponsored by the John Templeton Foundation under the aegis of its Humble Approach Initiative"—Pref.
Includes bibliographical references and index.
ISBN 1-890151-90-4 (hardcover : alk. paper)
1. Cosmology. 2.Eschatology. I. Ellis, George Francis Rayner.
II. Templeton Foundation.
QB981 .F175 2002
523.1—dc21
2002003108

Jacket photo: Orion Nebula/ CISCO/Subaru/ NAOJ

Printed in the United States of America
02 03 04 05 06 07 10 9 8 7 6 5 4 3 2 1

CONTENTS

PREFACE

The present book about events all the way at the end of time grew out of a symposium sponsored by the John Templeton Foundation under the aegis of its Humble Approach Initiative. The initiative is inherently interdisciplinary, sensitive to nuance, biased in favor of building linkages and connections, and promotes risk-taking discussion that leads to the generation of new ideas for writing, teaching, and research. It assumes a willingness to experiment on the part of all participants. Sir John Templeton has said that "humility is a gateway to greater understanding and open[s] the doors to progress" in all endeavors.[1] He believes that in their quest to comprehend ultimate reality, scientists, philosophers, and theologians have much to learn about and from one another. That those who gathered in Rome on 7–9 November 2000 did so in very considerable measure is indicated by the essays in this volume.

Their spirited conversation, chaired by Martin J. Rees, England's astronomer royal, took place in the Casina Pio IV, once a summer residence of Pope Pius IV. In 1922 the villa became the seat of the Pontifical Academy of the Nuovi Lincei, whose origins date to the founding of the Academy of the Lincei, the world's first scientific academy, by Prince Federico Cesi in 1603.[2]

1. John Marks Templeton, *Worldwide Laws of Life* (Radnor, Pa.: Templeton Foundation Press, 1997), 465.

2. As a result of Italy's political travails, there are now two scientific academies whose origins date to Prince Cesi: the Accademia dei Lincei of the Republic of Italy and the Pontifical Academy of the Nuovi Lincei.

Lincei is derived from the Latin word for lynx—the animal believed to have the sharpest vision of all, and the name was chosen because, according to Cesi, science should rely "on the acute observation of nature and its phenomena and on experimental work."[3] Galileo, who conducted his famous experiments with the pendulum on natural accelerated motion in 1602 and eight years later described his observations of the night sky in *Sidereus Nuncius (The Starry Message)*, was one of the academy's earliest members. The influence of his friend, Cesi, just eighteen years old at the founding of the academy and forty-five at his death in 1630, was so great and his lynx-like sight so good with respect to all manner of things that had he not died prematurely, the 1633 trial of the great astronomer and physicist would almost certainly never have occurred in the opinion of many scholars.

The audacious questions suggested by Galileo's astronomical discoveries continue to be debated as today's astronomers and physicists consider whether the universe will still be expanding 100 billion years from now and examine evidence that it may even be speeding up. The scientists, philosophers, and theologians who enjoyed the gracious hospitality of Bishop Marcelo Sanchez, director of the Pontifical Academy of Sciences, while participating in the Templeton symposium, discussed eschatology from a cosmic perspective. As they considered various scenarios for the long-range future, they benefited from the observations of James E. Peebles, Albert Einstein Professor of Science at Princeton University, Vera C. Rubin, senior researcher in the Department of Terrestrial Magnetism at the Carnegie Institution of Washington, and Allan R. Sandage, staff astronomer emeritus at The Observatories of the Carnegie Institution of Washington. Drs. Peebles and Sandage were presented with the inaugural Cosmology Prizes of the Peter Gruber Foundation at a ceremony held in the Vatican at the conclusion of the Far-Future Universe Symposium. It was a festive finale to a probing and provocative exchange of research findings, ideas, and opinions.

The Far-Future Universe: Eschatology from a Cosmic Perspective contains papers written by those at the symposium plus a few others either specially commissioned after the meeting or reprinted with permission from extant works. While some of them place the discussion in a present day scientific

3. Federico Cesi, et. al., *Lynceographi* (Roma, circa 1605), pars VI, particula 17, pp. 241–42. The manuscript is in the Archives Lincei, Pontifical Academy of Sciences, Vatican City.

setting, all of them consider issues that are themselves eternal, whatever the answers to our questions about the far future may turn out to be. My colleagues and I at the Templeton Foundation thank Martin Rees and John Barrow for their help in shaping a dynamic symposium and George Ellis for his thoughtful and patient editing of this volume. All of us connected with the venture are delighted that the Templeton Foundation Press has agreed to make these deliberations available to a wider audience.

Mary Ann Meyers
JANUARY 2002

PART I

ISSUES AND QUESTIONS

An Overview

[1]

INTRODUCTION

George F. R. Ellis

A main tension in human life is between intellect and emotion, specifically between impersonal rational analysis, which is driven by curiosity and the desire to understand both our universe and whatever life situations may face us on the one hand, and on the other, faith[1] and hope, which are driven by the need to make life choices in the face of uncertainty and adversity. Both capacities are needed in order to live a full life. Rationality, based on impartial analysis of repeated experience and carefully collected evidence, is what gives us our ability to plan sensibly and successfully in the face of reality and its inherent limitations, but hope is often needed in order to continue surviving and functioning in the face of desperate situations—to fight against the odds. Thus there are important roles for both rationality and hope in human life, but there is an ongoing tension between them, for rationality is based on logic and proof, while faith functions where there can be no proof.

1. We note here that even atheism is a faith.

This tension arises in particular when we consider what is the end of things, at both a personal and a collective level. This book is concerned with the latter issue: Will human life and all intelligence inevitably come to an end as the universe evolves, or is there some way in which they could survive until the end of time? What about the universe itself—does it end, or will it continue forever? If there is an end in either case, is that necessarily bad? Would continuation of life and/or the universe for eternity be a good thing?

We can engage with these questions in different ways: mainly on a scientific basis, obtaining rational impersonal answers based on simplified analytic models and repeatable experimental observations, with all the strengths and limitations that entails, or alternatively in terms of personal and communal faith and hope, based on wider aspects of our experience, and addressing other dimensions of understanding. In either case, when considering the far future we run into uncertainty because of an impossibility of proof—an inability to determine for sure what the final outcome will be, and so a lack of definite knowledge. The nature of last things is thus an area of ongoing speculation. There are various options that are not contradicted by any available evidence, and indeed may never be testable before the end of time.

The Far-Future Universe: Eschatology from a Cosmic Perspective examines these fascinating questions: the future of existence of life in the universe and the far future of the universe itself. Various aspects of the two questions interact with each other to some degree, and that interaction is explored in the essays here, which look at the scientific issues, the faith issues, and the ways in which they link with each other, as seen from various perspectives. The book attempts to clarify what can be said in these areas with reasonable certainty and what uncertainty remains. Indeed it is apparent that there is an impossibility of proof even with regard to many purely scientific issues, so even here there will remain areas of ongoing speculation.

The chapters that follow are divided into four main categories: cosmology and physics, biology and existence of life, issues of humanity that arise in the context of speculation about the far future, and metaphysics and theology. There is, however, no hard and fast separation among them; indeed, some major foci recur in the various essays—giving a rich resulting overview.

The introductory section comprises this overview chapter and chapter 2,

"Seeking the Future: A Theological Perspective" by George V. Coyne, S.J., which articulates some of the issues that arise in these discussions when seen from a faith point of view. From this vantage point, Coyne gives a larger perspective on the subject than a purely scientific stand is able to provide, including in his discussion issues of faith, goodness, and beauty. But he does so in an undogmatic way, emphasizing the uncertainties that arise and the need for humility in approaching them.

The first main category is cosmology and physics. In chapter 3, "The Far, Far Future," John Barrow presents an overview of the various physical possibilities and in particular the prospect of a "heat death" where the second law of thermodynamics prevails. He contrasts this prospect with cyclic possibilities and considers universes where a cosmological constant dominates in the far future. In doing so, he emphasizes the uncertainties in making very long-term predictions: for example, we do not know what changes there may be in the "constants" of physics, nor what the long-term future of the quantum vacuum is, and the outcome hinges on that.

In chapter 4, "Eternity: Who Needs It?," Paul Davies comments on the psychological desire for cyclical universes that has pervaded much of our thought, and gives a systematic summary of the different physical possibilities for the long-term history of the universe. He discusses the cyclic universe scenario in its various forms, universes with a finite duration, and eternal universes (including multiverses), and considers the conflict between two major human desires: "that the universe has an ultimate destiny or purpose, and that the universe will exist interestingly for eternity."

In chapter 5, "Time of the Universe," Michael Heller considers the point that any discussion of the far-future universe depends on assumptions about the nature of time—but time is one of the most mysterious features of physics, and many alternatives have been proposed. One particular basis for an alternative view of time is opened up by the concept of non-commutative geometries, where time is an emergent property, and Heller discusses this possibility. In this case, on the fundamental level, there is no distinction between singular and nonsingular states of the universe. Only at the Planck threshold do some states degenerate into singularities; it is from the perspective of the macroscopic observer that the universe had its beginning and will possibly have its end. The necessary condition for the far future to exist is the existence of global time. Its existence in the ensemble of all possible universes, however, is the exception rather than the rule. Thus in many universes, even the idea of "the far future" may not be well defined.

Finally, in chapter 6, "Living in a Multiverse," Martin Rees takes seriously the proposition that we may live in one of a multitude of universes, because this is one way of explaining the "fine tuning" of physical parameters that is required for life to exist. In this case, the idea of the "far future" takes on a different meaning, for then there will be numerous universes in the ensemble having each of the possible behaviors in the far distant future; and the passing away of life in one universe, or even of a universe as a whole, is rather immaterial when there are an infinite number of other universes unfolding in this ensemble of universes. After discussing whether such a hypothesis is testable, Rees literally brings us back to Earth, putting before us the current astrophysical understanding of the future fate of the observable universe and the structures in it. He suggests that

> In the entire domain that cosmologists explore—ten billion years of time, ten billion light years of space—the most crucial space-time location of all . . . could be here and now. . . . 21st century technology could destroy our species, and thereby foreclose the potentialities of our biophilic universe whose evolution has still barely begun. On the other hand, by prudence, and by spreading ourselves beyond the Earth, we could ensure sufficient diversity to safeguard life's potential for an infinite future. . . .

The second main category is biology and the existence of life. In chapter 7, "Exotic Genetic Materials and the Extent of Life in the Universe," Graham Cairns-Smith provides an alternative view on the difficult question of the origin of life, proposing that the original genetic material may have been crystalline minerals rather than organic molecules. This discussion relates to the possibility that formation of life may be taking place even now in our universe, and may continue some considerable way into the future; so the future of life in the universe may include the origin of new life. The more flexible the mechanisms are that lead to life, the more likely this is to be true; and this intriguing discussion points out the possibility of a considerable flexibility in this process.

Chapter 8 is a reprint of Freeman Dyson's influential 1979 article, "Time Without End: Physics and Biology in an Open Universe," which did much to revive the discussion of cosmic eschatology in recent times. In it he considers three main questions in the context of universes with negatively curved space sections: (1) Does the universe freeze into a state of permanent physical quiescence as it expands and cools? (2) Is it possible for life and intelligence to survive indefinitely? (3) Is it possible to maintain communication and transmit information across the constantly expanding distance between the galaxies? He tentatively answers them with a no, a yes, and a maybe. His answer to the second question involves the proposal that life may evolve into some other material embodiment than organic material—for example, evolving into transhuman life in a digital computer, thus silicon-based life.[2]

The resulting debate is picked up again in chapter 9, "Life in the Universe: Is Life Digital or Analogue?," also a reprint of an article by Dyson. He considers the possibilities of digital life, analogue life, and life based on quantum computation, concluding that life can survive forever with a finite supply of free energy, but only by reducing drastically the quality of life as the temperature goes down, and particularly by using a strategy of hibernation. This argument is based on classical physics; Glenn Starkman and Lawrence Krauss, however, elsewhere pointed out that quantization of energy imposes restrictions on what is possible, which they claim invalidates Dyson's analysis. Dyson responds to their arguments, and concludes, "In the closed and accelerating universes, we all agree that survival is impossible. In the decelerating universe, I say that survival is possible while Krauss and Starkman have not expressed an opinion. In the open universe, I say that survival is possible for analogue life but impossible for digital life."

Finally, in chapter 10, "Does Biology Have an Eschatology, and If So Does It Have Cosmological Implications?," Simon Conway Morris argues that a universe without an eschatological dimension is a universe that is incomplete. The argument examines the problem of navigating through protein and genetic hyperspace and the issue of the convergence of evolutionary paths, and ends up claiming that "biology and evolution possess an inherent structure that is not only consistent with the plenitude of the biosphere but more controversially is so arranged as to preordain the emergence of one (or more) sentient species." This argument, similar to the anthropic principle argument in cosmology, is seen as fitting naturally into a theological view with a strong eschatological component (thus linking into the fourth section of this book).

2. Thus mirroring in an intriguing way Cairns-Smith's discussion: Life could both start and end in a mineral form.

The third main category involves issues of humanity that arise in the context of speculation about the far future. In chapter 11, "Deep Time: Does It Matter?," Stephen Clark looks at the ethical and metaphysical effects of placing ourselves in the context of bygone and future ages. Specifically, he concentrates on the impact of possible futures, and addresses (1) the *doomsday argument*—that our future will be brief, (2) the *omega point argument*—that the future will be long and triumphant, and (3) the *presentist argument*—that all such stories are only metaphors for present-day experiences and desires. He places the discussion in a rich literary and theological context.

In chapter 12, "Games That End in a Bang or a Whimper," Steven Brams and Marc Kilgour consider whether human behavior may be expected to be rather different in the context of an infinite future as opposed to a finite future. Games theory models may be used to investigate this subject, the crucial difference being between *bounded games,* which end after a certain time or after a specific number of rounds have been played, and *unbounded games,* for which there is no such limit or bound. Unbounded outlooks encourage cooperative play, foster hope, and lead to more auspicious outcomes. These outcomes are facilitated by institutions that put no bounds on play—including reprisals—thereby allowing for a day of reckoning for those who violate established norms. The outcomes depend on beliefs about the future: cooperation can be sustained only if there is a sufficient level of hope. Brams and Kilgour conclude by relating their study to possible mechanisms that may help devise institutions that render destructive behavior unprofitable—which may be critical to the long-term survival of humanity.

In chapter 13, "Artificial Intelligence and the Far Future," Margaret Boden notes that AI will become more influential in the future. She asks: What is AI's potential for affecting our ideas about humanity? Will it support them, even enrich them—or undermine them? A key issue is whether far-future AI will support ways of thinking about people that have religious relevance, involving concepts of human freedom, uniqueness, emotion, self, and consciousness—including religious experience. She concludes that AI can be expected to have a positive influence in this regard and will enable a deepened understanding of humanity as such. She also emphasizes, however, the embodied nature of mind, which means that the proposal that intelligence will eventually be able to migrate to machines (vide Dyson above) is not the same as saying that intelligent life as such will be able to do so. She also argues that questions about whether far-future AI systems could "really" be conscious are incoherent and/or unanswerable.

Finally, in chapter 14, "Cosmic Eschatology vs. Human Eschatology," Owen Gingerich compares human timescales to the timescales of the cosmos, and puts forward pessimistic, optimistic, and realistic views on the future of humanity, stating "even ten million years seems to me unreasonably long for our species' survival." He then turns to theological reflections, querying traditional views of "eternity" and suggesting the idea of a "timeless eternity"—again showing how the question of the nature of time underlies the discussion. He finishes by emphasizing the human need for trust in the face of uncertainty.

The final category is an exploration of issues relating to metaphysics and theology. In chapter 15, "Cosmology and Religious Ideas about the End of the World," Keith Ward considers in turn the Eastern religions, Judaism and Islam, and Christianity. In the Eastern religions, either there is little concern with how the natural world will end or what it will be like in the far future (Taoism and Confucianism) or there is a cyclical world view (Hinduism and Buddhism). In Judaism and Islam, there are various planes of being, and a belief that this physical world has a future that is limitlessly better than its present and resurrection is conceived as incorruptible and beyond the reach of evil. Christianity has a similar view, and a vision of hope that is splendidly summarized at the end of section 3. Ward claims that modern knowledge of the universe adds considerable depth to the Christian vision, but compels a change in theological thinking that is rather less than that which occurred in the very first generation of Christian believers. Christianity is often taken to involve a view of the redemption of the cosmos itself from decay—that perhaps further evolutionary transformations of matter will occur beyond those that have already taken place. There is a tension between the view that what happens in the future of the physical universe is irrelevant and the hope that there will be sufficient time to express and understand forms of divine love and compassion in new ways. Nevertheless, the Christian faith is wholly consistent with the idea that space-time will have a temporal end—religious hope for an ultimate future is not confined to the future of this universe.

In chapter 16, "Cosmos and Theosis: Eschatological Perspectives on the Future of the Universe," Jürgen Moltmann emphasizes that theology does not acquire its eschatological horizons from the general observation of the world; it acquires them from its particular experience of God. In Christian eschatology we always find a combination of the contrasting ideas of an end and a beginning, a catastrophe and a new start. The possibilities that arise are annihilation of the world, transformation of the world, and deification of the world. Moltmann emphasizes that the theological foundation for an eschatological perspective on the far-future universe is to be found in the central importance for orthodox theology of the Resurrection of Christ. He conceives of two qualitatively differentiated eons, with different kinds of time: the time of this world is the time of the transitory world; the time of the future world is the time of a "world without end," the abiding and hence eternal world. The eschatological vision he puts forward is a universal transformation of the present world to a world of a new kind, where everything is to be "made new." He concludes with two questions: Is the universe contingent and is every event unique? Is the universe a closed system or an open one? As regards the latter, he argues a preference for an open universe because of its openness to the theological eschatology he describes.

In chapter 17, "Eschatology and Physical Cosmology: A Preliminary Reflection," Robert Russell addresses the great contrast between strong theological views (as represented by the first two articles in this section of this book), and scientific and humanistic views (as represented in the previous sections of this book). The question is, how may it be possible to reconcile them? As a thought experiment, while recognizing that there are other theological interpretations of these subjects, Russell takes up the extreme case of a literal interpretation of the idea of bodily resurrection of Jesus together with its eschatological implications regarding God's transformation of the creation into a "new creation," and argues how they might be compatible with present-day physics. He discusses physical cosmology, theological movements converging on resurrection, eschatology, and cosmology, and the relations between science and theology before turning to his central proposal: a set of guidelines for moving forward by revising eschatology in light of cosmology (thus scientific research programs help shape theological research programs), and cosmology in light of eschatology (the converse). He shows there could be significant interactions either way, and suggests how these interactions could be guided and shaped. The resultant discussion is provocative, as was intended: as Russell states clearly in the essay, it is a logical exploration of an extreme view, where there are indeed significant interactions either way.

Finally, in chapter 18, "Natures of Existence (Temporal and Eternal)," I make the point that the discussion about what happens in the far future depends on one's view of ontology, that is, what one considers as truly existing, for some forms of existence are more unchanging and eternal than others. Some may be conceived of as continuing forever unchanged, even while others decay away. This discussion provides a holistic view of ontology and of causality, first from a purely scientific viewpoint, and then with a possible extension to include issues of morality and theology. It then considers the way this viewpoint relates to eschatological questions, in terms of issues both of fact and of hope.

An appendix contains a reprint of "Olaf Stapledon," an essay by Stephen Clark, because it is in science fiction, and particularly in the writings of Stapledon, that many of the issues discussed here have been thought about and presented in considerable depth. Literature, science fiction in particular, remains a fascinating source of speculation on the far future of humanity and of the universe.

Taken together, these essays provide an intriguing and thought-provoking discussion of eschatological issues from radically different perspectives, and some thoughts about how to reconcile them even in their most extreme forms. Hopefully, they will encourage renewed interest and debate on this recurring subject.

[2]

SEEKING THE FUTURE

A Theological Perspective

George V. Coyne, S.J.

A glance at the program of the Far-Future Universe Symposium leaves one in a bit of a quandary. On paper the program looks quite synthetic and straightforward as it moves from cosmology and mathematics through biology to theological considerations. But when one looks at the content of the task of the symposium it is daunting. It appears to me that, with all of the scientific knowledge that we do solidly possess about the origins and evolution of the universe, we must also face a great deal of ignorance. And then when we compound this ignorance by trying to look into the future and by trying to draw some implications for philosophical and religious thought, what we are about to do must become clear to us. We are setting out on an adventure, one in which no one area of human thought and experience can claim proprietorship and one in which much that is tentative, at times even tendentious, will come to light. All of us are aware that our cultivated ignorance and our tentative proposals may hopefully bring us closer to the truth in this dialogue, but only if we exercise a degree of intellectual discipline, perhaps even more than is typical of our respective areas of inquiry.

But this is not only an intellectual adventure. It is rather a human adventure in which the emotional nature we possess must also have its say. When we human beings think about ourselves in comparison with all else we experience in the universe we recognize that we have much in common and yet that we are different. More than any other creatures we know we sense that we are open to an immense and rich array of possibilities in the way we exist and in the way we lead our lives. This is, perhaps, best described by saying that we are symbolic creatures. Like symbols we are always leaning toward a reality beyond ourselves, toward the far-future universe. We are never fully content with what we are now. We are directed to the other and to the future. Our very physical makeup, the constant tension in our flesh toward growth or toward disintegration, hints at this. Our need to love and to be loved, our unquenchable thirst to understand and to manage the environment in which we live, all speak to this great need.

Among the many symbolic, future-directed actions that express human nature, one of the most meaningful, and at the same time simple, ones is, perhaps, our gaze toward the sky in search of familiarity and understanding. That gesture of raising one's head and lifting one's eyes to gaze beyond immediate realities does not so much express a disorientation as it does an acknowledgement of insufficiency, of the need for something or someone out there, beyond oneself. Ancient mythologies, cosmologies, and cosmogonies bear witness to the immense power that drives us humans in our continuous search for deeper understanding. Modern science bears witness to this persistent journey. From time immemorial we have always sought this further understanding, many times in a person with whom we could converse, someone who shared our capacity to love and be loved and our desire to understand and to accomplish something of significance.

There is little doubt that science has been a dominating influence in setting the way we think today about the universe and ourselves. Nevertheless, we have become aware, in our attempt to unify our scientific knowledge with all that we have come to experience as human beings, that there is much in our experience that lies outside the domain of science. We experience a passionate desire to communicate and we sense that it is a call to love. Science, it is important to note, has brought us to this point; science today, more than ever in the past, throws open the doors to realities before which it senses that it is not totally competent, realities that require different approaches from those with which science itself is familiar. Science today is ever more human; it stimulates, provokes, questions us in ways that drive us beyond science in the search for satisfaction, while at the same time scientific data furnish the stimuli. In this context the best science, to its great merit, neither pretends nor presumes to have the ultimate answers. It simply suggests and urges us on, well aware that not all is within its ken. Freedom to seek further understanding, and not a dogmatic possession of that which is partially understood, characterizes the best scientists. Science, in fact, is a field where certainties lie always in the future; thus science is vital, dynamic, and very demanding of those who seek to discover the secrets of the universe and themselves.

In this book the reader will find an attempt to read the data of science more from a human than a scientific perspective without any presupposition, however, that the one precludes the other. In fact, it is altogether a very human endeavor to attempt to integrate rigorously scientific conclusions with those that are drawn from life's other experiences and from the very search for the meaning of life. However, we have no illusion, even those of us who are religious believers, that we possess the truth; quite the contrary, we are very aware that the truth is not to be possessed but to be contemplated. Contemplation is an end in itself and the object contemplated serves no other end than to bring the joy of contemplation to the observer. Our discussions about the far-future universe might bear witness to the fact that the search for truth brings with it many ways of knowing in addition to the commonly accepted ones. The truth itself requires that we neither exclude nor absolutize any possibility, if for no other reason than that our search be complete.

As many of you may know, the Vatican Observatory and the Center for Theology and the Natural Sciences at Berkeley, California have over the past decade sponsored a series of research conferences with an umbrella theme of "Scientific Perspectives on Divine Action."[1] Without prejudicing the work we are setting out to do, I would like to see us turn the tables a bit and emphasize in this symposium sponsored by the Templeton Foundation the

1. The books resulting from this series of meetings are references 1–5 listed at the end of this chapter.

theological perspectives on our scientific knowledge of the future of the universe. To do that difficult task with any hope of success will obviously require a grasp of the best that science can say and a theological stance that does not hold religious thought to be monolithic or stationary. Since the principal interest of theology is in the human being's relationship to God, I see two areas of utmost importance to such dialogue:

1. From our best scientific knowledge, what can we say about the degree of contingency in both the past and the future history of the universe, and especially of the human being in the universe?

2. Is the history of the human being linked inevitably to the physical history of the universe, or is there a transcendental aspect of the human? By physical, I mean all that science can investigate; by transcendental, I mean that which science cannot ultimately investigate. The distinction may not even be valid, but I think that the issue needs to be faced.

If we are to enter into a dialogue in some manner such as the above, then, before we sit down to the hard science, it may be helpful to have some idea of a theological stance, which, if not embraced by all, has a degree of accord with the stances of most others who, I believe, engage in dialogue at this level. I would like to do this by making three brief statements.

First, the religious believer is tempted by science to make God an "explanation." We bring God in to try to explain things that we cannot otherwise explain: "How did the universe begin?", "How did we come to be?" and all such questions. We sort of latch onto God, especially if we do not feel that we have a good and reasonable scientific explanation. She/He is brought in as the great God of the gaps. The truth is that the first moment in religious faith is to enter into a personal relationship with God. It is only in a second, and less important, moment that such a personal God can become a source of understanding. One gets the impression from certain religious believers that they fondly hope for the durability of certain gaps in our scientific knowledge of evolution, so that they can fill them with God. This is the exact opposite of what human intelligence is all about. We should be seeking for the fullness of God in creation not by using God but by using our brains.

It is not unusual for cosmologists to speak of the "mind of God." In most cases, it appears, this is taken to mean that ideal Platonic mathematical structure from which the shadow world we live in came to be. It would give us a unified theory and thus an understanding of all physical laws and the initial conditions under which they work. Would we also fundamentally understand life, ourselves, and our future? There appears to be no intentionality associated with the "mind of God" of the cosmologists. Can life and its future be understood without that intentionality? These are, I accept, pretentious questions that go beyond what a scientist would usually accept as a rational approach to questions about the world in which we live.

When scientists in their enthusiasm speak about the "theory of everything" and the "mind of God," they inevitably try to quantify what is not quantifiable: selflessness, graciousness, harmony, and so on. Musical scales can be carefully analyzed by mathematics; the beauty of a Mozart nocturne cannot. This is, of course, not to minimize science, nor, as a matter of fact, any of the other ways of knowing. It is simply to realize what a given discipline can or cannot do. That is precisely why our knowledge must be unifying. There will, of course, always be a tension between science and theology because of the transcendental (beyond reason) character of the latter, but considering the somewhat Platonic quest for the "mind of God" among cosmologists, that very tension could be the source of a quite creative dialogue.

At the same time, we must beware of a serious temptation toward oversimplification in such a dialogue. Within the culture of scientific cosmology God is essentially, if not exclusively, seen as an explanation and not as a person. God is the ideal mathematical structure, the theory of everything. God is Mind. It must remain a firm tenet of the reflecting religious person that God is more than that and that God's revelation of himself in time is more than a communication of information. Even if we discover the "mind of God" we will not have necessarily found God. The very nature of our emergence in an evolving universe and our inability to comprehend it, even with all that we know from cosmology, may be an indication that in the universe God may be communicating much more than information to us. If we are truly seeking a wholeness, a unification of our knowledge, then upon reflection we note a certain dynamism in our voyage to understand, a dynamism that draws us from knowledge, through the incompleteness of our knowledge, to wonder, and then to love.

My second remark on the theological perspectives on the far future is the following one. The scientific picture of the universe deals with the questions of origins, of how what we observe and experience today came to be. The theological notion of creation, and therefore of a God Creator, responds to the question of why there is anything in existence. Creation is not one of the ways whereby things originated as opposed to other ways that can be thought of, including, for instance, quantum cosmology and evolutionary biology. The claim that all things are created is a religious claim that all that exists depends for its existence on God. It says nothing scientifically of how things came to be, although beautiful stories are told in the Book of Genesis to elaborate on the dependence of all things for their existence upon God.

Thirdly, having opened the Pandora's box of the Bible, let us elaborate a bit upon it. The Bible is a collection of writings by various authors at various epochs using various literary genres. And so it best serves reason if one speaks of a specific book, or even part of a book, rather than of the Bible in general. It is clear for instance, that the overall intention of the authors of Genesis is to evoke religious faith, an adherence to the God of Abraham, Isaac, and Jacob, and not to teach. There is simply no scientific teaching in Genesis. In the Judeo-Christian tradition, the roots of religious belief reach to about two thousand years before Christ with the prophet Abraham. But modern science cannot be dated before the sixteenth or seventeenth centuries, roughly from the time of Galileo, and then through many others to Newton, with the discovery of the universal law of gravity, the differential calculus, and more. The modern science that speaks to religion today was born much later than the religion to which it speaks. It has to be recognized that the religious tradition is historically much longer and to a certain extent has that richness of the past that modern science does not.

Here is an example of that richness. The biblical account of creation in the Book of Genesis highlights the comment made by God after each act of creation: "And he saw that it was good." The Hebrew word used to express "good" has, in fact, a strong indication of something esthetically pleasing, so that, without betraying the original, one might translate the comment: "And he saw that it was beautiful." Thus, every creative act of God becomes a source of beauty and, as a sharing in God's creation, every human creation, the invention of a new scientific theory, for instance, is a source of beauty.

A study of the Old Testament shows that the first reflection of the Jewish people was that the universe was the source of their praise of the Lord who had freed them from bondage and had chosen them as his people. The Book of Psalms, written for the most part well before the Book of Genesis, bears witness to this: "The mountains and valleys skip with joy to praise the Lord"; "The heavens reveal the glory of the Lord and the firmament proclaims his handiwork." But if these aspects of the universe were to praise the Lord, they must be good and beautiful. Upon reflecting on their goodness and beauty, God's chosen people came to realize that these aspects must come from God—and so the stories in Genesis in which at the end of each day God declares that what he had created is good (beautiful). The stories of Genesis are, therefore, more about God than they are about the universe and its beginning.

They are not, in the first place, speaking of the origins of the created world. They are speaking of the beauty of the created world and the source of that beauty, God. The universe sings God's praises because it is beautiful; it is beautiful because God made it. In these simple affirmations we may even trace the roots of modern science in the West. The beauty of the universe invites us to know more about it and this search for knowledge discovers a rationality innate in the universe. The far-future universe must be seen in this light.

There are two implicit assertions in the Book of Genesis that set the faith of these people apart from their predecessors, the Canaanites, upon whose stories they rely. First, God is one and there is no other god; there is no struggle between God and some equal, even malevolent force. Second, everything else is not God, but depends for its beauty upon him. He made everything and declared it beautiful. It is very important to note that created things are first of all beautiful because God says that they are; it is only upon reflection in a second moment that they are seen as understandable, as having a rational structure. It is for this reason that we can gather to discuss the far-future universe. The future is forever drawing us forward in what we sense as an interminable journey to understand. Beauty is irresistible; beauty draws us to seek understanding.

It is unfortunate that, at least in America, creationism has come to mean some fundamentalistic, literal, scientific interpretation of Genesis. Judeo-Christian faith is radically creationist, but in a totally different sense. It is rooted in a belief that everything depends upon God, or better, all is a gift from God. The universe is not God and it cannot exist independently of God. The Judeo-Christian faith claims that neither pantheism nor naturalism is true.

Finally, if we confront what we know of origins scientifically with religious faith in God the Creator, in the senses described above, what results? I would claim that the detailed scientific understanding of origins has no bearing whatsoever on whether God exists or not. It has a great deal to do with my knowledge of God, should I happen to believe that God exists. Let me explain.

Take two rather extreme scientific views of origins: that of Stephen Gould of an episodic, totally contingent and, therefore, nonrepeatable evolutionary process, as contrasted to a convergent evolutionary process such as that of Christian de Duve, in which the interplay of chance, necessity, and opportunity leads inevitably to life and intelligence. In either case, it is scientifically tenable to maintain an autonomy and self-sufficiency of the natural processes in a natural world, so that recourse to God to explain the origins of all that exists is not required. It is not that chance in nature excludes God, or that destiny in nature requires God. In neither case is God required.

In the end it appears to me that one of the essential ingredients of wisdom is a serene openness to all of human experience. We scientists tend to be driven people with a passion to obtain solid scientific results. It is nice to see that we can also peacefully engage in the dialogue that is the stuff of this meeting.

REFERENCES

1. Russell, R. J., et al., eds., *Quantum Mechanics: Scientific Perspectives on Divine Action* (Vatican Observatory, Vatican City State/Center for Theology and the Natural Sciences, Berkeley, Calif., 2001).

2. Russell, R. J., Murphy, N., and Isham, C. J., eds., *Quantum Cosmology and the Laws of Nature* (Vatican Observatory, Vatican City State/Center for Theology and the Natural Sciences, Berkeley, California; University of Notre Dame Press, Notre Dame, Indiana., 1993).

3. Russell, R. J., Murphy, N., Meyering, T. C., and Arbib, M. A., eds., *Neuroscience and the Person* (Vatican Observatory, Vatican City State/Center for Theology and the Natural Sciences, Berkeley, Calif.; University of Notre Dame Press, Notre Dame, Ind., 1999).

4. Russell, R. J., Murphy, N., and Peacocke, A., eds., *Chaos and Complexity* (Vatican Observatory, Vatican City State/Center for Theology and the Natural Sciences, Berkeley, Calif.; University of Notre Dame Press, Notre Dame, Ind., 1995).

5. Russell, R., Stoeger, W., and Ayala, F., eds., *Evolutionary and Molecular Biology* (Vatican Observatory, Vatican City State/Center for Theology and the Natural Sciences, Berkeley, Calif.; University of Notre Dame Press, Notre Dame, Ind., 1998).

PART 2

COSMOLOGY/PHYSICS

[3]

THE FAR, FAR FUTURE

John D. Barrow

3.1. Past Predictions about the Far Future

The first attempts to predict the future of the universe in a manner that drew upon as much of known science as possible were those of Immanuel Kant in the progressive cosmology he published in 1755.[1] In the nineteenth century, the pioneers of thermodynamics identified the trend for closed systems to degenerate from order to disorder and Rudolf Clausius raised the specter of the "heat death" of the universe in 1850. It was first stated by the German physicist Hermann von Helmholtz in an article published in 1854 [36]. This pessimistic long-range forecast was seized upon with glee by many materialist philosophers and played an important role in the development of philosophies of progress.[2] Nor were physicists alone in extrapolating their science into the far future. Charles Darwin saw clearly that, in the long run,

1. Kant [25]; see also Barrow and Tipler [6], pp. 620–21.
2. A detailed history of these developments can be found in chapter 3 of [6].

evolution would continue leading to descendants whose appearance and nature were different from those of their ancestors. In the closing pages of *The Origin of Species* he wrote:

Judging from the past, we may safely infer that not one living species will transmit its unaltered likeness to a distant futurity.[12]

Haldane and Bernal, both ardent materialists, speculated on the detailed fate of intelligence in the far, far future. Haldane saw that the future looked bleak for life of any sort, but held that this sobering realization that life was doomed should nevertheless serve to convince us that

the use, however haltingly, of our imaginations upon the possibilities of the future is a valuable spiritual exercise.[3]

However, in pursuing those therapeutic considerations, it was important that we did not take too optimistic a view, or expect to find a comforting golden age awaiting us in the asymptotic realm. We must remember that

there are certain criteria which every attempt, however fantastic, to forecast the future should satisfy. In the first place, the future will not be as we should wish it. The Pilgrim Fathers were much happier in England under King James I than they would be in America under President Coolidge.[4]

Bernal wondered how life might develop so as to survive indefinitely into the future, in a disembodied form quite different from that displayed by any extant form of life. In *The World, the Flesh, and the Devil* (1929), he wonders if

finally, consciousness itself may end . . . becoming masses of atoms in space communicating by radiation, and ultimately resolving itself entirely into light . . . these beings . . . each utilizing the bare minimum of energy . . . spreading themselves over immense areas and periods of time. . . . The scene of life would be . . . the cold emptiness of space.[5]

He was reluctantly persuaded that the second law of thermodynamics would reduce everything to lifeless uniformity in the end, but held out a hope that some simple nonhuman form of life might sustain the processing of infor-

3. Haldane [19], p. 38.
4. Haldane [19], p. 37.
5. Bernal [10], p. 63.

mation over extremely long periods of time, staving off the evil day when life would become extinct, and far exceeding the survival capabilities of simple forms of life like ourselves:

The second law of thermodynamics which . . . will bring this universe to an inglorious close, may perhaps always remain the final factor. But by intelligent organizations the life of the Universe could probably be prolonged to many millions of millions of times what it would be without organisation.[6]

But the full context of the heat death could only be appreciated after the discovery of cosmological solutions of Einstein's general theory of relativity. Eddington and Jeans discussed the cosmological heat death in their widely read popular books in the 1930s and the idea can be found subsequently as a strong influence in popular works of contemporary fiction and science fiction. In 1931, Eddington wrote:

It used to be thought that in the end all the matter of the Universe would collect into one rather dense ball at uniform temperature; but the doctrine of the spherical space, and more especially the recent results as to the expansion of the Universe, have changed that. . . . It is widely thought that matter slowly changes into radiation. If so, it would seem that the Universe will ultimately became a ball of radiation growing ever larger, the radiation becoming thinner and passing into longer and longer wave lengths. About every 1,500 million years [this number would now be increased by about a factor 10 due to the more accurate determination of the Hubble constant] it will double its radius, and its size will go on expanding in this way in geometrical progression forever [14].

Nor were leading theologians silent about these prognostications. Scholars trained in science and theology, like Barnes and de Chardin, confronted the challenges head on. The most extensive study was that of William Inge, then Dean of St. Paul's Cathedral, London, whose influential book *God and the Astronomers* (1934), based upon lectures delivered in 1931–33, was entirely devoted to discussion of the heat death scenarios propounded by Jeans and Eddington in their influential talks and writings. Theologians like Inge actually welcomed the heat death of the cosmologists because a future in which all life must end was seen as a blow to the materialist conception of nature and humanity's self-sufficiency. Such a suicidal universe could not be an emotionally hospitable home for humankind. Inge's targets were the sup-

6. Bernal [10], p. 28.

porters of "modernist philosophy" whether they were atheistic materialists or theologians enamored of the idea of an evolving God, tuned to the needs of an ever-changing universe:

> The idea of the end of the world is intolerable only to modernist philosophy, which finds in the idea of unending temporal progress a pitiful substitute for the blessed hope of everlasting life, and in an evolving God a shadowy ghost of the unchanging Creator and Sustainer of the Universe. It is this philosophy which makes Time itself an absolute value, and progress a cosmic principle. Against this philosophy my book is a sustained polemic. Modernist philosophy is, I maintain, wrecked on the Second Law of Thermodynamics; it is no wonder that it finds the situation intolerable, and wriggles piteously to escape from its toils [*sic:* coils?] [24].

3.2. Modern Physical Eschatology

More recently, the application of all of known physics and astronomy to the question of the fate of the universe has been carried out most fully by Dyson in 1979 [13], who considered the long-term fate of matter and astronomical structures in an expanding universe, and by Barrow and Tipler in 1978 [5] and 1986 [6], who introduced consideration of the evolution of the anisotropy and inhomogeneity of the universe, proton decay, and other aspects of modern unified gauge theories, together with new forms of fast information processing, and looked at the heat death from the perspective of gravitational entropy stressed by Frautschi [18] in 1982. From these studies it became clear that predicting the future of an ever-expanding universe is at least as challenging as reconstructing its past history, and there have been many subsequent studies adding new details [30, 26] and speculative developments concerning the nature and possible consequences of indefinite information processing [32].

In order to include all the possible uncertainties in predicting the far cosmic future, we need to appreciate all the dynamical possibilities open to an expanding universe: the consequences of currently imperceptibly small corrections to our present theory of gravity becoming important in the long run, the large-scale topological constraints, the possibility of ultra-weak natural forces and lightweight elementary particles that might dominate the universal expansion in the far future, and the thermodynamic fate of matter and energy in all its forms. And if we want to predict the future of "life," then we must define it loosely enough to escape an entirely predictable fate along with all other atom-based structures. Defining "life" as information processing—requiring disequilibrium, and needing space and time to store information—enables us to ask what types of universe will allow this kind of minimal life to go on forever—processing an infinite amount of information. This question introduces the "problem of time" of general relativity into the larger problem. We need to ask what is the natural timescale on which information processing should be seen to occur in the far future. Perhaps it is simple comoving proper time as generally assumed, but if information processing takes global forms then it may be more appropriate to gauge it by means of a time that is intrinsically defined by the curvature of the universe, free of all reference to artifacts.[7]

One interesting consequence of these studies has been the realization that the traditional picture of the heat death of the universe requires drastic revision. Hawking's discovery that gravitational fields carry entropy means that it is possible for the maximum possible entropy of the universe to increase faster than the actual entropy in its matter and radiation. Thus the entropy of its matter and radiation can continue to increase inexorably, in accord with our expectations from the second law of thermodynamics, yet that entropy would always be getting further away from attaining the maximum possible entropy that the universe could have. Thus the universe gets further and further from thermal equilibrium and ultimate heat death even though its entropy is always increasing.

Within this complex of possibilities and inevitabilities there are a wealth of issues of great potential significance to philosophers and theologians alike:

- What is the response to a finite future for terrestrial, or any, forms of life?
- What if life can continue forever in some well-defined sense?
- How do we feel about the infamous infinite replication paradoxes [17, 31] of infinity?
- Is the block space-time picture of the future reconcilable with free will?

The problem changes if our universe is going to recollapse in the far future, heading ultimately for a "big crunch" in finite time. Observations can only tell us whether our visible fraction of the universe may undergo such a

7. Barrow and Tipler [6], ch. 10. This discussion poses the problem of what the structure of space-time must be like in order that an infinite number of bits of information can be processed to the future.

fate [2]. We are tantalizingly close to the divide separating indefinite expansion from collapse in the future and small fluctuations from place to place over large scales in the universe could see local regions of very large scale undergoing collapse while the rest of the universe carries on expanding forever. Interestingly, the currently favored inflationary universe theory leads us to expect that such an inhomogeneous state of affairs may exist throughout the universe at all times. If a region of (or an entire) universe undergoes collapse there is the possibility that it might bounce back into a state of expansion, so producing an oscillating version of the ancient stoic and Eastern conception of the cyclic reborn universe.

3.3. The Ups and Downs of Oscillating Universes

If our expanding universe of stars and galaxies did not appear spontaneously out of nothing at all, then from what might it have arisen? One option that has an ancient pedigree is that it had no beginning. It has always existed. A persistently compelling picture of this sort is one in which the universe undergoes a cyclic history, periodically disappearing in a great conflagration before reappearing phoenix-like from the ashes [16, 6]. This scenario has a counterpart in modern cosmological models of the expanding universe. If we consider closed universes that have an expansion history that expands to a maximum and then contracts back to zero, then there is a tantalizing possibility that this episode of cosmic history might continue to repeat itself into the future. Suppose the universe reexpands and repeats this behavior over and over again. If this can happen then there is no reason why we should be in the first cycle. We could imagine an infinite number of past oscillations and a similar number to come in the future. However, we are ignoring the fact that a singularity arises at the start and the end of each cycle. It could be that repulsive gravity stops the universe just short of the point of infinite density or some more exotic passage occurs "through" the singularity, but this is pure speculation.

This speculation is not entirely unrestrained, though. Let us assume that one of the central principles governing nature, the second law of thermodynamics, which tells us that the total entropy (or disorder) of a closed system can never decrease, governs the evolution from cycle to cycle.[8] Gradually, or-

8. The big assumption here is that nothing counter to the second law of thermodynamics occurs at the moments when the universe bounces.

dered forms of matter will be transformed into disordered radiation and the entropy of the radiation will steadily increase. The result is to increase the total pressure exerted by the matter and radiation in the universe and so to increase the size of the universe at each successive maximum point of expansion.[9] As the cycles unfold they get bigger and bigger! Intriguingly, the universe expands closer and closer to the critical state of flatness that we saw as a consequence of inflation. If we follow it backwards in time through smaller and smaller cycles it need never have a beginning at any finite past time although life can only exist after the cycles get big enough and old enough for atoms and biological elements to form.

For a long time this sequence of events used to be taken as evidence that the universe had not undergone an infinite sequence of past oscillations because the build up of entropy would eventually make the existence of stars and life impossible (see e.g., [20]), and the number of photons that we measure on average in the universe for every proton (about one billion) gives a measure of how much entropy production there could have been. However, we now know that this measure does not need to keep on increasing from cycle to cycle. It is not a gauge of the increasing entropy. Everything goes into the mixer when the universes bounces and then the number of protons that there are compared with photons gets set by processes that occur early on. One problem of this sort might be that of black holes. Once large black holes form, like those observed at the centers of many galaxies, including the Milky Way, they will tend to accumulate in the universe from cycle to cycle, getting ever more massive until they engulf the universe, unless they can be destroyed at each bounce or become separate "universes" that we can neither see nor feel gravitationally. Smolin [29] has proposed an adventurous scheme in which black hole collapses bounce back to produce new expanding universes in which the values of the physical constants are slightly shifted. In the long run, this could lead to the population of universes being dominated by those that maximize the production of black holes. Thus small shifts in the constants should reduce the production of black holes in our universe. However, it is possible that such universes do not permit observers and the real prediction of the scenario is that we should be in a universe that maximizes the production of black holes given that observers can exist.

9. This was first pointed out by R. C. Tolman [33, 34]. Recently, a detailed reanalysis was given by Barrow and Dabrowski [4].

A curious postscript to the story of cyclic universes was recently discovered by Mariusz Dabrowski and me [4]. We showed that if Einstein's cosmological constant exists, then no matter how small a positive value it takes its repulsive gravitational effect will eventually cause the oscillations of a cyclic universe to cease. The oscillations will get bigger and bigger until eventually the universe becomes large enough for the cosmological constant to dominate over the gravity of matter. When it does so it will launch the universe off into a phase of accelerating expansion from which it can never escape unless the vacuum energy creating the cosmological constant stress were to decay mysteriously away in the far future. Thus the bouncing universe can eventually escape from its infinite oscillatory future in the presence of a positive cosmological constant. If there has been a past eternity of oscillations we might expect to find ourselves in the last ever-expanding cycle so long as it is one that permits life to evolve and persist.

Another means by which the universe can avoid having a beginning is to undergo the exotic sequence of evolutionary steps created by the eternal inflationary history [27]. There seems to be no reason why the sequence of inflations that arises from within already inflating domains should ever have had an overall beginning. It is possible for any particular domain to have a history that has a definite beginning in an inflationary quantum event, but the process as a whole could just go on in a steady fashion for all eternity, past and present.

3.4. The Speculative Futures Market

Any astronomer entering the futures market has to grapple with the following questions:

- Will there be an "end" or asymptote of any sort(s)?
- Will it be of "us" in particular, of life in general, of matter, of space or time, or of the entire universe?
- Will the end be sudden or gradual?
- Can there be cycles of behavior or some form of natural or artificial selection of universes over long periods of time?
- Will something always be changing in an ever-expanding universe or are we headed for stasis?
- What is the impact of the apparent existence of the cosmological constant, or cosmic vacuum energy, in the universe?

An end that is gradual could result from a number of subtle slow changes in the nature of the universe over eons of cosmic time. The constants of nature, whose current values are in many cases quite finely "tuned" for the existence of atom-based life, may be very slowly changing [37]. Eventually, they may drift out of the narrow window that permits stable atoms and stars to exist. If other dimensions of space exist then any change in the size of those extra dimensions can induce changes at a similar rate in the values of the "constants" that we have found to define the nature of physics in our three-dimensional space.

We have begun to appreciate how major features of physics like the matter-antimatter balance in space or the number of fundamental forces of nature may arise by a process of symmetry breaking, which produces a result that is very different in different parts of the universe. As the universe expands and ages, different domains, possessing very different fundamental physical properties, will eventually meet, with dramatic results at their boundary.

We have detected evidence for the existence of many black holes formed by the collapse of short-lived massive stars and of giant black holes in the centers of galaxies where they grow by cannibalizing the stars and gas around them. Eventually, a very large fraction of the material in the universe may end up inside black holes, but it will not stay there forever. Hawking [22] showed that black holes will slowly evaporate their mass by quantum particle production, radiating a thermal spectrum of radiation and relativistic particles back into space. What happens in the explosive final throes of this evaporation process is still a mystery. Superficially, it appears to leave a singular hole in space and time just like that resulting from a big crunch. If it does then the universe will become peppered by these holes in space-time—places where the laws of nature may break down. Anything can emerge from them—photons, particles, stars, even whole universes.

All manner of dramatic sudden ends of us or of the universe around us can be contemplated given what we know to be possible from current theories of high- and low-energy physics. The vacuum state in which the universe currently resides is generally assumed to be the lowest possible energy state, reached following a sequence of symmetry breakings that dropped it further and further downstairs in the past. But what if there are more symmetry breakings to come? What if the next drop is imminent with a significant probability that the whole universe suddenly makes the transition? Then the new vacuum may have very different properties to our own. All particles could be massless, for example. We would just disappear in the twinkling of an eye.[10] If inflation left us lodged on a shallow ledge in the potential landscape then we might suddenly find ourselves nudged over the brink and on the way to another lower minimum. That nudge might be supplied by very high energy events in the universe.

If collisions between stars or black holes generated cosmic rays of sufficiently high energies they might be able to initiate the transition to the new vacuum in a region of space [23, 35]. This picture of the vacuum landscape is a speculative one. We do not know the overall form of the landscape well enough to be able to tell whether we are already at the ground level or whether there are other vacuums into which the states of matter in our locale can fall, either accidentally or deliberately. As one contemplates this radical possibility of an unannounced change in some of the basic properties of the forces of nature, it is tempting to portray it as the ultimate extension of the idea of punctuated equilibrium promoted by Niles Eldredge and Stephen Jay Gould [15]. They proposed that the course of biological evolution by natural selection on the earth proceeded by a succession of slow changes interspersed by sudden jumps rather than as a steady ongoing process. Indeed, we can characterize it as a movement through a landscape with many hills and valleys in which a force is dragging someone along. The pattern of change under these circumstances is for a slow climb up each hill but when the top is reached there will be a sudden jump across to the side of the next hill and another spell of steady hill climbing [1].

The way in which our form of biochemical life relies on rather particular coincidences between the strengths and properties of the different forces of nature means that any change of vacuum state would very likely be catastrophic for us. It would leave us in a new world where other forms of life might be possible, but there is no reason why they should be just a small evolutionary step away from our own biochemical forms.

If the universe follows this lead there may be a shock for some in eons to come. As with the puzzle of why the lambda force should come into play so

10. Recently, there seems to have been some public worry in the United States that a planned sequence of high-energy particle collisions at a national laboratory might induce just such a catastrophe.

close to our time, so we might regard it as unlikely that the epoch at which the fall "downstairs" could occur should be close to the time of human existence in the universe—unless, of course, there is a link with lambda, or the presence of life can do something to precipitate the great fall downstairs. Prophets of doom, do not despair. Things may be even worse than you thought.

In particle physics we are used to the dogma that all broken symmetries ultimately get restored if we go to high-enough energies. However, we don't believe that *all* symmetries ultimately get broken if we go to low-enough energies and temperatures. Maybe that belief is wrong. If all symmetries ultimately break at low-enough temperatures then, in the far future, when the universe has cooled sufficiently, the U(1) symmetry of electromagnetism might break down, giving all photons a positive mass. Again, the results would be dramatic. Equally sudden might be an end that results from us hitting a space-time singularity locally or being zapped by an incoming gravitational shock wave.

The most remarkable possibility of all that could happen in an ever-expanding universe with an infinite future to wait is that it could quantum tunnel back into "nothing." At present, many quantum cosmologists are looking for a way to explain the universe as having tunneled out of "nothing" in some sense. Now in most scenarios for this happening it is vastly more probable for something to appear from nothing than for nothing to appear from something. But with an infinite time to wait eventually any process with a finite probability of happening will happen.

3.5. The Future of the Vacuum

The vacuum energy of the universe that manifests itself as a cosmological constant in Einstein's general theory of relativity may prevent the universe from having a beginning, influence its early inflationary moments, and may be driving its expansion today. But its most dramatic effect may still be to come: its domination of the universe's future. The vacuum energy that manifests itself as Einstein's cosmological constant stays constant while every other contribution to the density of matter in the universe—stars, planets, radiation, black holes—is diluted away by the expansion. If the cosmic vacuum energy has recently started accelerating the expansion of the universe and that expansion will continue forever, as observations imply, then its domination will grow overwhelming in the future. The universe will continue expanding and accelerating forever. The temperature will fall faster, and the stars will exhaust their reserves of nuclear fuel and implode to form dense dead relics of closely packed cold atoms or concentrated neutrons, or large black holes. Even the giant galaxies and clusters of galaxies will eventually follow suit, spiraling inward upon themselves as the motions of their constituent stars are gradually slowed by the outward flow of gravitational waves and radiation. All their stars will be swallowed up in great central black holes, growing bigger until they have consumed all the material within reach. Ultimately, all these black holes will evaporate away by the Hawking evaporation process.

The most fascinating thing about the cosmic vacuum energy is that, ultimately, it wins out over all other forms of matter and energy in the struggle to determine the shape of space and the rate of expansion of the universe. No matter what the structure of the universe in its earlier days, before the vacuum energy comes to dominate, just as all ancient roads led to Rome, so all ever-expanding universes approach a very particular accelerating universe, called the de Sitter universe after Willem de Sitter, the famous Dutch astronomer who discovered it was a solution of Einstein's general theory relativity in 1917. It is distinguished by being the most symmetrical possible universe.

This property of an accelerating universe, that it loses all memory of how it began, is sometimes called the "cosmic no hair property." This curious terminology is chosen to capture the fact that all the accelerating universes become the same: they retain no individual distinguishing features (hairstyles, metaphorically speaking). This inexorable slide toward the same future state signals that there is a loss of information taking place when the universe starts accelerating. The expansion is so fast that the information content of signals sent across the universe gets degraded as fast as possible. Everything looks smoother and smoother; all differences in the rate of expansion from one direction to another are expunged at a rapid rate; no new condensations of matter can appear out of the cosmic matter distribution; local gravitational pull has lost the last battle with the overwhelming repulsion of the lambda force.

This effect of accelerating cosmological expansion has important consequences for any consideration of "life" in the far future. If life requires information storage and processing to take place in some way then we can ask whether the universe will always permit these things to occur. When the vacuum energy is not present, and so the expansion does not ultimately accelerate, Dyson [13] and Barrow and Tipler ([6], chap. 10) showed that there are a range of possibilities open for this rather basic form of life to perpetuate itself. It can store information in elementary-particle states that are vastly better information storage repositories than those used for storing data in our present computers. To continue to process information indefinitely, living systems need to create and sustain deviations from perfect uniformity in the temperature and energy of the universe.[11] This may always be possible when the accelerating vacuum energy is not present, although it probably requires that "life" make use of the gravitational energy differences sustaining the differences in the expansion of the universe from one direction to another. The energy density supporting these differences falls off far more slowly than that in any ordinary form of matter. Tiny deviations in the way in which the universe is expanding from one direction to another can be exploited to make radiation cool at slightly different rates in different directions. The gradient in temperature thereby created can then be used to do work or process information. This does not of course mean that life in any shape or form *will* survive [11] forever, let alone that it *must* survive forever, merely that it is logically and physically possible given the known laws of physics in the absence of a vacuum energy permeating the universe.

However, as Barrow and Tipler showed (see [6], p. 668), if the vacuum energy exists then everything changes—for the worse. All evolution heads inevitably for a state of uniformity characterized by the accelerating universe of de Sitter. Information processing cannot continue forever: it must die out. There will be less and less utilizable energy available as the state of the material universe is driven closer and closer to a state of uniformity. If the vacuum energy exists but there is insufficient matter in the universe to

11. The absolute minimum amount of energy required to process a given amount of information is determined by the second law of thermodynamics. If ΔI is the number of bits of information processed, the second law requires $\Delta I \leq \Delta E / kT ln2 = \Delta E / T(\text{ergs}/K)(1.05 \times 10^{16})$, where T is the temperature in degrees Kelvin, k is Boltzmann's constant, and ΔE is the amount of free energy expended. If the temperature operates at a temperature above absolute zero ($T>0$, as required by the third law of thermodynamics), there is a minimum amount of energy that must be expended to process a single bit of information. This inequality is due to Brillouin.

reverse its expansion into contraction before the vacuum energy gets a grip on the expansion and begins to accelerate it,[12] then the universe seems destined for a lifeless far future. Eventually, the acceleration leads to the appearance of communication barriers. We will be unable to receive signals from sufficiently remote parts of the universe. It will be as if we are living inside a black hole. The part of the universe that can affect us (or our descendants) and with whom we (they) may be in contact will be finite. To escape this claustrophobic future we would need the vacuum energy to decay. We think it must stay constant forever, but maybe it is slowly eroding imperceptibly away. Or, maybe one day it will decay suddenly into radiation and ordinary forms of matter and the universe will be left to pick up the pieces and slowly use gravity to aggregate matter and process information again. But the decay may not be so benign. We have seen that it could herald a slump into an even lower energy state for the universe, with a sudden change in the nature of physics accompanying it. It is even possible for the vacuum to decay into a new type of matter that is even more gravitationally repulsive than the lambda force. If its pressure is even more negative then something very dramatic can lie in the future. The expansion can run into a singularity of infinite density after a finite time.[13]

3.6. Varying Constants

When we assess the long-term cosmological outlook it is important that we know precisely what things cannot change, no matter how long we wait. Physicists are in the habit of designating such invariant quantities "constants of nature." They are presumed to be the same forever. Various stringent observational and experimental limits exist to support these presumptions and the standard models of particle physics and big bang cosmology are developed with this constancy assumed. However, if one or more of these constants were to be changing, even by an amount that was for all practical pur-

12. The current observations are indicating that this is not the case in our universe. It appears to be destined to keep expanding forever, locally, and if the eternal inflation scenario is true it will continue expanding globally as well.

13. This must lie at least about 30 billion years to the future. It should be noted that it is possible for us to encounter a singularity in the future without this lambda energy decay, even if the expansion appears to be going to carry on forever. There could be a gravitational shock wave travelling toward us at the speed of light that hits us without warning.

poses insignificant today, it could radically alter our picture of the universe's long-range future. Until quite recently there had been no positive evidence that any of the traditional constants of nature are anything other than constant. Theories had been developed to explore the consequences of variations in Newton's gravitation constant, G, but these were largely to provide a theoretical context that placed limits on the allowed variations by using observational data. However, a series of detailed observational studies [37, 38, 39] have produced tantalizing evidence that the fine structure was smaller, by about seven parts in a million, at red shifts between about one and three.

The theoretical investigation [7, 8] of this situation has revealed a very unusual feature of universes with varying constants, like the fine structure constant, α, or G. If the universe is flat and has zero cosmological constant then the value of α will remain constant during the early radiation era of the universe but begin to change during the dust era so long as the curvature of the universe or a vacuum energy does not come to affect the expansion. When either the curvature or a vacuum energy controls the expansion, any variation in α ceases. Thus we see that it might be essential for life that our universe not have curvature or vacuum energy that is too *small*.[14] For, the smaller these effects, the longer the value of α keeps increasing in size. Eventually, it will become too large for atoms to exist and the possibility for life as we know it will be lost. There will be a niche of cosmological history during which atom-based life can exist [9]. The existence of a very small level of spatial curvature, which produces an open universe, or a non-zero vacuum energy will turn off the increase in α and allow atoms to exist for vastly longer periods.

These considerations show that a full understanding of the constancy of the traditional constants of nature is needed to make reliable long-range forecasts about the future of the universe and the forms of matter that will be able to exist within it.

3.7. A Theory That, If True, Cannot Be Original

There is one last line of speculation that must not be forgotten. In science we are used to neglecting things that have a very low probability of occur-

14. We appreciate that life seems to require that neither the curvature nor the vacuum energy must be too *big* or they will come to dominate the expansion so early in the history of the universe that neither galaxies nor stars will be able to form [6]. Consequently, in such universes we would not expect atom-based life to emerge.

ring even though they are possible in principle. For example, it is permitted by the laws of physics that my desk rise up and float in the air. All that is required is that all the molecules "happen" to move upward at the same moment in the course of their random movements. This is so unlikely to occur, even over the 15-billion-year history of the universe, that we can forget about it for all practical purposes. However, when we have an infinite future to worry about all of these fantastically improbable physical occurrences will eventually have a significant chance of occurring. An energy field sitting at the bottom of its vacuum landscape will eventually take the fantastically unlikely step of jumping right back up to the top of the hill. An inflationary universe could begin all over again for us. Yet more improbably, our entire universe will have some minutely small probability of undergoing a quantum transition into another type of universe. Any inhabitants of universes undergoing such radical reform will not survive. Indeed, the probability of something dramatic of a quantum-transforming nature occurring to a system gets smaller as the system gets bigger. It is much more likely that objects within the universe, like rocks, black holes, or people, will undergo such a remake before it happens to the universe as a whole. This possibility is important, not so much because we can say what might happen when there is an infinite time in which it can happen, but because we can't. When there is an infinite time to wait then *anything* that can happen eventually *will* happen. Worse (or better) than that, it will happen infinitely often.

Globally, the universe may be self-reproducing but that will merely provide other expanding regions with new beginnings. Perhaps some of their inhabitants will master the techniques needed to initiate these local inflations to order and engineer their outcomes. For us there is a strange symmetry to existence. The universe may once have appeared out of the quantum vacuum, retaining a little memory of its energy. Then in the far future that vacuum energy will reassert it presence and accelerate the expansion again, this time perhaps forever. Globally, the self-reproduction may inspire new beginnings, new physics, new dimensions, but, along our world line, in our part of the universe, there will ultimately be sameness, starless and lifeless forever, it seems.

I would like to thank Mary Ann Meyers and the Templeton Foundation for their invitation to participate in this symposium, the participants for many stimulating discussions, and George Coyne and the staff of the Pontifical Academy of Sciences for their unfailing hospitality.

REFERENCES

1. Bak, P., *How Nature Works* (Oxford University Press, Oxford, and Copernicus, New York, 1996).

2. Barrow, J. D., *Impossibility* (Oxford University Press, Oxford, 1998).

3. Barrow, J. D., "Life, the Universe, and Almost Everything," *Physics World, 12,* 31–35 (1999).

4. Barrow, J. D., and Dabrowski, M., "Oscillating Universes," *Mon. Not. Roy. Astron. Soc., 275,* 850–62 (1995).

5. Barrow, J. D., and Tipler, F. J., "Eternity Is Unstable," *Nature, 276,* 453 (1978).

6. Barrow, J. D., and Tipler, F. J., *The Anthropic Cosmological Principle* (Oxford University Press, Oxford, 1986).

7. Barrow, J. D., Sandvik, H., and Magueijo, J., "The Behaviour of Varying-Alpha Cosmologies," *Phys. Rev. D., 65,* 063504 (2002).

8. Barrow, J. D., Sandvik, H., and Magueijo, J., "A Simple Varying-Alpha Cosmology," *Phys. Rev. Lett., 88,* 031302 (2002).

9. Barrow, J. D., Sandvik, H., and Magueijo, J., "Anthropic Reasons for Non-zero Flatness and L," *Phys. Rev. D., 65,* 1235XX (2002).

10. Bernal, J. D., *The World, the Flesh, and the Devil,* 2nd ed. (Indiana University Press, Bloomington, 1969, 1st ed., 1929), 63.

11. Clark, S. R. L., *How to Live Forever* (Routledge, London, 1995).

12. Darwin, C., *On the Origin of Species by Means of Natural Selection,* 2nd ed. (Murray, London, 1860), 489.

13. Dyson, F., "Life in an Open Universe," *Rev. Mod. Phys., 51,* 447 (1979).

14. Eddington, A. S., "The End of the World: From the Standpoint of Mathematical Physics," *Nature, 127,* 447–53 (1931).

15. Eldredge, N., *Macro-evolutionary Dynamics* (McGraw-Hill, New York, 1989).

16. Eliade, M., *The Myth of the Eternal Return* (Pantheon, New York, 1934).

17. Ellis, G. F. R., and Brundrit, G. B., "Life in the Infinite Universe," *Q. J. Roy. Astron. Soc., 20,* 37 (1979).

18. Frautschi, S., "Entropy in an Expanding Universe," *Science, 217,* 593 (1982).

19. Haldane, J. B. S., *The Last Judgement* (Harper Bros., New York, 1927), 38.

20. Harrison, E. R., *Cosmology* (Cambridge University Press, Cambridge, 1981), 299–300.

21. Harrison, E. R., "The Natural Selection of Universes," *Q. J. Roy. Astron. Soc., 36,* 193 (1995).

22. Hawking, S. W., "Black Hole Explosions," *Nature, 248,* 30 (1974).

23. Hut, P., and Rees, M. J., "How Stable Is Our Vacuum?" *Nature, 302,* 508 (1983).

24. Inge, W., *God and the Astronomers* (Longmans Green, London, 1934), 28.

25. Kant, I., *Universal Natural History and Theory of the Heavens,* trans. W. Hastie (Greenwood Pub., New York, 1968), 59–70.

26. Krauss, L., and Starkman, G., "Life, the Universe, and Nothing: Life and Death in an Ever-Expanding Universe," *Astrophys. J., 531,* 22–31 (2000).

27. Linde, A., "The Self-Reproducing Inflationary Universe," *Sci. Amer.* 5, November, 48–55 (1994).

28. Sandvik, H. B., Barrow, J. D., and Magueijo, J., "A Simple Varying-Alpha Cosmology," *Phys. Rev. Lett., 88,* 031302 (2002).

29. Smolin, L., "Did the Universe Evolve?" *Class. Q. Gravity, 9,* 173 (1984).

30. Starobinskii, A. A., "Future and Origin of Our Universe: Modern View," *Grav. Cosmol., 6,* 157–63 (2000).

31. Tipler, F. J., "A Brief History of the Extraterrestrial Intelligence Concept," *Q. J. Roy. Astron. Soc., 22,* 133 (1981).

32. Tipler, F. J., *The Physics of Immortality* (Doubleday, New York, 1995).

33. Tolman, R. C., "On the Problem of the Entropy of the Universe as a Whole," *Phys. Rev., 37,* 1639–1771 (1931).

34. Tolman, R. C., "On the Theoretical Requirements for a Periodic Behaviour of the Universe," *Phys. Rev., 38,* 1758 (1931).

35. Turner, M. S., and Wilczek, F., "Is Our Vacuum Metastable?" *Nature, 298,* 633 (1982).

36. von Helmholtz, H., "On the Interaction of the Natural Forces," reprinted in *Popular Scientific Lectures,* ed. M. Kline (Dover, New York, 1961).

37. Webb, J. K., Flambaum, V. V., Churchill, C. W., Barrow, J. D., and Drinkwater, M. J., "Evidence for Time Variation of the Fine Structure Constant?" *Phys. Rev. Lett., 82,* 884–87 (1999).

38. Webb, J. K., Murphy, M., Flambaum, V. V., Dzuba, V., Barrow, J. D., Churchill, C. W., Prochaska, J., and Wolfe, A., "Further Evidence for Cosmological Evolution of the Fine Structure Constant," *Phys. Rev. Lett., 87,* 091301 (2001).

39. Webb, J. K., Murphy, M., Flambaum, V. V., Dzuba, V., Barrow, J. D., Churchill, C. W., Prochaska, J., and Wolfe, A., "Possible Evidence for a Variable Fine Structure Constant from QSO Absorption Lines—I. Motivations, Analysis, and Results," *Mon. Not. Roy. Astron. Soc., 327,* 1208 (2001).

[4]

ETERNITY

Who Needs It?

Paul Davies

4.1. Introduction

The famous adage "It is better to journey interestingly than arrive" well captures an exquisite tension that pervades the human fascination with time and eternity. Death is anathema to most people, yet eternal life can seem pointless. This tension applies equally to human existence, or to the "life" of the universe as a whole. Indeed, it resurfaces in scientific cosmology, and the strong emotive appeal that attaches to certain cosmological models in various cultures.

Mercea Eliade [9] has argued that all ancient cultures are founded on the myth of the eternal return. There are good reasons from evolutionary psychology to expect cyclicity to be deeply embedded in the human psyche. Survival in the premodern era depended crucially on people being in harmony with the cycles and rhythms of nature: the diurnal, lunar, and annual cycles; menstrual cycles; and other biorhythms. It was natural that these cultures should develop cyclic cosmologies. Cyclicity pervades the creation myths of many contemporary cultures, from the Dreaming of the Australian Aborigines to the cycles within cycles of Hinduism. Notions of "the wheel of fortune" and popular expressions like "what goes around comes around" attest to the strong appeal of cyclicity even in modern Western society.

A decisive break with cyclicity came with Judaism, and its emphasis on God as Creator, who made the universe at a specific moment in the past, and then supervised its unidirectional unfolding. In Judaism, Christianity, and Islam time is linear, not circular, and God is revealed through history in a specific sequence (e.g., Creation, Fall, Incarnation, Redemption, Judgment). In other words, there is a cosmic story to tell, a story that has a beginning, a middle, and an end.

4.2. The Scientific Cosmological Options

Science adopted the Judeo-Christian world view when it developed into its modern form in the seventeenth century. The flux of linear time became the foundation on which Newton's mechanics was built (his theory of fluxions being what is now known as the differential calculus). Curiously, Newton's cosmology was static and eternal, but sustained by occasional divine maintenance.

Science clung to static, eternal cosmologies until the twentieth century, and the discovery that the universe is expanding. Since then, a variety of cosmological models has been advanced consistent with this basic fact. Their implications for the ultimate fate of the universe differ radically. Here is a broad-brush summary:

1. The universe begins at a finite time in the past and expands forever, degenerating according to the second law of thermodynamics, approaching a state of near-featureless equilibrium (the "heat death" of nineteenth-century cosmology) at very late times.

2. The universe begins at a finite time in the past and terminates at a finite time in the future (e.g., by collapsing to a final "big crunch" singularity, or being overwhelmed by a cosmic catastrophe such as quantum vacuum decay). Entropy rises throughout, but the universe (at least in its present form) is extinguished before a final equilibrium state is reached.

3. Scenario 2, but with the "arrow of time" reversing at some stage, perhaps near the point of maximum expansion in a recontracting model, returning the universe toward a final, relatively ordered state similar or even identical to the initial big bang state.

4. A cyclic universe, in which expansion and contraction is followed by a "big bounce" into another cycle of expansion and contraction. The universe continues to pulsate in this manner indefinitely. Suboptions include the following:

a. Information about the pre-bounce physical state survives the bounce, so the universe continues to evolve according to the laws of thermodynamics. The cycles grow larger with each bounce (as pointed out by Tolman [25] as long ago as the 1930s), perhaps approaching something like our own universe after very many cycles [16].

b. The bounce represents such an extreme physical state that information is effectively reprocessed [27], perhaps at random. Possibly the physical laws themselves are reprocessed. In that case the evolution of one cycle need not be correlated with that of previous or successive cycles. It is moot whether terms like *prior* or *successive* are meaningful in this case, as temporal continuity is irrelevant here. One might as well say that the cycles are parallel rather than sequential.

c. The cyclic model is combined with time reversal in some way. For example, the arrow of time might flip in successive cycles, switching (more plausibly) at the bounce rather than near the midpoint of expansion.[1] The universe could then return after two cycles to its initial state.

5. A steady state universe, which has no beginning or end, but continues to expand indefinitely. New matter is continually created to fill in the gaps left by the receding galaxies. This injection of low-entropy matter effectively "refuels" the universe forever, combining eternity with everlasting novelty. Variations on this theme include Hoyle's C field cosmology [13], which is asymptotically steady state, or models in which evolutionary episodes are embedded within an overall steady state arrangement.

6. Multiverse cosmologies.[2] The general idea in these models is that what

1. I published a theory like this myself, see [4].

2. For a review of this concept, see Rees [20].

we have hitherto taken to be "the universe" is but a specific "bubble" in a much larger system in which other "bubbles" of space, perhaps with very different physical conditions, exist at very great distance from each other. While each bubble might go through a life cycle of birth, evolution, and perhaps death, the assemblage as a whole might be in something like a steady state. Thus the multiverse is eternal, although individual components are not. Models of this sort include Linde's chaotic cosmology [18] and Smolin's "cosmic Darwinism" [23]. In the latter model, one "universe" can spawn others through a sort of "budding" process, so the situation is reminiscent of biological organisms—as the parent universe ages, so new universes can be born, ad infinitum.

Although astronomical observation offers the possibility of distinguishing among these various models, cosmology has (at least until recently) been notorious for the disputatious nature of the observational results. In this atmosphere, it is perhaps understandable that proponents and opponents of various models should bring rather more emotional or theological baggage to bear than is usual in science. For example, Fred Hoyle made it clear that for him the steady state theory had a strong appeal because it eliminated the big bang [14], which in the 1950s had become identified in some circles (e.g., Pope Pius XII) with the Judeo-Christian notion of creation. Hoyle (at that stage) found the idea of special divine creation unpalatable, and welcomed the possibility of a universe that has no definite beginning or end as a means of divesting scientific cosmology of its theological roots. In the same vein, many scientists (e.g., Robert Jastrow [15]) took exactly the opposite view, proclaiming their belief in the big bang as science's version of the biblical creation myth. There is no doubt that, even today, theists feel more comfortable with the big bang theory, even though the doctrine of creation *ex nihilo* was never intended to invest huge significance in the originating cosmic event as such.

4.3. Cyclical Universes

From my own experience lecturing to the public on model cosmologies, I have been struck by the number of people in the audience who find the cyclic universe congenial. This model most accords with ancient cosmological myths, and with the Hindu system. I have been unable to determine whether the appeal of this model is due to the recent popularity of Eastern philosophy, or whether it represents a deep-seated uneasiness with linear time that predates Western scientific thought. At the same time, other people find the notion of endless cycles of repetition unpalatable.

Here one must distinguish between strictly cyclic cosmologies, in which time is closed into a circle, and those in which only the gross features of the universe are repeated. There is a popular misconception that strict cyclicity means that we are doomed to repeat the same events endlessly, and pointlessly, a concept (roughly) dramatized in the Hollywood movie *Groundhog Day.* However, this conflates two quite separate notions of the arrow of time. On the one hand, there is the asymmetry of the physical world with respect to time: the fact that later states are objectively distinct from earlier states (e.g., entropy is greater, people are older). On the other hand, there is the psychological impression that time is flowing or passing, a concept that has no existence in known physics, and is dismissed by many philosophers as an illusion, or linguistic confusion (see, for example, "The Myth of Passage" by Donald Williams [28]). In a cyclic universe, the (former) arrow of time must reverse at some stage so as to return the state of the universe to its earlier form.[3] This will entail various physical processes "running backwards," for example, entropy falling, people getting younger. However, it is utterly wrong to characterize this process as "*time* running backward" (since time doesn't "run" at all). Most physicists identify the psychological impression that time is flowing as in some way connected with brain processes and memory. In a universe in which the arrow of time is reversed, brain processes would reverse too, and an observer would never see crazy reversed sequences like broken eggs assembling themselves and heat flowing spontaneously from cold to hot bodies. So the notion that things "keep going around" the same way forever and ever in a strictly cyclic universe is simply a misconception about the flow of time versus the asymmetry of the world in time.[4] Once this point is appreciated, the status of the strictly cyclic models is seen to be very similar to that of the finite time bang-to-crunch model. In both cases there is a finite duration in which things can happen, projects get accomplished, and so on.

3. The cyclic universe theory itself seems to keep coming around. Versions of it have been published by, for example, Gold [11], Hawking [12], Price [20], Gell-Mann and Hartle [10], and Shulman [22].

4. See, for example, my book [5], and for a more popular account, [6].

4.4. *Finite Duration Universes*

How appealing is a universe of finite duration? Some people react with horror to the notion that the entire universe might be obliterated, even though the annihilation may take place in the very far future, long after their personal deaths. It can often be a comfort when confronting bodily death that "life goes on" afterward, and that one's sacrifices and accomplishments might endure, or that one's aspirations might be achieved vicariously through one's children. There can be a feeling that one's life is ultimately pointless if the universe as a whole is doomed. Perhaps the most famous expression of this sentiment that "life must be pointless and ultimately meaningless if the universe will die" is Bertrand Russell's assessment [21]:

> All the labours of the ages, all the devotion, all the inspiration, all the noonday brightness of human genius are destined to extinction in the vast death of the solar system, and . . . the whole temple of man's achievement must inevitably be buried beneath the debris of a universe in ruins.

Russell wrote these words in his book *Why I Am Not a Christian*. His reasoning seems to be that if the universe will not endure forever, then there cannot be a God. Peter Atkins, another militant atheist, uses the same reasoning in support of his attack on religion [1]:

> We have looked through the window on to the world provided by the Second Law, and have seen the naked purposelessness of nature. The deep structure of change is decay; the spring of change in all its forms is the corruption of the quality of energy as it spreads chaotically, irreversibly and purposelessly in time. All change, and time's arrow, point in the direction of corruption. The experience of time is the gearing of the electrochemical processes in our brains to this purposeless drift into chaos as we sink into equilibrium and the grave.

Many theists concur with this position. It is no coincidence that some Christian fundamentalists challenge the second law of thermodynamics, with its prediction of cosmic degeneration and decay. These fundamentalists reason that a universe that can undergo renewal and be sustained for eternity is more God-friendly than one that decays. However, some notable atheists (e.g., Friedrich Engels[5]) have expressed similar distaste for the cosmic heat death, and sought ways in which the laws of nature might operate

5. See Barrow and Tipler [3].

to circumvent it. These negative sentiments about a doomed universe are more normally applied to the ever-expanding cosmological models, which I shall consider soon.

In the case of a universe that collapses to a big crunch in, say, 100 billion years, there is some subtlety about the physical meaning to be attached to "the end." The definition of time in simple cosmological models refers to an averaged-out state of matter. Using this so-called cosmic time, the big crunch universe does indeed endure for a finite time. However, as the final singularity is approached, the temperature rises without limit, and there is a possibility that an infinite amount of activity may occur in the final stages. This is also true if the universe undergoes departures from strict uniformity (as is realistic). Depending on the technical details, the distribution of matter and the shape of space can undergo infinitely complex changes of escalating rapidity. Tipler [24] has pointed out that it is possible for an infinite amount of information to be processed in the finite cosmic time available. In this sense, the recontracting model still offers a type of limitless potential.

4.5. *Eternal Models*

Turning now to the standard ever-expanding models, these have infinite longevity but are subject to the heat death. The specific details of the final state of such a universe depend on assumptions made about the nature of matter. The general trend in all cases is for the universe to cool and approach thermodynamic equilibrium as it continues to expand. However, the classical heat death predicted in the nineteenth century is modified due to the fact that the universe is expanding, which has the effect of slowing the approach to equilibrium. The specific details (proton decay, black hole evaporation, etc.) have been extensively discussed elsewhere,[6] and I shall not repeat them here, except to say that in a standard picture the very far future of the universe would be characterized by a dilute and ever-diminishing soup of extremely low-energy photons, neutrinos, and gravitons moving virtually freely through a slowly expanding space. Normal matter, if it exists at all, might consist of the occasional electron and positron drifting slowly through this invisible soup.

Following a trailblazing paper by Freeman Dyson [8], a number of analy-

6. For a review see, for example, my book *The Last Three Minutes* [7] and the work cited therein.

ses have been made of the far future of an ever-expanding universe, and the challenges faced by sentient beings attempting to eke out an existence amid dwindling thermodynamic resources. Dyson discovered that, significantly, a sufficiently resourceful community that husbanded free energy might enjoy infinite longevity, albeit by "living it down" and spending ever-longer periods hibernating. Strictly, Dyson showed that information processing might possess no limits. Whether "life as we know it" demands more than merely manipulating informational bits is another matter.

Is Dyson's scenario a prospect that can inspire hope or comfort? It should be pointed out that the timescale for known sources of energy to run out is immense. For example, the time required for a solar mass black hole to evaporate by the Hawking process is a staggering 10^{66} years! But we are used to big numbers in this game. Psychologically, however, there seems to be a crucial difference between the prospect that our descendants will survive for a huge length of time, and life going on literally forever. There is no doubt that the image of an unobserved and almost empty universe growing steadily more cold and dark for all eternity is profoundly depressing (for this writer, at least). Better surely that this moribund and redundant universe were decently and properly "killed off." Recent evidence for a non-zero cosmological constant, which acts as a sort of universal repulsive force, serves only to makes matters worse, for if this force exists then the universe will actually accelerate in its expansion in the future, thereby speeding the approach to a state of dark emptiness.

This sorry fate for our beloved cosmos might not seem so bad if there are other universes springing into existence to take its place. This, I suspect, is the reason why the various multiverse models are currently proving so popular, because they capture something of the perpetual novelty factor of the old (and now discredited) steady state universe. Consider, for example, Smolin's theory [23] that the collapse of stars to black holes might create entire new universes, as space balloons out on the far side of the hole, eventually detaching itself from the "mother" universe to constitute a separate entity enjoying an independent existence. This whole process of universes begetting other universes could continue ad infinitum. Smolin's theory projects into cosmology the same emotive appeal of "life going on" among our descendants, even as we ourselves die. Our mother universe may be destined for a fate of useless eternal expansion into dark emptiness, but her children, her children's children . . . would embark on better things. Note that in principle our descendants could decamp into a newborn universe via a wormhole (the ultimate in emigration) and enjoy another lease of life, or even create their own baby universe in the lab for future habitation. So even if the universe we now see is doomed, conscious beings might survive forever. In other words, life and mind might transcend actual universes! (If so, there is the obvious problem of deciding whether such cosmic engineers had long ago entered our universe from another, and if not, why not.)

Although the prospect of the multiverse enjoying infinite longevity is more cheering, there remains the theological and philosophical problem of whether eternal life (of whatever sort) serves any actual purpose. A friend of mine once remarked that from what he had heard of paradise (by which I took him to mean joyful inactivity for eternity), he wasn't all that interested. For him, it is the challenges and goals of (mortal) life that make it worthwhile. Weinberg [26] has famously written that "the more the universe seems comprehensible, the more it also seems pointless." But if the universe has a point, and its goal is achieved, its continued existence would seem to be unnecessary and gratuitous. Can a universe (or multiverse) that endures forever *have* a point?

The answer to this conundrum hinges on whether perpetual novelty is a possibility. If the universe (or multiverse) is capable of achieving an infinite number of states, then there will be an inexhaustible supply of possibilities open to it. In a similar vein, if mental states can likewise never be exhausted, a set of conscious beings could think an infinite number of thoughts and have an infinite number of experiences. There is then the possibility of an infinitely interesting journey and eternally deferred arrival. From the viewpoint of human sentiment, this would appear to be the best bet. Whether the real universe (or multiverse) lives up to our hopes is a matter for science to decide.

In favor of the perpetual novelty ideal is the famous theorem of Kurt Gödel, which proves that the realm of mathematics contains an inexhaustible supply of novelty—theorems that are true but cannot be proved to be true from what has come before. If we suppose that, however much our understanding of the physical world may change, the fundamental processes of nature are mathematical in form, then the universe can share this property of inexhaustibility.[7]

7. Freeman Dyson has discussed Gödel's incompleteness theorem as a message of hope for a universe of perpetual novelty.

There is still an uncomfortable conundrum lurking here. There are two ways in which a universe might experience perpetual novelty—progressive and steady state. In the progressive case, there is an arrow of time that points in the direction of greater advancement, so that in some sense the universe gets better and better with time (or more and more complex, or the experiences of sentient beings become progressively richer, more pleasant, etc.). This "onward and upward" scenario means accepting that we human beings are situated in time, curiously and perhaps bafflingly, at the (manifestly imperfect) near beginning of an infinitely improving cosmic trajectory. If we accept the simplest case of a single universe model, with the big bang marking the beginning of time, then humans occupy a temporal niche close to the start of the great journey, whereas an infinite number of other sentient beings enjoy a vastly more developed state of affairs in the unlimited future.

The alternative—that there is some sort of steady state (e.g., a multiverse of many big bangs with no beginning or end to the ensemble as a whole)—is that the multiverse is going "nowhere fast," scanning through all possible states at random with no directionality, destination, or goal. Individual universes (or communities, or lives) may have a strong sense of purpose, but there is no ultimate purpose to the assemblage as a whole. A randomly selected universe (say ours) finds all manner of sentient beings having all manner of experiences, but no systematic trend exists. This runs counter to both the traditional Judeo-Christian linear time model of history as a process *toward* something (i.e., a divine plan enacted through history), and the Hindu and pagan concepts of cyclicity, but at least it has the virtue of "sublime pointlessness" rather than nihilistic absurdity.

4.6. The Fundamental Tension

I believe there is no plausible cosmological model on offer at this time that simultaneously satisfies two widespread but seemingly conflicting desires: that the universe has an ultimate destiny or purpose, and that the universe will exist interestingly for eternity. This fundamental tension between progress and eternity pervades the history of religion, and is solved in classical Christian theology, for example, by placing God (and perhaps paradise) outside of time altogether.[8] That is, eternity is to be understood not as an infinite duration of time, but as "not in time at all."

8. See, for example, Pike [19].

The theme of this essay is that human beings will find certain cosmological models congenial or repugnant according to their culture or personality. Perhaps future generations of humans, or perhaps already existing alien beings, might feel completely differently about cosmology. There may be communities (even on Earth) that rejoice in the prospect of the cosmic heat death and eternal empty darkness, or thrill to the prospect of big crunch obliteration (which has a sort of clean finality at least). It is also possible we shall discover that time is not a fundamental physical quantity at all, but an emergent property of some deeper structure [2]. Since the foregoing "tension" between time and eternity arises from our species' particular temporal anxieties, it may be that the discovery that time is only a secondary phenomenon will induce a wholly radical change of world view, along with our hopes and fears.

REFERENCES

1. Atkins, P., "Time and Dispersal: The Second Law," in *The Nature of Time*, eds. R. Flood and M. Lockwood (Basil Blackwell, Oxford, 1986), 98.

2. Barbour, J., *The End of Time* (Phoenix Press, London, 1999).

3. Barrow, J. D., and Tipler, F. J., *The Anthropic Cosmological Principle* (Oxford University Press, Oxford, 1986).

4. Davies, P. C. W., "Closed Time as an Explanation of the Black Body Background Radiation," *Nature, 240,* 3 (1972).

5. Davies, P. C. W., *The Physics of Time Asymmetry* (University of California Press, Berkeley, 1974).

6. Davies, P. C. W., *About Time* (Penguin, London, 1995; Simon and Schuster, New York, 1995).

7. Davies, P. C. W., *The Last Three Minutes* (Basic Books, New York, 1994; Weidenfeld and Nicolson, London, 1999).

8. Dyson, F., "Time without End: Physics and Biology in an Open Universe," *Rev. Mod. Phys., 51,* 447 (1979).

9. Eliade, M., *The Myth of the Eternal Return,* trans. W. R. Trask (Pantheon Books, New York, 1954).

10. Gell-Mann, M., and Hartle, J., "Time Symmetry and Asymmetry in Quantum Mechanics and Quantum Cosmology," in *Physical Origins of Time Asymmetry,* eds. J. J. Halliwell et al. (Cambridge University Press, Cambridge, 1994), 311.

11. Gold, T., "The Arrow of Time," *Amer. J. Phys., 30,* 403 (1962).

12. Hawking, S. W., "The Arrow of Time in Cosmology," *Phys. Rev., D 32,* 2489 (1985).

13. Hoyle, F., "A New Model for the Expanding Universe," *Mon. Not. Roy. Astron. Soc., 108,* 372 (1948); *109,* 365 (1949).

14. Hoyle, F., *Frontiers of Astronomy* (Harper, New York, 1955).

15. Jastrow, R., *God and the Astronomers* (Norton, New York, 1992).
16. Landsberg, P. T., and Park, D., "Entropy in an Oscillating Universe," *Proc. Roy. Soc. Lond., A 346*, 485 (1975).
17. Linde, A., *Particle Physics and Inflationary Cosmology* (Gordon and Breach, New York, 1991).
18. Pike, N., *God and Timelessness* (Routledge and Kegan Paul, London, 1970).
19. Price, H., *Time's Arrow and Archimedes' Point* (Sydney University Press, Sydney, 1993).
20. Rees, M. J., *Before the Beginning* (Helix Books, Cambridge, Mass., 1998).
21. Russell, B., *Why I Am Not a Christian* (Allen and Unwin, New York, 1957), 107.
22. Shulman, L., "Opposite Thermodynamic Arrows of Time," *Phys. Rev. Letts., 83*, 5414, (1999).
23. Smolin, L., *The Life of the Cosmos* (Oxford University Press, Oxford, 1997).
24. Tipler, F. J., *The Physics of Immortality* (Doubleday, New York, 1994).
25. Tolman, R., *Relativity Thermodynamics and Cosmology* (Clarendon Press, Oxford, 1934).
26. Weinberg, S., *The First Three Minutes* (Harper and Row, New York, 1988).
27. Wheeler, J. A., "Law without Law," in *Quantum Theory and Measurement* (eds. Wheeler, J. A. & Zurek, W. H., Princeton University Press, 1983).
28. Williams, D., "The Myth of Passage," *Journal of Philosophy, 48*, 27 (1951).

[5]

TIME OF THE UNIVERSE

Michael Heller

5.1. Introduction

Time is certainly one of the most mysterious aspects of the Universe.[1] On the one hand, it seems to be quasi-nonexistent: we can observe and measure changes of things in time, but we can neither observe nor measure the flux of time itself. On the other hand, time appears to be more real than anything else: the arrow of time inexorably points to the future, and we are unable, even if we wished so, either to stop it, or to act against it. This "time constraint" is so strong that we exhibit a natural tendency to regard the existence in time as an ontological necessity without which the being of the Universe would be hard to imagine. For many people it is a surprise to learn that some results of modern science strongly suggest that the tyranny of old Chronos is not as absolute as they believed. In this essay, I present and try to

1. If I speak of the particular Universe we inhabit, I capitalize it; if I speak of possible universes (see below), I use the lowercase "u."

substantiate a view of time based on recent cosmological investigations. In contemporary cosmology, time is no longer regarded as something a priori with respect to the Universe or its laws, but rather as a part of the "big game" played by physical forces in the arena that we call the Universe.

Another common belief concerning time is that it should be global, that is, that it should cover the entire history of the Universe: from the big bang to the possible end. However, within the conceptual framework of general relativity there are serious reasons to distrust this belief, and to ask the following question: Is the existence of global time a "generic property" for cosmological models, or something "rather exceptional"? However, to make this question meaningful we must first determine a "population" of universes (cosmological models) to which this question is to be addressed. Within the paradigm of relativistic cosmology, such a "population" is determined by the Einstein field equations: each solution of this system of equations can be interpreted as a certain universe,[2] and the set of all such solutions is a highly structured mathematical space. This space is sometimes called an *ensemble of universes* (for details see [7], pp. 70–76). The above question concerning the existence of global time, when asked in reference to the ensemble of universes, has a negative answer: the existence of global time is not a generic property in the ensemble; only a zero-measure subset of universes admits global time. Roughly speaking, this means the following: if we decided to choose a world from among all possible worlds—members of the ensemble of universes at random—there would be practically zero probability to pick up a universe admitting the existence of global time.

Any discussion concerning the far future of the Universe presupposes that the far future idea as referred to the entire Universe is meaningful, that is, that there exists a global time with which we are able to measure (or at least to estimate) the distance from now to the future, and that this distance is sufficiently great. The sense in which such a future might exist or not exist is not at all obvious. The aim of the present essay is to discuss this problem.

In section 2, I briefly present Hawking's theorem, which formulates the necessary and sufficient condition for the existence of global time in a cos-

2. Strictly speaking, only solutions with "cosmological initial or boundary conditions" deserve to be called universes; other solutions can represent stars or other objects. However, for the sake of simplicity we shall not enter into such details.

mological model. There are strong reasons to believe that global time is only a macroscopic phenomenon—one should ask about its roots in the fundamental level. In section 3, I prepare the ground for dealing with this problem by introducing the reader to some elementary material concerning C*-algebras and non-commutative geometry. In section 4, I briefly present a model unifying general relativity and quantum mechanics based on a certain non-commutative C*-algebra. In this model, on the fundamental level there is no time (in the usual sense), but the model shows how time emerges when the Universe undergoes a "phase transition" from the non-commutative regime to the commutative one. In section 5, I consider the problem of the beginning and end of the Universe. By using methods of non-commutative geometry, it has been possible to prove several theorems throwing new light on the nature of classical singularities, such as the ones representing the beginning and the end in the standard cosmological model. It has been demonstrated that on the fundamental, non-commutative level there is no difference between singular and non-singular states of the Universe. It is when passing through the Planck threshold that some states degenerate into singularities. The very concepts of the beginning and the end become meaningful only for the macroscopic observer living in a global time that extends from the far past to the far future.

5.2. *Global Time*

We certainly live in a Universe in which there exists a global time (or at least a "sufficiently global time") covering the world's history from the big bang to the present epoch. Moreover, there are strong reasons to believe that we owe our own existence to the fact that this global time exists in our Universe: we are made out of carbon chemistry, and to produce carbon a long history of a few star generations is required. We are then fully entitled to ask: What are the sufficient and necessary conditions for the existence of global time in a world model?

It was Leibniz who had a correct intuition as far as this question was concerned. In his view, time is nothing but the successive order of events (in contrast to the views of Newton and Clarke, who claimed that time was prior with respect to events), and the order of events is determined by causal relations (for details see [18], pp. 42–50). In the theory of general relativity there is a beautiful theorem that puts this ideology into strict mathematical form. The theorem is due to Hawking [5], and it asserts that *space-time admits a global time function if and only if it is stably causal.* Let me explain the technical terms appearing in this theorem.

First of all, how to mathematically describe global time? Imagine a clock (for instance, a vibrating particle) present in the Universe for its earliest moments. It traces a curve in space-time. Indications of this clock (its subsequent vibrations) can be represented as a monotonically increasing function on this curve, which represents motion at less than the speed of light, because the clock has mass. This is a time function for this clock. Now consider a family of such clocks filling all of space-time, and assume there exists a single space-time function that is a time function for all these clocks. Technically, such a function is called a *global time function;* we can regard it as a scientific synonym of global time. The theorem asserts that such a global time exists in a given space-time if and only if this space-time is stably causal. What does this mean? In general relativity, Leibniz's causal relations ordering events have the form of the so-called light cones. The existence of a limiting velocity of physical signals, the velocity of light, implies that not all events in space-time can be causally related with each other. The structure of light cones determines which events can causally influence other events, and which cannot. Curves along which causal influences can be propagated are called *causal curves.* Space-time is said to be *stably causal* if it contains no closed causal curves (technically, the space-time is *strongly causal*), and a slight perturbation of the gravitational field (which is mathematically represented by the so-called *Lorentz metric*) cannot produce closed causal curves. The appearance of such curves would lead to temporal loops, and this would ruin the idea of global time.[3]

In general relativity, the gravitational field is strictly connected with the metric properties of space-time, that is, with those aspects of space and time that can be measured. Were space-time not stably causal, any slight perturbation of the gravitational field could result in arbitrarily great differences of space and time measurement results. In other words, measurement results within any "box of errors" could differ by any value. And since measurement

3. We should notice that the existence of global time, understood as a global time function, does not imply that there should be a unique "present" in the Universe, that is, that all clocks in the Universe should indicate the same hour. This would require stronger causality conditions [6].

errors are unavoidable, in such a situation no measurement would give reliable results. The very possibility of doing physics, in the way we are so successfully doing it, would be in jeopardy. Here we can observe the strange interaction of good physics with beautiful mathematics. The elegant Hawking theorem relates them to each other: time, causality, and the very possibility of physics.

5.3. Wonderful Algebras

The above "story of global time" is based entirely on classical physics, that is, without considering any quantum effects. We know, however, that sufficiently close to the initial singularity (which is the mathematical counterpart of the big bang), and sufficiently deep into the strata of the Universe (below the so-called *Planck scale*), quantum effects dominate the scene, and in such a case the picture that emerges is drastically different from what we are accustomed to in classical physics. Because of the Heisenberg uncertainly principles, space-time becomes fuzzy, and the history of a particle cannot be represented as a well-defined curve in space-time. Below the Planck scale the world's stuff can be thought of as a collection of quantum fluctuations, a set of states that are continuously created and annihilated.

These ideas come from quantum mechanics, but we are pretty sure that on the fundamental level (below Planck's scale) quantum physics should be unified with the theory of gravity. The best classical candidate for a theory of gravity is Einstein's general theory of relativity. What we need is a world model in which space-time of general relativity emerges out of quantum fluctuations when the evolution passes through the Planck threshold. Today we do not have such a commonly accepted model, although many attempts are being undertaken to create it. In the following, I shall present a beautiful mathematical structure that seems to have the required unification properties. Even if this structure will not be used in the final unification, it discloses fascinating possible interactions between such important physical concepts as space, time, dynamics, causality, and probability—the concepts that will certainly be involved in the unification process. This mathematical structure is called a C*-algebra.[4]

4. For the purposes of the present essay the definition of C*-algebra is not indispensable; it can be found in any textbook of functional analysis or advanced quantum mechanics.

C*-algebras are very well known in quantum mechanics. The standard mathematical formulation of this physical theory is in terms of states of a given quantum system, which are represented as vectors in a space, called *Hilbert space.* There exists, however, another formulation of quantum mechanics, namely the one in terms of C*-algebras. This formulation is even more satisfactory than the standard one. Why? Because the measurable properties of quantum systems (called *observables*) are elements of this algebra, and measurable properties in physics are always more fundamental than purely theoretical ones.

It came as a nice surprise that C*-algebras can also be used to formulate Einstein's general theory of relativity. Usually, one formulates general relativity in terms of space-time geometry, but Robert Geroch [4] demonstrated that one can *equivalently* formulate it in terms of the family of all smooth functions on space-time, and this family is trivially a C*-algebra. In other words, one can forget about space-time itself and deal, instead, exclusively with the C*-algebra of smooth functions.[5]

This algebra of smooth functions has a simple property: the multiplication of two functions does not depend on the order of factors, that is, if f, g are smooth functions on space-time, then $f \cdot g = g \cdot f$; we say that multiplication is *commutative,* and algebras that have this property are said to be *commutative algebras.* It turns out that many important features of geometric spaces depend on this simple property. This becomes evident if we get rid of it. Algebras deprived of this property are called *non-commutative algebras.* It was Alain Connes's idea [1] to treat these algebras in the same way as algebras of smooth functions, and assume that they also define some spaces. The idea worked, and these new spaces are now suitably called *non-commutative spaces.* The geometric theory that has resulted from Connes's seminal work is termed *non-commutative geometry.* A special role is played in this new geometric theory by those spaces that are defined in terms of non-commutative C*-algebras.

Non-commutative spaces are powerful generalizations of ordinary (commutative) spaces. Many spaces that so far were regarded as untractable or pathological ones now surrender elegantly to the machinery of non-commutative geometry. A striking property of non-commutative spaces is their

5. The idea of Geroch was further developed in [8, 9, 16, 17].

global character. In the conceptual framework of non-commutative geometry all local concepts, such as the concepts of point and its neighborhood, are in principle meaningless.

Let us collect our results. C*-algebras can be used:

1. to formulate quantum mechanics;
2. to formulate general relativity;
3. to define a non-commutative space.

In the light of these results, it seems natural to look for a C*-algebra that would do all these three things together. In such a way, quantum mechanics and general relativity would be suitably generalized and unified within the theoretical framework of a non-commutative space. Nonlocality would be the prize of unification. In the following, I shall assume that such a C*-algebra has been found (we shall denote it by A), and presuppose that it correctly models physics below the Planck threshold. It is remarkable that even without knowing the exact form of the algebra A one can deduce from its very existence so many important facts concerning the hypothetical fundamental level.[6] It is worthwhile to notice that recently some deep connections have been found between non-commutative geometry and such popular candidates for the final unification as superstring theory and M-theory [3, 19].

5.4. *Emergence of Time*

In non-commutative geometry there is no concept of time as a succession of instants (since instants are local concepts), but the concept of state of a physical system remains valid. This is so because the state concept is nonlocal: it is the entire system that can be in this or in another state. With every C*-algebra there can be associated another algebra, called a *von Neumann algebra.* Roughly speaking, a von Neumann algebra is a C*-algebra together with a distinguished state.[7] This state fulfills two functions.

First, it allows us to define a parameter t, which *imitates time.* It should be emphasized that it is not ordinary time; it only imitates time (technically, t is

6. In fact, a candidate for such an algebra has been suggested in [11, 12, 14, 15]. Since, however, the model proposed in these works is certainly not a final one, I do not want to involve the present considerations in its details.

7. For the precise definition see, for example, [20].

a parameter of a one-parameter group). It is like time since by using the parameter t we can write down the dynamical equation that describes the behavior of the system, but it is not time in the usual sense since the parameter t depends on the state: if the system goes to another state, the counting of the parameter t changes, and with it the entire dynamical regime changes as well.

Second, the state distinguished by the von Neumann algebra can be regarded as a *generalized probability* (generalized with respect to probability we meet in the standard probability calculus; strictly speaking, it is a *generalized probability measure*). By definition, a state is a functional on the von Neumann algebra that is positive and normed to unity, just as is each probability measure. In the context of non-commutative geometry one cannot, in general, speak about the probability of individual events, and the lack of the usual concept of time prevents one from connecting with probability a feeling of an "uncertain expectation." If one likes imaginary pictures, one could possibly think of this generalized probability as a "field of global potentialities," which, however, has a certain degree of reality.

As we can see, von Neumann algebras are both "dynamical objects" and "probabilistic objects." In the non-commutative regime every dynamics is probabilistic, and every probability has a dynamical aspect. It is only after passing through the Planck threshold that dynamics and probability split and become independent concepts. Quantum mechanics can be regarded as an intermediary stage at which a certain connection between dynamics and probability is still preserved. As it is well known, dynamical equation of quantum mechanics (the Schrödinger equation) describes the evolution of probabilities: at every time instant various measurement results (of a given measurable quantity) are possible, each with a strictly determined probability. However, at the instant of measurement this "field of possibilities" collapses to the single value, the one that has been measured. This effect, widely discussed in quantum mechanics, is known under the name of the *collapse of the wave function* (or *reduction of the state vector*). So far there have been serious difficulties in providing a satisfactory explanation of this effect; in the framework of the non-commutative model it can be done naturally [15].

But what about time? How does it emerge out of non-commutative era? Mathematics can handle this problem beautifully. If we glue together with the help of a certain equivalence relation some elements of the original algebra A, then the picture becomes more coarse; it looks as if some averaging processes were at work and, as the result of them, various "*t* parameters" merge together and become state independent. The usual time is born [11].

5.5. Story of the Beginning and End

The beginning and end of the Universe in classical cosmology (i.e., the one that does not take into consideration quantum gravity effects) are known under the technical names of the *initial* and *final singularities.* Several important theorems have been proved stating that such singularities are unavoidable under very general conditions, which are expected to be satisfied in any nonquantum universe [6]. But what if we do take into account quantum gravity effects? Although the feeling that the future theory of quantum gravity (or some other final theory) will remove singularities is rather common, there are some working models that testify to the opposite. Anyway, two answers—"yes" and "no"—are open possibilities.

How to mathematically describe singularities in a strict way is a difficult problem. The fact that some physical magnitudes (such as matter density and space-time curvature) go to infinity as we approach the singularity compels us to regard singularities not as situated in space-time but rather as kinds of space-time boundaries, and if we try to extend standard geometric methods to these "singular boundaries," the methods break down. This is an unmistakable sign that the standard methods should be generalized. If so, why not try new methods of non-commutative geometry? It has turned out they do indeed work.

Surprisingly enough, all kinds of singularities (even the strongest ones) appearing in general relativity surrender to the analysis in terms of non-commutative geometry [10, 13]. Moreover, this analysis was able to disclose the way the singularities originate.[8] As we already know, on the level of non-commutative geometry, the concept of localization is in principle meaningless, but it can, to some extent, be replaced by the concept of state. It has been shown that in the non-commutative regime there is no difference between singular and nonsingular states: all states are on equal footing. However, if we change from the non-commutative description of the Universe to its usual commutative description, ordinary space-time, with its

8. These results are independent of the non-commutative unification model briefly presented in the preceding section.

well-localized events, emerges and some states degenerate into singularities.

If we connect these strict mathematical results with the proposal (sketched in the preceding section) to model the fundamental level of physics with the help of non-commutative geometry, a new interpretative possibility is open. As we have seen, two mutually exclusive "yes" and "no" answers were given to the question of whether the future quantum gravity theory would be able to remove singularities from our cosmological scenario; now the third possibility enters the scene: the fundamental level of physics (below the Planck scale) is atemporal and aspatial, and the question of the existence of the initial or final singularities, as referred to this era, is meaningless. From the non-commutative perspective, everything is "regular," albeit drastically different from the post-Planck epoch. It is only when the Universe undergoes the "phase transition" from non-commutative geometry to commutative geometry that space-time appears together with its singular boundaries. Consequently, from the perspective of the macroscopic observer, who is always situated in space-time, it looks as if the Universe had a beginning (initial singularity) in its finite past and would possibly have an end (final singularity) in its final future. The notions themselves of the beginning and end become meaningful only when the macroscopic observer has at his disposal a global time that extends (or flows) from the past to the future.

Let us further explore our perspective as macroscopic observers. We look for the Planck era "in two directions": backward in time till we reach the Planck density $\rho_{Pl} = 10^{93}$ g.cm^{-3}, and deeper and deeper in space till we reach the Planck distance $L_{\rm Pl} = 10^{-33}$ cm. From our macroscopic point of view, these two directions are distinct, but on the fundamental level there is no time and no space; consequently, these two directions are the same. In this sense, the beginning and the end of the Universe are always and everywhere.

5.6. Results and Speculations

In the preceding sections, I have presented some new developments concerning the time problem as it is involved in recent cosmological and physical investigations. I did so in the order that seemed to me most suitable for the reader to follow the logical intricacies of these modern physical theories and models. Let us now try to compose, from the pieces of information obtained in this way, a more coherent cosmic scenario.

Let us begin with the fundamental level. There is no space and no time. Neither the beginning nor the end has any meaning. The Universe simply is. Not only all physical forces are fully unified, but also the whole is maximally self-contained on the conceptual level. Such seemingly different concepts as dynamics, causality, and probability (and some others of which I had no chance to speak) are but various aspects of the same mathematical structure. Here the question arises: Why was this "perfect state" unstable and the Universe had to pass through the Planck threshold? I think this question is badly posed. We are inclined to think of such a "primordial epoch" as situated in the beginning. However, as we have seen, it also can be thought of as existing even now at distances below the Plank scale. If so, the fundamental level still exists,[9] and it did not pass through the Planck threshold. The existence of the threshold and what happens on our side of it is just a part of the non-commutative game.[10]

We know the mechanism, operating in the Planck epoch, that leads to the appearance of a (state independent) time in the post-Planck epoch (it was described in section 4). But we need something more—time should be global. There are some hints suggesting that the global character of time is connected with the non-commutative origin of entropy and the second law of thermodynamics [2].

The non-commutative model of the fundamental level, explored in this essay, is very attractive from the conceptual point of view but at the present stage of its development has only a hypothetical character. A lot more has to be done to change it into a fully competitive scenario of the cosmic genesis. The solid story begins on our side of the Planck threshold. The story has begun and is open toward the far future.

9. The word "still" is not adequate as far as an atemporal regime is concerned, but we are now speaking from our temporal point of view.

10. A few technical remarks could help the reader to better understand these speculations. The mathematical structure of the non-commutative regime is shaped by a C^*-algebra A. The algebra A has a so-called *center,* that is, the subset of elements of A, which commute with all other elements of A. The transition to commutative geometry consists essentially in restricting the algebra A to its center (or to a subset of it). But the center is always in A; it belongs to the very essence of the non-commutative regime. We are beings of the center.

REFERENCES

1. Connes, A., *Non-commutative Geometry* (Academic Press, New York, London, 1994).

2. Connes, A., and Rovelli, C., "Von Neumann Algebra Automorphisms and Time-Thermodynamics Relation in Generally Covariant Quantum Theories," *Class. Q. Grav.*, *11*, 2899–917 (1994).

3. Fröhlich, J., Grandjean, O., and Recknagel, A., "Supersymmetric Quantum Theory and (Non-commutative) Geometry," *Commun. Math. Phys.*, *193*, 527–94 (1998). hep-th/9612205.

4. Geroch, R., "Einstein Algebras," *Commun. Math. Phys.*, *26*, 271–75 (1972).

5. Hawking, S. W., "The Existence of Cosmic Time Functions," *Proc. Roy. Soc. Lond.*, *A 308*, 433–35 (1968).

6. Hawking, S. W., and Ellis, G. F. R., *The Large Scale Structure of Space-Time* (Cambridge University Press, Cambridge, 1973).

7. Heller, M., *Theoretical Foundations of Cosmology* (World Scientific, Singapore, London, 1992).

8. Heller, M., "Einstein Algebras and General Relativity," *Int. J. Theor. Phys.*, *31*, 277–88 (1992).

9. Heller, M., and Sasin, W., "Sheaves of Einstein Algebras," *Int. J. Theor. Phys.*, *34*, 387–98 (1995).

10. Heller, M., and Sasin, W., "Non-commutative Structure of Singularities in General Relativity," *J. Math. Phys.*, *37*, 5665–71 (1996).

11. Heller, M., and Sasin, W., "Emergence of Time," *Phys. Lett.*, *A 250*, 48–54 (1998).

12. Heller, M., and Sasin, W., "Non-commutative Unification of General Relativity and Quantum Mechanics," *Int. J. Theor. Phys.*, *38*, 1619–42 (1999).

13. Heller, M., and Sasin, W., "Origin of Classical Singularities," *Gen. Rel. Grav.*, *31*, 555–70 (1999).

14. Heller, M., Sasin, W., and Lambert, D., "Groupoid Approach to Non-commutative Quantization of Gravity," *J. Math. Phys.*, *38*, 5840–53 (1997).

15. Heller, M., Sasin ,W., and Odrzygozdz, Z., "State Vector Reduction as a Shadow of Non-commutative Dynamics," *J. Math. Phys.*, *41*, 5168–79 (2000).

16. Landi, G., and Marmo, G., "Lie Algebra Extensions and Abelian Monopoles," *Phys. Lett.*, *B 195*, 429–34 (1987).

17. Landi, G., and Marmo, G., "Einstein Algebras and the Kaluza-Klein Monopole," *Phys. Lett.*, *B 210*, 68–72 (1988).

18. Mehlberg, H., *Time, Causality, and Quantum Theory*, vol. 1: *Essay on the Causal Theory of Time* (Reidel, Dordrecht, Boston, London, 1980).

19. Seiberg, N., and Witten, E., "String Theory and Non-commutative Geometry," *J. High Energy Phys.*, *9909*, 32 (1999). hep-th/9908142.

20. Sunder, V. S., *An Invitation to von Neumann Algebras* (Springer, New York, Berlin, Heidelberg, 1987).

[6]

LIVING IN A MULTIVERSE

Martin J. Rees

6.1. A Biophilic Universe?

If we ever established contact with intelligent aliens, how could we bridge the "culture gap"? One common culture would be physics and cosmology. Other intelligent life would, like us, be made of atoms, and we'd all trace our origins back to the same "genesis event"—the so-called big bang, which happened about 13 billion years ago. We'd all share the potentialities of a (perhaps infinite) future.

But our existence (and that of the aliens, if there are any) depends on our universe being rather special. Any universe hospitable to life—what we might call a biophilic universe—has to be "adjusted" in a particular way. The prerequisites for any life—long-lived stable stars, stable nuclei such as carbon, oxygen, and silicon that are able to combine into complex molecules, and so on—are sensitive to the physical laws and to the size, expansion rate, and contents of the universe. If the recipe imprinted at the time of the big bang had been even slightly different, we could not exist.[1]

1. See references [1, 10, 11].

Our universe evolved from a simple beginning—a big bang—specified by quite a short recipe, but this recipe seems rather special. Different "choices" for some basic numbers would have a drastic effect, precluding the hospitable cosmic habitat in which we emerged. For example, we could not exist in a world where gravity was much stronger. In an imaginary "strong-gravity" world, stars (gravitationally-bound fusion reactors) would be small; gravity would crush anything larger than an insect. But what would preclude a complex ecosystem even more would be the limited time. The mini-sun would burn faster, and would have exhausted its energy before even the first steps in organic evolution had got under way. A large, long-lived, and stable universe depends quite essentially on the gravitational force being exceedingly weak.

The nuclear fusion that powers stars depends on a delicate balance between two forces: the electrical repulsion between protons, and the strong nuclear force between protons and neutrons. If the nuclear forces were slightly stronger than they actually are relative to electric forces, two protons could stick together so readily that ordinary hydrogen would not exist, and stars would evolve quite differently. Some of the details are still more sensitive. For instance, carbon—crucial for all life—wouldn't be so readily produced in stars were it not for some seeming "tuning" in the properties of its nucleus, which depend even more sensitively on the nuclear force.

Even a universe as large as ours could be very boring: it could contain just black holes, or inert dark matter, and no atoms at all. Even if it had the same ingredients as ours, it could be expanding so fast that no stars or galaxies had time to form; or it could be so turbulent that all the material formed vast black holes rather than stars or galaxies—an inclement environment for life. And our universe is also special in having three spatial dimensions. A four-dimensional world would be unstable; in two dimensions, nothing complex could exist.

6.2. Three Interpretations of the Apparent "Tuning"

If our existence depends on a seemingly special cosmic recipe, how should we react to the apparent fine tuning? There seem to be three lines to take: we can dismiss it as happenstance; we can acclaim it as the workings of providence; or (my preference) we can conjecture that our universe is a specially favored domain in a still vaster multiverse. Let's consider them in turn.

6.2.1. HAPPENSTANCE (OR COINCIDENCE)

Maybe a fundamental set of equations, which some day will be written on T-shirts, fixes all key properties of our universe uniquely. It would then be an unassailable fact that these equations permitted the immensely complex evolution that led to our emergence.

But I think there would still be something to wonder about. It's not guaranteed that simple equations permit complex consequences. To take an analogy from mathematics, consider the beautiful pattern known as the Mandelbrot set. This pattern is encoded by a short algorithm, but has infinitely deep structure: tiny parts of it reveal novel intricacies however much they are magnified. In contrast, you can readily write down other algorithms, superficially similar, that yield very dull patterns. Why should the fundamental equations encode something with such potential complexity, rather than the boring or sterile universe that many recipes would lead to?

One hard-headed response is that we couldn't exist if the laws had boring consequences. We manifestly are here, so there's nothing to be surprised about. But shouldn't we still wonder why the unique recipe for the physical world permits consequences as interesting as those we see around us (and which, as a by-product, allowed us to exist)?

6.2.2. PROVIDENCE OR DESIGN

Evidence of design in the cosmos was a traditional theme of "natural theology." Two hundred years ago, the Cambridge theologian William Paley introduced the famous metaphor of the watch and the watchmaker—adducing the eye, the opposable thumb, and so on as evidence of a benign creator. This line of thought fell from favor, even among most theologians, in post-Darwinian times. We now view any biological contrivance as the outcome of prolonged evolutionary selection and symbiosis with its surroundings.

But Paley would have reacted differently if he'd known about the providential-seeming physics that led to galaxies, stars, planets, and the ninety-two elements of the periodic table. The seemingly biophilic features of basic physics and chemistry can't be as readily dismissed as the old claims for design in living things: biological systems evolve in symbiosis with their environment, but the basic laws governing stars and atoms are given, and nothing biological can react back on them to modify them. A modern counterpart of Paley, John Polkinghorne, interprets our fine-tuned habitat as "the creation of a Creator who wills that it should be so" [8].

6.2.3. A SPECIAL UNIVERSE DRAWN FROM AN ENSEMBLE, OR MULTIVERSE

If one doesn't believe in providential design, but still thinks the fine tuning needs some explanation, there is another perspective—a highly speculative one, however. There may be many "universes" of which ours is just one. In the others, some laws and physical constants would be different. But our universe wouldn't be just a random one. It would belong to the unusual subset that offered a habitat conducive to the emergence of complexity and consciousness. The analogy of the watchmaker could be off the mark. Instead, the cosmos maybe has something in common with an "off the shelf" clothes shop: if the shop has a large stock, we're not surprised to find one suit that fits. Likewise, if our universe is selected from a multiverse, its seemingly designed or fine-tuned features wouldn't be surprising.

6.3. Are Questions about Other Universes Part of Science?

Science is an experimental or observational enterprise, and it's natural to be troubled by assertions that invoke something inherently unobservable. Some might regard the other universes as being in the province of metaphysics rather than physics. But I think they already lie within the proper purview of science. It is not absurd or meaningless to ask "Do unobservable universes exist?", even though no quick answer is likely to be forthcoming. The question plainly can't be settled by direct observation, but relevant evidence can be sought, which could lead to an answer.

There is actually a blurred transition between the readily observable and the absolutely unobservable, with a very broad gray area in between. To illustrate this point, one can envisage a succession of horizons, each taking us further than the last from our direct experience.

6.3.1. LIMIT OF PRESENT-DAY TELESCOPES

There is a limit to how far out into space our present-day instruments can probe. Obviously there is nothing fundamental about this limit: it is constrained by current technology. Many more galaxies will undoubtedly be revealed in the coming decades by bigger telescopes now being planned. We would obviously not demote such galaxies from the realm of proper scientific discourse simply because they haven't been seen yet. When ancient navigators speculated about what existed beyond the boundaries of the then-known world, or when we speculate now about what lies below the oceans of Jupiter's moons Europa and Ganymede, we are speculating about something "real"—we are asking a scientific question. Likewise, conjectures about remote parts of our universe are genuinely scientific, even though we must await better instruments to check them.

6.3.2. LIMIT IN PRINCIPLE AT PRESENT ERA

Even if there were absolutely no technical limits to the power of telescopes, our observations are still bounded by a horizon, set by the distance that any signal, moving at the speed of light, could have traveled since the big bang. This horizon demarcates the spherical shell around us at which the red shift would be infinite. There is nothing special about the galaxies on this shell, any more than there is anything special about the circle that defines your horizon when you're in the middle of an ocean. On the ocean, you can see farther by climbing up your ship's mast. But our cosmic horizon can't be extended unless the universe changes so as to allow light to reach us from galaxies that are now beyond it.

If our universe were decelerating, then the horizon of our remote descendants would encompass extra galaxies that are beyond our horizon today. It is, to be sure, a practical impediment if we have to await a cosmic change taking billions of years, rather than just a few decades (maybe) of technical advance, before a prediction about a particular distant galaxy can be put to the test. But does that introduce a difference of principle? Surely the longer waiting time is a merely quantitative difference, not one that changes the epistemological status of these faraway galaxies.

6.3.3. NEVER-OBSERVABLE GALAXIES FROM "OUR" BIG BANG

But what about galaxies that we can never see, however long we wait? It's now believed that we inhabit an accelerating universe. As in a decelerating universe, an accelerating universe contains galaxies so far away that no signals from them have yet reached us. But if the cosmic expansion is accelerating, we are now receding from these remote galaxies at an ever-increasing rate; if their light hasn't yet reached us, it never will. Such galaxies aren't merely unobservable in principle now—they will be beyond our horizon forever. But if a galaxy is now unobservable, it hardly seems to matter whether it remains unobservable forever, or whether it would come into view if we waited a trillion years. (And I have argued, under (ii) above, that the latter category should certainly count as "real.")

6.3.4. GALAXIES IN DISJOINT UNIVERSES

The never-observable galaxies in (iii) would have emerged from the same big bang as we did. But suppose that, instead of causally disjoint regions emerging from a single big bang (via an episode of inflation), we imagine separate big bangs. Are space-times completely disjoint from ours any less real than regions that never come within our horizon in what we'd traditionally call our own universe? Surely not—so these other universes too should count as real parts of our cosmos.

This step-by-step argument (those who don't like it might dub it a slippery slope argument!) suggests that whether other universes exist or not is indeed a scientific question. So how might we answer it?

6.4. Scenarios for a Multiverse

At first sight, nothing seems more conceptually extravagant—more grossly in violation of Ockham's razor—than invoking multiple universes. But this concept follows from several different theories (albeit all speculative).

Linde [7], Garriga and Vilenkin [4], and others have performed computer simulations depicting an "eternal" inflationary phase where many universes sprout from separate big bangs into disjoint regions of space-times. Guth [5] and Smolin [13] have, from different viewpoints, suggested that a new universe could sprout inside a black hole, expanding into a new domain of space and time inaccessible to us. And Randall and Sundrum [9] suggest that other universes could exist separated from us in an extra spatial dimension; these disjoint universes may interact gravitationally, or they may have no effect whatsoever on each other. In the hackneyed analogy where the surface of a balloon represents a two-dimensional universe embedded in our three-dimensional space, these other universes would be represented by the surfaces of other balloons: any bugs confined to one, and with no conception of a third dimension, would be unaware of their counterparts crawling around on another balloon: other universes would be separate domains of space and time. In some of these scenarios, we couldn't even meaningfully say whether the other universes existed before, after, or alongside our own, because such concepts make sense only insofar as we can impose a single measure of time, ticking away in all the universes.

Fahri and Guth [3] and Harrison [6] have even conjectured that universes could be made in the laboratory, by imploding a lump of material to make a small black hole. Is our entire universe perhaps the outcome of some experiment in another universe? If so, the theological arguments from design could be resuscitated in a novel guise. Smolin [13] speculates that the daughter universe may be governed by laws that bear the imprint of those prevailing in its parent universe. If that new universe were like ours, then stars, galaxies, and black holes would form in it; those black holes would in turn spawn another generation of universes; and so on, perhaps ad infinitum.

Parallel universes are also invoked as a solution to some of the paradoxes of quantum mechanics, in the "many worlds" theory, first advocated by Everett and Wheeler in the 1950s. This concept was prefigured by Stapledon, as one of the more sophisticated creations of his Star Maker [14]:

> Whenever a creature was faced with several possible courses of action, it took them all, thereby creating many . . . distinct histories of the cosmos. Since in every evolutionary sequence of this cosmos there were many creatures and each was constantly faced with many possible courses, and the combinations of all their courses were innumerable, an infinity of distinct universes exfoliated from every moment of every temporal sequence.

None of these scenarios has been simply dreamed up out of the air: each has a serious, albeit speculative, theoretical motivation. However, one of them, at most, can be correct. Quite possibly none is: there are alternative theories that would lead to just one universe.

Firming up any of these ideas will require a theory that consistently describes the extreme physics of ultra-high densities, how structures on extra dimensions are configured, and so on. But consistency is not enough: there must be grounds for confidence that such a theory isn't a mere mathematical construct, but applies to external reality. We would develop such confidence if the theory accounted for things we can observe that are otherwise unexplained.

At the moment, we have an excellent physics framework, called the standard model, that accounts for almost all subatomic phenomena that have been observed. But the formulas of the "standard model" involve numbers which can't be derived from the theory but have to be inserted from experiment. Perhaps, in the twenty-first-century theory, physicists will develop a theory that yields insight into (for instance) why there are three kinds of neutrinos, and the nature of the nuclear and electric forces. Such a theory would thereby acquire credibility. If the same theory, applied to the very beginning of our universe, were to predict many big bangs, then we would have as much reason to believe in separate universes as we now have for believing inferences from particle physics about quarks inside atoms, or from relativity theory about the unobservable interior of black holes.

6.5. Universal Laws, or Mere Bylaws?

"Are the laws of physics unique?" is a less poetic version of Einstein's famous question "Did God have any choice in the creation of the Universe?" The answer determines how much variety the other universes—if they exist—might display. If there were something uniquely self-consistent about the actual recipe for our universe, then the aftermath of any big bang would be a rerun of our own universe. But a far more interesting possibility (which is certainly tenable in our present state of ignorance of the underlying laws) is that the underlying laws governing the entire multiverse may allow variety among the universes. Some of what we call "laws of nature" may in this grander perspective be local bylaws, consistent with some overarching theory governing the ensemble, but not uniquely fixed by that theory.

As an analogy, consider the form of snowflakes. Their ubiquitous six-fold symmetry is a direct consequence of the properties and shape of water molecules. But snowflakes display an immense variety of patterns because each is molded by its micro-environments: each flake grows in response to the fortuitous temperature and humidity changes during its growth. If physicists achieved a fundamental theory, it would tell us which aspects of nature were direct consequences of the bedrock theory (just as the symmetrical template of snowflakes is due to the basic structure of a water molecule) and which are (like the distinctive pattern of a particular snowflake) the outcome of accidents. The accidental features could be imprinted during the cooling that follows the big bang—rather as a piece of red-hot iron becomes magnetized when it cools down, but with an alignment that may depend on chance factors.

The cosmological numbers in our universe, and perhaps some of the so-called constants of laboratory physics as well, could be "environmental accidents," rather than uniquely fixed throughout the multiverse by some final theory. Some seemingly "fine-tuned" features of our universe could then only be explained by "anthropic" arguments, which are analogous to what any observer or experimenter does when he or she allows for selection effects in his or her measurements: if there are many universes, most of which are not habitable, we should not be surprised to find ourselves in one of the habitable ones!

The entire history of our universe could just an episode of the infinite multiverse; what we call the laws of nature (or some of them) may be just parochial bylaws in our cosmic patch. Such speculations dramatically enlarge our concept of reality. Putting them on a firm footing must await a successful fundamental theory that tells us whether there could have been many big bangs rather than just one, and (if so) how much variety they might display. Such a theory must unify quantum theory (which governs the microworld) with gravity, the force that dominates the large-scale universe. For most natural phenomena, such unification is superfluous: quantum theory is only crucial in the microworld of atoms, where gravity is too weak to be significant; conversely, gravity is only important on the scale of stars and planets, where the intrinsic "fuzziness" owing to quantum effects can be ignored. But right back at the beginning, everything was squeezed so densely that quantum effects could shake the entire universe.

Will such a theory—reconciling gravity with the quantum principle, and transforming our conception of space and time—be achieved in coming decades? The smart money is on superstring theory, or M-theory, in which each point in our ordinary space is actually a tightly-folded origami in six or seven extra dimensions. There is still a daunting gap between the intricacies of ten or eleven dimensions and anything that we can observe or measure. It has not yet predicted anything new—experimental or cosmological—caused by the extra dimensions. But many are willing to bet on superstrings almost for aesthetic reasons. Edward Witten has said, "Good wrong ideas are extremely scarce, and good wrong ideas that even remotely rival the majesty of string theory have never been seen."

6.6. Testing Multiverse Theories Here And Now

We may one day have a convincing theory that accounts for the very beginning of our universe, tells us whether a multiverse exists, and (if so) whether some so-called laws of nature are just parochial bylaws in our cosmic patch. But while we're waiting for that theory—and it could be a long wait—the analogy of the "ready made clothes shop" can already be checked. It could even be refuted: this would happen if our universe turned out to be even more specially tuned than our presence requires. Let me give two examples of this style of reasoning, applied to two different scientific issues.

First, Boltzmann argued that our entire universe was an immensely rare "fluctuation" within an infinite and eternal time-symmetric domain. There are now many arguments against this hypothesis, but even when it was proposed one could already have noted that fluctuations in large volumes are far more improbable than in smaller volumes. So, it would be overwhelmingly more likely that we would (if Boltzmann were right) be in the smallest fluctuation compatible with our existence: indeed a solipsistic universe where only one brain existed, complete with memories, would be more likely than any other interpretation of our experience. If we weren't prepared to be solipsists, then whatever our initial assessment of Boltzmann's theory, its probability would plummet as we came to realize the extravagant scale of the cosmos.

Second, even if we knew nothing about how stars and planets formed, we would not be surprised to find that our Earth's orbit was fairly close to circular: had it been highly eccentric, water would boil when the earth was at perihelion, and freeze at aphelion—a harsh environment unconducive to our emergence. However, a modest orbital eccentricity is plainly not incompatible with life. If it had turned out that the earth moved in a much more nearly perfect circle, then we could rule out a theory that postulated anthropic selection from orbits whose eccentricities had a "Bayesian prior" that was uniform in the range 0–1.

We could apply this style of reasoning to the important numbers of physics (for instance, the cosmological constant L) to test whether our universe is typical of the subset that that could harbor complex life [15]. The methodology requires us to decide what values are compatible with our emergence. It also requires a specific theory that gives the relative Bayesian priors for any particular value. For instance, in the case of L, are all values equally probable? Are low values favored by the physics? Or are there a finite number of discrete possible values? With this information, one can then ask if our actual universe is "typical" of the subset in which we could have emerged. If it is a grossly atypical member even of this subset (not merely of the entire multiverse) then we would need to abandon our hypothesis.

As another example of how "multiverse" theories can be tested, consider Smolin's conjecture [13] that new universes are spawned within black holes, and that the physical laws in the daughter universe retain a memory of the laws in the parent universe: in other words there is a kind of heredity. Smolin's concept is not yet bolstered by any detailed theory of how any physical information (or even an arrow of time) could be transmitted from one universe to another. It has, however, the virtue of making a prediction about our universe that can be checked.

If Smolin were right, universes that produce many black holes would have a reproductive advantage, which would be passed on to the next generation. Our universe, if an outcome of this process, should therefore be near optimum in its propensity to make black holes, in the sense that any slight tweaking of the laws and constants of physics would render black hole formation less likely. (I personally think Smolin's prediction is unlikely to be borne out, but he deserves our thanks for presenting an example that illustrates how a multiverse theory can in principle be vulnerable to disproof.)

These examples show that some claims about other universes may be refutable, as any good hypothesis in science should be. We cannot confidently assert that there were many big bangs—we just don't know enough about the ultra-early phases of our own universe. Nor do we know whether the underlying laws are "permissive": settling this issue is a challenge to twenty-first-century physicists. But if they are, then so-called anthropic explanations would become legitimate—indeed they'd be the only type of explanation we'll ever have for some important features of our universe.

What we've traditionally called "the universe" may be the outcome of one big bang among many, just as our solar system is merely one of many planetary systems in the galaxy. Just as the pattern of ice crystals on a freezing pond is an accident of history, rather than being a fundamental property of water, so some of the seeming constants of nature may be arbitrary details rather than being uniquely defined by the underlying theory. The quest for exact formulas for what we normally call the constants of nature may consequently be as vain and misguided as was Kepler's quest for the exact numerology of planetary orbits. And other universes will become part of scientific discourse, just as "other worlds" have been for centuries. Nonetheless (and here I gladly concede to the philosophers), any understanding of why anything exists—why there is a universe (or multiverse) rather than nothing—remains in the realm of metaphysics.

6.7. The Long-Range Future

Let us now refocus on our own universe—our "cosmic oasis" in (perhaps) an infinite multiverse. What is its long-range future? The Sun will continue to shine for six billion years—several times longer than it has taken for the earth's biosphere (including us) to evolve from the first multicellular organisms. Intelligently controlled modifications lead to far faster changes than natural selection does. Even if life is now unique to Earth it could eventually "take over" the cosmos. The entire galaxy, extending for a hundred thousand light-years, could be "greened" in less time than it took for us to evolve from the first primates. There is plenty of time for life to spread through the entire galaxy, and even beyond.

We are far from the culmination of evolution—the emergence of structure, intelligence, and complexity is still near its cosmic beginnings. Wormholes, extra dimensions, and quantum computers open up speculative scenarios that we cannot yet even envision. The future may lie in artifacts, created by us, that develop via their own directed intelligence. Diffuse living structures, freely floating in interstellar space, would think in slow motion, but may nonetheless come into their own in the long-range future.

Six billion years from now, when the Sun dies, the galaxies will be more widely dispersed, and will be intrinsically somewhat fainter because their stellar population will have aged, and less gas will survive to form bright new stars. But what might happen still further ahead? We can't predict what role life will eventually carve out for itself. In the long term the universe may, in a sense, become alive. Entities descended from us may be able to "create" other universes within black holes. Our universe has the potential to harbor a teeming complexity of life far beyond what we can even conceive.

But we can give long-range projections for a universe that continues to be governed just by physical forces. It seems fated to continue expanding, and even—if current evidence is borne out—accelerating. Eventually, even the slowest-burning stars would die, and all the galaxies in our local group—our Milky Way, Andromeda, and dozens of smaller galaxies—would merge into a single system. Most of the original gas would by then be tied up in the dead remnants of stars—some would be black holes, others would be cold neutron stars or white dwarfs.

Still further ahead, events far too rare to be discernible today could come into their own—stellar collisions, for instance. Stars are so thinly spread through space that collisions between them are immensely infrequent (fortunately for our Sun), but their number would mount up. The terminal phases of galaxies would be sporadically lit up by intense flares, each signaling an impact between two dead stars.

Eventually, even black holes would decay. The surface of a black hole is made slightly fuzzy by quantum effects, and it consequently radiates. This effect would be important in our present universe if it contained mini-holes the size of atoms: such black holes would erode away, emitting radiation and particles; the smaller they would get, the more powerful and energetic the radiation would be, and they would eventually disappear in an explosion. But it seems unlikely that such mini-holes exist. They could form only in the hyperdense early universe, and even there this would be unlikely unless conditions were far more turbulent than theory suggests. And to form them today, a kilometer-sized asteroid (or something of similar mass) would need to be compressed to the size of an atomic nucleus. The evaporation of black holes, being a quantum process, is far less important for big holes: the time it would take for a hole to erode away depends on the cube of its mass. The lifetime of a hole with the mass of a star is 10^{66} years. Even black holes as heavy as a billion suns—such as those that lurk in the centers of galaxies—would erode away in less than 10^{100} years.

6.8. *The Asymptotic Future of Life*

Cosmologists have produced an immense speculative literature on the ultra-early universe. In contrast, cosmic futurology has been left to science-fiction writers. I can myself claim to have made one of the first scientific contributions, back in 1968, when I wrote a short paper entitled "The Collapse of the Universe: An Eschatological Study" [10]. Many cosmologists then suspected that we might live in a bounded universe that would end in a "big crunch," and I calculated what might then happen after the cosmic expansion had come to a halt and the universe had started to recollapse. During the countdown to the crunch, galaxies would merge together, and individual stars would accelerate to almost the speed of light (rather as the atoms speed up in a gas that is compressed); eventually these stars would be destroyed in massive explosions because the heat irradiating their surfaces (the blueshifted radiation from other stars) would be hotter than their interiors.

Eleven years later, Freeman Dyson made the subject scientifically respectable: he published in *Reviews of Modern Physics* a fascinating and detailed article called "Time Without End: Physics and Biology in an Open Universe" [2].[2] The evidence for an ever-expanding universe was then less clear than it is now. But Dyson already had his prejudices: he wouldn't countenance the big crunch option because it "gave him a feeling of claustrophobia." He discussed the prognosis for intelligent life. Even after stars have died, he asked, can life survive forever without intellectual burnout? Energy reserves are finite, and at first sight this might seem to be a basic restriction. But he showed that this constraint was actually not fatal. As the universe expands and cools, lower-energy quanta of energy (or, equivalently, radiation at longer and longer wavelengths) can be used to store or transmit information. Just as an infinite series can have a finite sum (for instance $1 + ½ + ¼ + \ldots = 2$), so there no limit to the amount of information processing that could be achieved with a finite expenditure of energy. Any conceivable form of life would have to keep ever cooler, think slowly, and hibernate for ever longer periods. But there'd be time to think every thought. As Woody Allen once said, "Eternity is very long, especially toward the end."

Dyson imagined the endgame being spun out for a number of years so large that to write it down you'd need as many zeros as there are atoms in all the galaxies we can see. At the end of that time, any stars would have tunneled into black holes, which would then evaporate in a time that is, in comparison, almost instantaneous.

In the twenty plus years since Dyson's article appeared, our perspective has changed in two ways, and both make the outlook more dismal. First, most physicists now suspect that atoms don't live forever. In consequence, white dwarfs and neutron stars will erode away, maybe in 10^{36} years—the

2. Reprinted as chapter 8 of this book.

heat generated by particle decay will make each star glow, but as dimly as a domestic heater. By then our local group of galaxies will be just a swarm of dark matter and a few electrons and positrons. Thoughts and memories will only survive beyond the first 10^{36} years if downloaded into complicated circuits and magnetic fields in clouds of electrons and positrons—maybe something that would resemble the threatening alien intelligence in *The Black Cloud,* the first and most imaginative of Fred Hoyle's science-fiction novels, written in the 1950s.

Dyson was optimistic about the potentiality of an open universe because there seemed be no limit to the scale of artifacts that could eventually be constructed. He envisioned the observable universe getting ever vaster; moreover, many galaxies, whose light hasn't yet had time to reach us, would eventually come into view, and therefore within range of possible communication and "networking"; gravity tends to slow the recession of distant galaxies even though they move ever farther away, so the disruptive effect of the expansion would become less important. But it now seems that the expansion isn't slowing down: some repulsive force or "antigravity" seems to be pushing galaxies apart at an accelerating rate. The long-term future is then more constricted. Galaxies will fade from view even faster: they will get more and more red-shifted—their clocks, as viewed by us, will seem to run slower and slower, and freeze at a definite instant, so that even though they never finally disappear we would see only a finite stretch of their future. The situation is analogous to what happens if cosmologists fall into a black hole: from a vantage point safely outside the hole, we would see our falling colleagues freeze at a particular time, even though they would experience, beyond the horizon, a future that is unobservable to us.

Our own galaxy, Andromeda, and the few dozen small satellite galaxies that are in the gravitational grip of one or other of them, will merge together into a single amorphous system of aging stars and dark matter, and the universe will look ever more like an "island system" (the kind of universe that was originally proposed by Laplace). In an accelerating universe, everything else disappears beyond our horizon; if the acceleration is due to a fixed "lambda," this horizon never gets much farther away than it is today. So there's a firm limit—though of course a colossally large one—to how large any network or artifact can ever become. This translates into a definite limit on how complex anything can get. One important recent development has been to quantify this limit. Space and time cannot be infinitely divided. The inherent "graininess" of space sets a limit to the intricacy that can be woven into a universe of fixed size.

Even if the problem of limited energy reserves could be surmounted, there would be a limit to the variety and complexity. The best hope of staving off boredom in such a universe would be to construct a time machine and, subjectively at least, exhaust all potentialities by repeatedly traversing a closed time loop.

These long-range projections involve fascinating physics, most of which is quite well understood. But readers who are really concerned about what happens in zillions of years should be mindful of some uncertainties. First, we can't be absolutely sure that the regions beyond our present horizon are like the parts of the universe we see. On the ocean there could be something extraordinary just beyond the horizon. Likewise, a universe that's decelerating, but seemingly not enough ever to stop, could eventually be crushed by denser material not yet in view. Even if that didn't happen, the trend toward greater smoothness on larger scales may not continue indefinitely. There could be a new range of structures, on scales far larger than the part of the universe that we have so far observed. John Barrow and Frank Tipler have pointed out that a new source of energy—so-called shear-energy—would become available if the universe expanded at different rates in different directions.

Second, we don't know what may eventually happen to "quintessence"—the mysterious energy in space that drives the accelerating cosmic expansion. This residual energy could convert into some new kinds of particles. If this conversion happened smoothly, it would not lift the terminal gloom. However, the residual energy could decay in bubbles, whose surfaces would crash together, giving rise to energy concentrations where it's even possible that atoms could be regenerated. This revived universe would be patchy, with "islands" of revived activity separated by vast voids. To quote Paul Steinhardt, would future beings "figure out that their origin was the isotropic universe we see around us today? Would they ever know that the universe had once been alive and then died, only to be given a second chance?"

A more disconcerting prospect is that empty space could be vulnerable to a catastrophic transfiguration. Very pure water can "supercool" below its freezing point, but it suddenly freezes when a speck of dust is put into it. In an analogous way, our present "vacuum" may be merely meta-stable, and could then transform to a quite different universe, governed by different laws, and perhaps with a large negative lambda that would cause everything to implode, rather than to accelerate outward.

There are periodic scares that this kind of transition could be induced artificially by high-energy particle collisions in accelerator experiments. It's reassuring, however, that far more energetic collisions—involving cosmic ray particles—have been happening naturally for billions of years, without tearing the fabric of space.

Another route to armageddon could be the conversion of ordinary atomic nuclei into hypothetical particles called strangelets: these contains a third kind of quark, additional to the kinds that make up normal protons and neutrons. If strangelets can exist as stable objects, and if their electric charge is negative (which is thought very unlikely), they could attract other nuclei and then, by contagion, convert their surroundings, and eventually the entire earth, into so-called strange matter. This eventuality was taken seriously enough by those in charge of the Brookhaven accelerator that they commissioned an expert assessment on whether experiments that crash together very heavy nuclei could trigger such a catastrophe.

Our remote descendants are likely to have an eternal future (unless they come within the clutches of a black hole). Nonetheless, it is worth remarking that the alternative fate—being snuffed out in a big crunch—could be an enriching experience. We've seen that events happen ever more slowly in the ever-expanding universe; the total number of discrete events or "thoughts" could then be bounded, even in an infinite future. John Barrow and Frank Tipler have emphasized that, in a collapsing universe, the converse is possible: there can be an infinite number of "happenings" within a finite time [1]. Cosmologists are used to the idea that a lot could happen in the initial instants after the big bang: as we extrapolate back to more extreme densities, time must be measured by a series of progressively smaller, more robust, and faster-ticking clocks. An infinite set of numbers can add up to a finite sum. Likewise, in the final instants before the crunch we could not only see our entire earlier life flash by, but could experience an infinite number of new events.

Even if our own universe is destined for perpetual expansion, the big crunch scenario could perhaps be played out elsewhere in the multiverse, within a universe where the global curvature is different but the microphysical laws are the same as in ours (and so equally conducive to life). Most universes governed by different microphysics would be less propitious for life, but some could allow even greater complexity than our own. It is beyond most biologists' capacities to envision the variety of alien life forms that could exist elsewhere in our universe; still less can we imagine the potentialities that could unfold in universes governed by other "well-tuned" laws.

6.9. Earth in a Cosmic Perspective

An iconic image from the 1960s was the first photograph from space, showing our home planet of land, oceans, and clouds: its fragile beauty contrasts with the stark and sterile moonscape on which the astronauts left their footprints. We now know that other stars are not mere "points of light"—many are orbited by retinues of planets. Within twenty years we'll be able to hang on our walls another poster that will have even more impact—a telescope image of another earth, orbiting some distant star.

But will this planet have a biosphere? If so, will it harbor life that is advanced or intelligent? Evolution of intelligence may require such an improbable chain of events that it hasn't emerged around even one of the trillion billion stars within range of our telescopes. In that case, our Earth, though tiny, would thus acquire deepened significance—not just for what it is now, but for its cosmic potential.

Our 13-billion-year-old universe is still just beginning. Our actions during this century could initiate the spread of life beyond the solar system—through the galaxy, and eventually even beyond. Or they could—if life is indeed now unique to the earth—stunt the future of our entire cosmos. Snuffing out advanced life while it was still restricted to our Earth could foreclose the "creative" potential of the entire cosmos billions of years hence: it would be a cosmic disaster, not "just" a terrestrial one.

Twenty-first-century technology confronts us with many lethal prospects. We could face global environmental catastrophe. But I rate even higher the threat that we could be wiped out by lethal, "engineered" airborne viruses, or by rogue nanomachines that replicate catastrophically. The risk stems from accelerating technical advances. There are, frighteningly, many ways in which small dissident groups, or accidents in small laboratories, could trigger global catastrophe. Any of these catastrophes could arise from malign intent—or, even more worryingly, simply from technical misadventure. (The kind of accelerator-induced catastrophe mentioned earlier cannot be absolutely dismissed, but would rate far lower on my personal "risk list" than those stemming from bio- or nanotechnology.)

On grounds of prudence, governments should put the brakes on risky areas of genetic engineering and nanotechnology. The same caution should be exercised by physicists. If an experiment could destroy the world, the acceptable level of risk is surely less than one chance in a trillion. Theoretical arguments cannot offer adequate comfort at this level: they can never be firmer than the assumptions on which they are based, and only recklessly overconfident theorists would stake odds of a trillion to one on the validity of their assumptions. We can't sleep completely soundly unless we are sure that essentially identical events—for example, equally energetic collisions of cosmic ray particles—have already happened naturally without any disaster ensuing. An extreme "precautionary principle" should prevail.

But even if all nations imposed strict regulations on potentially threatening experiments, the prospects for effective enforcement are no better than in the case of the drug laws. And even a single infringement could trigger global disaster—hence, extreme pessimism seems to me the only rational stance.

Though it would be little consolation to those on Earth, for whom the 50-50 risk remains, it would be feasible to ensure that the long-term cosmic potential of life isn't quenched. Once intelligent life, in real or encoded form, has spread beyond the earth, no terrestrial disaster could thereafter foreclose its cosmic potential—life would have "tunneled through" its era of maximal jeopardy.

Development of "space habitats"—feasible before the end of the twenty-first century—would offer an insurance. When self-sustaining communities have been established away from the earth—on the Moon, on Mars, or freely floating in space—our species will be invulnerable to any global disaster on Earth, and whatever potential it has for the five-billion-year future could not be snuffed out. This is perhaps the only reason for giving priority to programs of manned space flight. (These need not involve an Apollo-style venture, but could be privately funded—perhaps, indeed, becoming the province of wealthy adventurers prepared to accept high risks to boldly explore the far frontier.)

New technologies may offer another option: downloading our blueprint into inorganic memories that could be launched into the cosmos (perhaps able to duplicate themselves). Such concepts confront us with profound issues about the limits of information storage, and philosophical implications of identity.

Our Earth may have cosmic importance, as the one place from which life could spread through the universe. This realization raises the stakes from the earth to the entire cosmos. It may be less stunning than the "immediate" risk to those already living on Earth, but for me, and perhaps for others (especially for those without religious belief), this cosmic perspective strengthens the imperative to cherish life on Earth.

This new century, on this planet may be a defining moment for the cosmos. In the entire domain that cosmologists explore—ten billion years of time, ten billion light-years of space—the most crucial space-time location of all (apart from the big bang itself) could be here and now. Through malign intent, or through misadventure, the technology of the twenty-first century could destroy our species, and thereby foreclose the potentialities of our biophilic universe whose evolution has still barely begun. In contrast, by prudence, and by spreading ourselves beyond the earth, we could ensure sufficient diversity to safeguard life's potential for an infinite future, in a cosmos that is even vaster and more diverse than previously believed.

REFERENCES

1. Barrow, J., and Tipler, F., *The Anthropic Cosmological Principle* (Oxford University Press, Oxford, 1988).

2. Dyson, F. J., "Time without End: Physics and Biology in an Open Universe," *Rev. Mod. Phys.*, *51*, 447–60 (1979).

3. Fahri, E. H., and Guth, A. H., "An Obstacle to Creating a Universe in a Laboratory," *Phys. Lett.*, *B 183*, 149 (1987).

4. Garriga, J., and Vilenkin, A., "Many Worlds in One," *Phys. Rev.*, *D 64*, 043511 (2001); gr-qc/0102010.

5. Guth, A. H., *The Inflationary Universe* (Addison-Wesley, 1996), especially 245–52; astro-ph/0101507.

6. Harrison, E. R., "The Natural Selection of Universes Containing Intelligent Life," *Q. J. Roy. Ast. Soc.*, *36*, 193 (1995).

7. Linde, A., "The Self-Reproducing Inflationary Universe," *Sci. Amer.*, *5*, 32–39 (1994).

8. Polkinghorne, J., *Quarks, Chaos, and Christianity* (SPCK Triangle Press, London, 1994).

9. Randall, L., and Sundrum, R., "An Alternative to Compactification," *Phys. Lett.*, *83*, 4690 (1999).

10. Rees, M., "The Collapse of the Universe: An Eschatological Study," *Observatory, 89*, 193 (1969).

11. Rees, M., *Just Six Numbers* (Basic Books, New York, 1999).

12. Rees, M., *Our Cosmic Habitat* (Princeton University Press, Princeton, 2001).

13. Smolin, L., *The Life of the Cosmos* (Oxford University Press, New York, 1996).

14. Stapledon, O., *Star Maker* (Penguin, New York, 1972; first published Methuen, London, 1937), 251.

15. Weinberg, S. W., Conference Summary in "Relativistic Astrophysics," ed. J. C. Wheeler and H. Martin (AIP Publications, New York), 893–910; astro-ph/0104482.

PART 3

BIOLOGY

[7]

EXOTIC GENETIC MATERIALS AND THE EXTENT OF LIFE IN THE UNIVERSE

A. Graham Cairns-Smith

7.1. Introduction

That all life on Earth is so similar in its biochemistry tells us something about the history of life on Earth, but not how life must always be constituted. Even life on Earth may have started with exotic genetic materials, by which I mean materials very different from DNA structurally, although like DNA in abstract essentials, in being able to store replicable information that can influence its surroundings so as to enhance its own survival and propagation, and hence be subject to evolution through natural selection. Materials that might *most easily* support such an evolution seem not to be organic polymers at all, but inorganic crystals, and so perhaps life on Earth started

Note: I would like to thank John Barrow, Paul Davies, George Ellis, and Hyman Hartman for helpful discussions.

with the evolution of information in such materials (minerals) [3]. Exotic genetic materials should also be considered for designs of artificial life forms [4], and they may be of interest in thinking about the extent of life in the universe—of life in unearthly places, now and in the far future.

Many of the detailed features of life on Earth appear to be universal—in the use of particular amino acids and nucleotides, in the near universality of a protein manufacturing code, and so on. But there is no good reason to suppose that the so-called molecules of life are some kind of biological invariant. There is a more incidental and sufficient explanation for the unity of central biochemistry now observed: it is that all life now on Earth descended from a last common ancestor in which precisely these features were present *and were already fixed,* already locked in—presumably through mutual dependence [3]. Yet if the last common ancestor was itself the product of a Darwinian evolution, then there was a time when its components were not fixed. How could this be?

By means of a naive analogy we can try to imagine the general situation within which a step-by-small-step Darwinian process could give rise to a fixed interdependence of components. Think of an arch of stones built without mortar:

FIG. 7.1

This is a crude model for any system consisting of mutually dependent components. So we might ask how such an arch could be produced in small steps—say by touching only one stone at a time. One answer would be by building a heap first

FIG. 7.2

and then removing stones to create the arch. I have suggested that the strong interdependence of components in today's central biochemistry is itself evidence for there having been earlier biochemical control systems whose components did not lean together so critically, and that created a kind of scaffolding which allowed more sophisticated systems to be built up piece by piece, after which that scaffolding became redundant and was discarded [3].

We have then a picture of a fluid stage of early biochemical evolution followed by a freezing in of one highly interdependent system. We would not know directly what materials had been used during such a more fluid era, especially during its earliest stages. But fortunately there are some things that we can say about evolving systems in general, and chemical constraints that we can apply to limit possibilities.

7.2. Streams and Genes: The Genetic View of Life

Living things are open systems operating within open systems, needing maintained supplies of energy and materials, and the means to dispose of waste products and corpses. Sherrington once likened living things to eddies in a stream of energy [18]. But they are well-controlled "eddies" and the most characteristic part of that control resides ultimately in stable structures that we call genes. While the environment for an organism must be dynamic, and most of the organism itself may be highly dynamic, the genes represent an essential core of control that is stable over the short and medium term: incredibly stable between generations as Schrödinger so vividly saw [15]. To emphasize the contrast between the essentially dynamic activities and the essentially stable genetic aspect of a living thing, we might extend the eddy analogy by likening an organism to a particular system of eddies produced by a stone of a particular shape placed in a stream. The shape of the stone is the organism's inherited genetic information. The system of eddies is the phenotype of the organism. So the phenotype can be thought of as part of the environment—that part which is under genetic control—and there is no sharp boundary between the phenotype and the rest of the environment [2].

The genetic view of life puts the genetic memory at the center of our understanding—as opposed to the more visible, dynamic aspects of living things. Yes, organisms are open systems and as such are "self-organizing" in many interesting ways.[1] This is part of the story of how life works. But being

1. I refer here to the spontaneous appearance of interestingly complex structures in simple open systems: for example, convection cells predictably appear in a uniformly heated pan of liquid. Living things can be said to be often "self-organizing" in this sort way. See [3] (ch. 3) and [2] (ch. 5) on meanings and ambiguities of the term "organization."

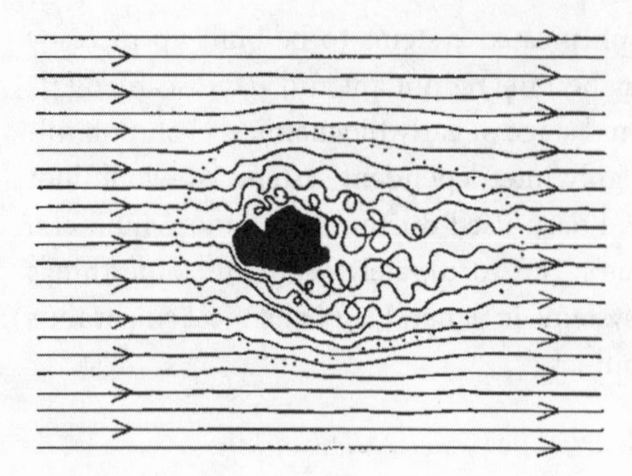

FIG. 7.3 *A metaphor for the relationship between genetic information, phenotype, and environment.*

an open system is too general a feature of the universe to be a defining feature of life. Open systems are not just common, but universal parts of our physical world. On a long-enough timescale everything is an open system within an open system. But we should not say that everything is "alive." The defining feature of living things is that they have a set of recipes for success in survival and reproduction, a long-term memory put there through the process of evolution by natural selection. In that long process the recipes were adjustable, for sure, but first and foremost they had to be remembered. For that they had to be stable in themselves and in repeated copies of copies between generations.

In considering the possible extent of life in the universe we should think of both "streams" and "genes." Perhaps we feel that the only suitable place for life in the universe is an earth-like planet with a maintained supply of photons from its star. But even for life on Earth now this is not the only energy stream that is important. The earth itself is a nuclear reactor. Radioactive heating drives the earth's geology, continually renewing surface mineral resources necessary for life as we know it, quite apart from making active open systems such as deep-sea hydrothermal vents. Life forms remote from stars are perfectly feasible.

In a discussion following the symposium meeting, John Barrow suggested another possibility for a nonstellar energy source arising from the kind of thing that is thought to have happened at Oklo, in the West African Republic of Gabon, where geological processes in the past apparently created a uranium-235 fission reactor [13]. Much smaller bodies than the earth might thus provide suitable life environments, at least for a time.

On the question of "genes" we may also be inclined to think too narrowly. Perhaps we think that genes must be DNA-like (or RNA-like) molecules. No one has yet made any other organic polymer that can replicate more than trivial amounts of information. This is evidently a difficult thing to do. But then organic molecules are not the only possible stores of replicable information to be found in nature.

7.3. Exotic Genetic Materials and Genetic Takeover

If we try to look beyond the last common ancestor to the very beginning of the whole bioevolutionary process that took place on Earth, it is hard to see anything like DNA as the holder of the very earliest hereditary information. Neither DNA nor RNA are remotely plausible as molecules that might have been produced on a prevital earth.[2] Their nucleotide units are too difficult to make and activate, and nucleic acids themselves are too delicate and too dependent on subsidiary systems in their operation to be anything other than *products* of evolution through natural selection. Yet how can this be so if a genetic material is an essential *prerequisite* for evolution?

I used an arch analogy to provide, perhaps, a hint of how we might see a way through this dilemma. "Genetic takeover" is another somewhat more specific idea that might help us out. It is a speculative scheme for early evolution, for the evolution of the central control machinery itself. In its simplest form we imagine a first genetic material, say a mineral, able to operate in an unevolved environment. This was later replaced by another quite different and more efficient material (say RNA). This second material was too elaborate to have started *de novo* on the primitive earth and only became possible for the first time within evolved micro-environments (phenotypes) created by the original genes. Then the second genetic material gradually ousted the first.

A genetic takeover would be analogous, in some ways, to the replacement of one technology by another in a civilization: the quill pen by the word processor for writing, the horse by the internal combustion engine as a power unit—one can think of many such examples. Replacements like this may be gradual, in the sense that a background technology had gradually evolved to make possible the inception of some particular new "high tech"

2. See [3], pp. 56–59. See also Robert Shapiro [16, 17].

system. And the perfection of a new system, and the final redundancy of an old one, may be gradual too. But none of this is to say that a new technology is necessarily some kind of elaboration of an old one. At least as often it is a replacement of one thing with something else that works altogether differently. (Word processors did not evolve through the gradual elaboration of quill pens, for example.) Among other things one expects that a "high tech" replacement will be made of different materials.

In its simplest form, as described above, the genetic takeover idea may be represented by a pair of overlapping lines:

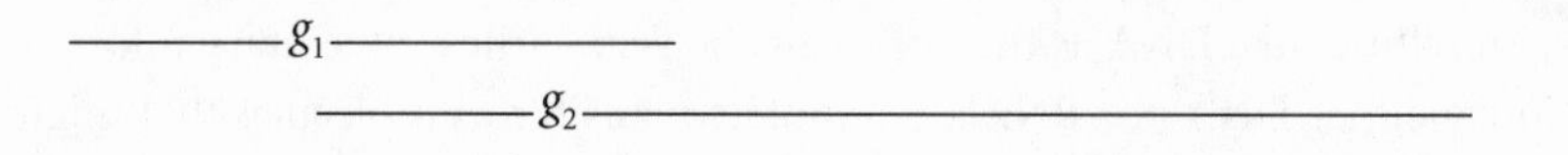

First there are organisms that have one (primitive) genetic material, g_1, the line representing the long-term passage of information between generations, written in this material. Then a new replicating material is discovered within the environment created by evolved phenotypes of a g_1 life form to make organisms that now have two different kinds of genes in them, controlling different aspects of the activities through which together they contrive their survival and propagation. The new genetic material need not be "compatible" with the old one. Indeed, rather the reverse: part of its virtue might be that it holds information of a new kind and replicates it in a new way. Mutual usefulness would be a sufficient explanation for why the old and new kinds of genes should stay together.

Finally, according to this simplest form of genetic takeover, the "high tech" g_2-based subsystems had evolved far enough to be able to make organisms that no longer needed the old "low tech" systems at all.

I will come shortly to more specific suggestions about g_1. But first I would like to take the general idea further and suggest that during the earliest stages of the evolution of life on Earth many, perhaps dozens, of types of genetic material came and went, nucleic acid being the final outcome of a distant "trying out" process through which the genetic material itself could be said to have evolved. So the picture might have been more like this:

———g_1——— ———g_4——— ———g_8———
———g_2——— ———g_6——— ———RNA———
———g_3——— ———g_7———
———g_5——— ———g_9—

The first step might be imagined as g_1 replicating some single simple feature that enhanced its survival and propagation—some exceedingly simple "function", f_1:

$$[g_1 \rightarrow f_1]$$

There is perhaps an example of this in submarine illite clays. These clays occur as tiny flexible laths a few microns wide and a few nanometers thick. The laths are made up of stacks of, perhaps, three or four unit layers with a mica-like structure. Such laths are commonly found attached to the grains of sandstones, including oil-bearing sandstones beneath the sea [14]. It seems that the illite crystallites grow by sideways extension of the laths, through atoms being added exclusively to their edges. Thus the laths tend to maintain a constant thickness. Typically this is of about three or four mica layers. The thickness, then, is a replicating feature, although variations can result occasionally from accidents of growth. Now we might ask: Is there a selective advantage in being about three or four mica layers thick? Perhaps. The illites grow slowly from "nutrient" solutions flowing within the pores of the sandstone. Crystals that grow too thick would clog the pores and reduce or stop the nutrient flow. They would not propagate well. Crystals that were too thin would be liable to break, or break away from their attachment to the sand grains, and that too would clog the pores downstream. (This actually happens if too much pressure is applied in trying to move oil through submarine sandstones.) So perhaps to keep local nutrient solutions flowing it is better to be about three or four units thick!

Of course, the amount of information involved here is *tiny*, and my explanation may not be correct in this case. But notice that even in such a case one can start talking like a biologist in seeking explanations in terms not only of thermodynamic stability, or availability of cations, or ease of formation and so on, but also in terms of whether or not a structural feature is adapted to its own survival and propagation. Evolution through natural selection can start at very low levels indeed given appropriate materials. We may call these genetic materials, even if only of the most modest sort.

The imagined next step would be a collaboration of different materials replicating different survival promoting features ("functions"):

$$[g_1 \rightarrow f_1;\ g_2 \rightarrow f_2;\ g_3 \rightarrow f_3 \text{ etc.}]$$

I will not attempt to identify the various early functions (although see [3] for somewhat more detailed speculations) except to remark that for mineral crystal genes such functions might have included the manipulation of locally produced organic molecules. There is now a considerable literature on the activities of clay and similar minerals in adsorbing organic molecules in an orderly way in the spaces between layers or at their edges, as well as in catalyzing organic chemical reactions.[3] And there are reasons for thinking that such activities could be functional in the sense discussed above of promoting survival and propagation: if you were a clay gene, it would be a smart thing to surround yourself with organic molecules that would assist your replicative growth, or in some way protect you when local conditions, such as pH, changed for the worse. Organic acids, for example citric acid, are known actively to assist the crystallization of clay minerals by transporting otherwise insoluble cations such as aluminum [19].

Multifunctional genetic materials would be another kind of advance:

$$[..........G_x \rightarrow f_n, f_{n+1} \text{ etc. }].$$

When eventually organic molecules could be synthesized consistently by well-organized mineral-genetic assemblages, then replicating organic polymers would become a possibility. An RNA-like polymer able to replicate and twist into different pieces of machinery according to the information it contains would be such a multifunctional genetic material.

Then we can follow something like the "RNA world" story [10, 9]. G_x (RNA) comes also to operate indirectly by controlling the synthesis of other molecules that are not themselves able to reproduce, but are specialists at making micromachinery ($Y =$ protein).

$$[..........G_x \rightarrow f_n, f_{n+1} \text{ etc. }] \searrow \searrow Y(f_p, f_{p+1} \text{ etc.})]$$

Finally an indirect acting, multifunctional genetic system is established (G_z = DNA) and the mineral scaffolding has finally been kicked away.

3. Reviewed in [5]. See also many papers in the journal *Origins of Life and Evolution of the Biosphere* giving accounts of research in which possible roles of clay and other similar minerals in the origin of life have been investigated.

$$[\ G_z \rightarrow G_x \rightarrow f_n,\ f_{n+1}\ \text{etc.}\ ] \searrow \searrow Y(f_p,\ f_{p+1}\ \text{etc.})]$$

This is "life as we know it." The structural and catalytic functions f_n, f_{n+1}, and so on are carried out by RNA directly; f_p, f_{p+1}, and so on are the far more numerous functions carried out by thousands of different kinds of proteins, each kind folded in its particular way as determined by its unique amino acid sequence, in turn controlled by a DNA sequence from which it is derived. It is a wonderfully sophisticated system, but surely far off the ground in evolutionary terms.

7.4. Crystal Gene Candidates

In seeking possible mineral genetic materials we should think about microcrystalline strong-bonded materials with crystal structures possessing some superimposed, randomlike variability (broadly comparable to the variability of covalently bonded stacking sequences in DNA molecules) so that an individual crystal can be unique. Many such "defects" affect physicochemical properties of crystalline materials—shapes and sizes of crystals, ability to adsorb small molecules, catalytic effects, and so on. The critical question is whether any such "defect information" is stable and capable of replicating through the processes of crystal growth. The answer is that while most stable "defects" do not replicate, there are several that do, and some that do so spectacularly.

Here is an example. The unit layers in micas and mica-like clays (referred to earlier) have quite a complex sandwich structure with seven planes of atoms within them. These composite unit layers have negative charges and are stacked on top of each other in the crystal via intermediate cations. However, the unit layers have a somewhat asymmetric structure. The patterns of their top and bottom planes of oxygen atoms, although the same, do not lie directly over each other within a unit layer. This offset gives a directionality to the layer as a whole, a direction that can be represented by a little arrow. The question then arises as to how the "arrows" in successive layers stacked on top of each other are arranged.

Often they all lie the same way. This can be formally represented as follows:

$$\rightarrow \rightarrow \rightarrow \rightarrow \rightarrow \rightarrow \rightarrow$$

This example shows how seven successive layers are stacked on top of each other. Often, too, they stack with an alternating sequence of directions:

→ ↖ → ↖ → ↖

These represent the two most common "regular polytypes" of mica. More usually there are some irregularities—"disordered polytypes"—which could in principle hold information as a particular irregularity of stacking (somewhat as in DNA). Furthermore, there are cases where some apparently disordered feature occurs at absolutely regular intervals. For example, the following were observed in samples of biotite mica [1].

→→→→→→→ ↖→→→→→→→ ↖→→→→→→→ ↖

→→↙→↖↖→→→↙→↖↖→→→↙→↖↖→→→↙→↖↖→

It looks very much as if such long-range features result from a copying process taking place during crystal growth.[4]

Another similar form of permutation, found in clays and elsewhere, is where a material consists of a stack of chemically different layers, which may be arranged in more or less irregular sequences. Barium ferrite is a particularly striking example. Here long-range repetitions of irregular stacking features have frequently been observed over distances of one hundred nanometers and more [12]. At Glasgow and Paisley we have studied the crystal growth of this material and have suggested a copying mechanism to account for the long-range repeats [20]. According to this suggestion, an initial plate consisting of a randomly stacked sequence of layers grows by sideways addition of atoms so that the initial stack sequence is preserved. Like illite the layer stacks are tough and flexible. From the micromorphology of early products, crystal growth appears to proceed unevenly in different directions to make flexible branching fronds ("seaweed growth") in such a way that different parts of the same layer stack come to overlap and then fuse together to make larger pieces of crystal within which the stacking sequence in the initial plate is repeated, perhaps many times. Barium ferrite itself proba-

4. A screw dislocation growth mechanism may produce such repetitions, as evidenced by characteristic spiral surface features on crystals [8]. However, such features are notably absent in cases where very long repeats have been found. An alternative copying mechanism has been suggested for these (see [20]).

bly only grows at high temperatures (our experiments were done at about 1300° C), but we are inclined to think that "seaweed growth" provides a general mechanism for the really long repeats found in a number of materials, including the mica polytypes discussed above.

Clay and micas can crystallize at ordinary temperatures, even if only slowly. Rapid synthesis of mica calls for at least hydrothermal conditions of a few hundred degrees Celsius. This fits with the idea that the earliest stages of evolution might have taken place in oceanic hydrothermal systems [6, 11].

7.5. Long-Term Survival and Different Kinds of Panspermia

There are many kinds of organisms that can survive difficult conditions through a state of so-called "suspended animation": plant seeds and bacterial spores are examples. These might be said to be potential forms of life maintaining one of the two essential prerequisites for a fully living system—evolved information. The other key requirement is that there exist open systems to which this information refers, but these do not need to be permanent—they only need to be recurrent.

Perhaps we can think of ways in which organisms might even evolve a competence in space travel, overcoming not only "drought" and "famine" but low temperatures, immense timescales, and cosmic radiation. Suppose there were a pair of planets somewhat like Mars and the earth exchanging fragments of material resulting from meteorite impacts. This exchange might create selection pressures in favor of micro-organisms that could hitch a lift and survive these journeys (because periodically "Mars" would be a better place to live than "the earth" and *vice versa*)—that is, there would be a selection pressure for space travel, including the ability to survive through long periods of shut-down. Thus, perhaps, spores might evolve that could survive for possibly millions of years and be able to "come to life" in the rare event of landing in a suitable place.

We might imagine then a kind of primitive panspermia, in which the spores in question were of organisms in which the genetic memory and its essential control machinery were inorganic, that is to say not made of organic molecules. Information held as (say) a stacking sequence of extensive layers, as in mica, would be less easily destroyed by cosmic radiation. This would be to imagine a primitive life form, but an evolved one all the same. Such a form would be storing the information that would "bring it back to life" when suitable conditions were encountered: it would be set perhaps to incorporate and manipulate organic molecules and so possibly shorten the initiation stages for the evolution of higher forms.

Then one can think of kinds of "advanced panspermia" depending on genetic takeovers. The first kind would arise from late takeovers driven by the selection pressures of spore space travel, using perhaps new, more robust organic polymers, or reverting to solid state devices.

Another possibility for advanced panspermia is nearer to home. Appalled by the difficulties of imagining nucleic acid arising from a primordial soup on the primitive earth, Crick and Orgel introduced the idea of "directed panspermia" [7]. This is a way of explaining the origin of life on the earth—that it was put there by spacemen. Crick and Orgel were not too serious about the idea, but it isn't so crazy when we realize that we humans might be just such spacemen in the near future. We are surely near to designing artificial organisms: and I mean real organisms, not just computer simulations. Perhaps we could design ones that would survive far into the future.

When we imitate nature, for example when we make flying machines or thinking machines, it is usual that certain general principles will be analogous, but the materials used are seldom the same. The airliner splays the control surfaces of its wings as it comes in to land. It is like a swan splaying its feathers, except that airplanes don't have feathers. I expect that when we come to design free living, evolving machines, for whatever purposes, we will most likely choose other materials than "the molecules of life" and, among these, other genetic materials [4].

7.6. Summary and Conclusion

The major part of this chapter has been looking backward. I started with the speculation that the unity of biochemistry resulted from an elaborate frozen accident of early evolution: all forms of life now on Earth have a similar biochemistry because their central control systems became fixed during early evolution as a result of the irreversible loss of earlier control systems. The suggestion here was that what we might call "low tech" starter systems, not requiring the collaboration of many components, were replaced by a more efficient "high tech" system consisting of elaborately interdependent components.

Such a perspective naturally leads to the idea that similar mechanisms may be operating now. New life should still be emerging in different parts of the universe. There should be forms of life that are at different stages of evolution. Among advanced forms there should be different "high tech" systems that had happened to have been frozen in—increasing the chances that some of these biochemical systems might be better adapted than ours to persist into the far future. And then when it comes to forms of life invented by humankind, among desirable features to be built into some of the specifications might be hyperlongevity, the ability to last within far future conditions. All such ideas depend on seeing a strong distinction between "life as we know it" and "life in general."

REFERENCES

1. Baronnet, A., and Kang, Z. C., "About the Origin of Mica Polytypes," *Phase Transitions, 16/17*, 477–93 (1989).

2. Cairns-Smith, A. G., *The Life Puzzle* (Toronto University Press, Toronto, 1971), 63–64.

3. Cairns-Smith, A. G., *Genetic Takeover and the Mineral Origins of Life* (Cambridge University Press, Cambridge, 1982).

4. Cairns-Smith, A. G., "The Chemistry of Materials for Artificial Darwinian Systems," *Int. Rev. Phys. Chem., 7*, 209–50 (1988).

5. Cairns-Smith, A. G., "The Origin of Life: Clays," in *Frontiers of Biology*, vol. 1, eds. D. Baltimore, R. Dulbecco, F. Jacob, and R. Levi-Montalcini (Academic Press, New York, 2001), 169–92.

6. Corliss, J. B., "Hot Springs and the Origin of Life," *Nature, 347*, 624 (1990).

7. Crick, F., and Orgel, L. E., "Directed Panspermia," *Icarus, 19*, 341–46 (1973).

8. Frank, F. C., *Phil. Mag., 42*, 1014 (1951).

9. Gesteland, R. F., Cech, T. R., and Atkins, J. F., *The RNA World*, 2nd ed. (Cold Spring Harbor Press, Cold Spring Harbor, N.Y., 1999).

10. Gilbert, W., "The RNA World," *Nature, 319*, 618 (1986).

11. Holm, N. G., ed., *Marine Hydrothermal Systems and the Origin of Life* (Kluwer Academic Publications, Dordrecht, 1992).

12. Kohn, J. A., Eckart, D. W., and Cook, C. F., "Crystallography of the Hexagonal Ferrites," *Science, 172*, 519–25 (1971).

13. Maurette, M., "The Oklo Reactor," *Annual Reviews of Nuclear and Particle Science, 26*, 319 (1976). Described in Barrow, J. D., *Impossibility: The Limits of Science and the Science of Limits* (Oxford University Press, Oxford, 1998), 187.

14. McHardy, W. J., Wilson, M. J., and Tait, J. M., "Electron Microscope and X-ray Diffraction Studies of Filamentous Illitic Clay from Sandstones of the Magnus Field," *Clay Minerals, 17*, 23–39 (1982).

15. Schrödinger, E., *What Is Life?* (Cambridge University Press, Cambridge, 1944), chapter 4.

16. Shapiro, R., "Prebiotic Ribose Synthesis: A Critical Analysis," *Orig. Life Evol. Bio.*, *18*, 71–85 (1988).

17. Shapiro, R., "The Pre-biotic Role of Adenine: A Critical Analysis," *Orig. Life Evol. Bio.*, *25*, 83–98 (1995).

18. Sherrington, C., *Man on His Nature* (1937 Gifford Lecture; Cambridge University Press, Cambridge, 1940), chapter 3.

19. Siffert, B., "Clay Synthesis: The Role of Organic Complexing Agents," in *Clay Minerals and the Origin of Life*, eds. A. G. Cairns-Smith and H. Hartman (Cambridge University Press, Cambridge, 1986), 75–78.

20. Turner, G., Stewart, B., Baird, T., Peacock, R. D., and Cairns-Smith, A.G., "Layer Morphology and Growth Mechanisms in Barium Ferrites," *J. Cry. Gr.*, *158*, 276–83 (1996).

[8]

TIME WITHOUT END

Physics and Biology in an Open Universe

Freeman J. Dyson

Quantitative estimates are derived for three classes of phenomena that may occur in an open cosmological model of Friedmann type. (1) Normal physical processes taking place with very long time-scales. (2) Biological processes that will result if life adapts itself to low ambient temperatures according to a postulated scaling law. (3) Communication by radio between life forms existing in different parts of the universe. The general conclusion of the analysis is that an open universe need not evolve into a state of permanent quiescence. Life and communication can continue forever, utilizing a finite store of energy, if the assumed scaling laws are valid.

Lecture I. Philosophy

A year ago Steven Weinberg published an excellent book, *The First Three Minutes* (Weinberg, 1977), explaining to a lay audience the state of our

Note: This material was originally presented as four lectures, the "James Arthur Lectures on Time and Its Mysteries" at New York University, Autumn 1978. The first lecture is addressed to a general audience, the other three to an audience of physicists and astronomers. Published as F. J. Dyson, *Reviews of Modern Physics, 51, no. 3,* 447–60 (July 1979).

knowledge about the beginning of the universe. In his sixth chapter he describes in detail how progress in understanding and observing the universe was delayed by the timidity of theorists.

"This is often the way it is in physics—our mistake is not that we take our theories too seriously, but that we do not take them seriously enough. It is always hard to realize that these numbers and equations we play with at our desks have something to do with the real world. Even worse, there often seems to be a general agreement that certain phenomena are just not fit subjects for respectable theoretical and experimental effort. Alpher, Herman and Gamow (1948) deserve tremendous credit above all for being willing to take the early universe seriously, for working out what known physical laws have to say about the first three minutes. Yet even they did not take the final step, to convince the radio astronomers that they ought to look for a microwave radiation background. The most important thing accomplished by the ultimate discovery of the 3°K radiation background (Penzias and Wilson, 1965) was to force all of us to take seriously the idea that there *was* an early universe."

Thanks to Penzias and Wilson, Weinberg and others, the study of the beginning of the universe is now respectable. Professional physicists who investigate the first three minutes or the first microsecond no longer need to feel shy when they talk about their work. But the end of the universe is another matter. I have searched the literature for papers about the end of the universe and found very few (Rees, 1969; Davies, 1973; Islam, 1977 and 1979; Barrow and Tipler, 1978). This list is certainly not complete. But the striking thing about these papers is that they are written in an apologetic or jocular style, as if the authors were begging us not to take them seriously. The study of the remote future still seems to be as disreputable today as the study of the remote past was thirty years ago. I am particularly indebted to Jamal Islam for an early draft of his 1977 paper, which started me thinking seriously about the remote future. I hope with these lectures to hasten the arrival of the day when eschatology, the study of the end of the universe, will be a respectable scientific discipline and not merely a branch of theology.

Weinberg himself is not immune to the prejudices that I am trying to dispel. At the end of his book about the past history of the universe, he adds a short chapter about the future. He takes 150 pages to describe the first three minutes, and then dismisses the whole of the future in five pages. Without any discussion of technical details, he sums up his view of the future in twelve words:

"The more the universe seems comprehensible, the more it also seems pointless."

Weinberg has here, perhaps unintentionally, identified a real problem. It is impossible to calculate in detail the long-range future of the universe without including the effects of life and intelligence. It is impossible to calculate the capabilities of life and intelligence without touching, at least peripherally, philosophical questions. If we are to examine how intelligent life may be able to guide the physical development of the universe for its own purposes, we cannot altogether avoid considering what the values and purposes of intelligent life may be. But as soon as we mention the words value and purpose, we run into one of the most firmly entrenched taboos of twentieth-century science. Hear the voice of Jacques Monod (1970), high priest of scientific rationality, in his book *Chance and Necessity:*

"Any mingling of knowledge with values is unlawful, forbidden."

Monod was one of the seminal minds in the flowering of molecular biology in this century. It takes some courage to defy his anathema. But I will defy him, and encourage others to do so. The taboo against mixing knowledge with values arose during the nineteenth century out of the great battle between the evolutionary biologists led by Thomas Huxley and the churchmen led by Bishop Wilberforce. Huxley won the battle, but a hundred years later Monod and Weinberg were still fighting Bishop Wilberforce's ghost. Physicists today have no reason to be afraid of Wilberforce's ghost. If our analysis of the long-range future leads us to raise questions related to the ultimate meaning and purpose of life, then let us examine these questions boldly and without embarrassment. If our answers to these questions are naive and preliminary, so much the better for the continued vitality of our science.

I propose in these lectures to explore the future as Weinberg in his book explored the past. My arguments will be rough and simple but always quantitative. The aim is to establish numerical bounds within which the destiny of the universe must lie. I shall make no further apology for mixing philosophical speculations with mathematical equations.

The two simplest cosmological models (Weinberg, 1972) describe a uniform zero-pressure universe which may be either closed or open. The closed universe has its geometry described by the metric

$$ds^2 = R^2[d\psi^2 - d\chi^2 - \sin^2\chi d\Omega^2], \tag{1}$$

where χ is a space coordinate moving with the matter, ψ is a time coordinate related to physical time t by

$$t = T_0 (\psi - \sin\psi), \tag{2}$$

and R is the radius of the universe given by

$$R = cT_0 (1 - \cos\psi). \tag{3}$$

The whole universe is represented in terms of the coordinates (ψ,χ) by a finite rectangular box

$$0 < \psi < 2\pi, 0 < \chi < \pi. \tag{4}$$

This universe is closed both in space and in time. Its total duration is

$$2\pi T_0, \tag{5}$$

where T_0 is a quantity that is in principle measurable. If our universe is described by this model, then T_0 must be at least 10^{10} years.

The simple model of a uniform zero-pressure open universe has instead of (1) the metric

$$ds^2 = R^2[d\psi^2 - d\chi^2 - \sinh^2\chi d\Omega^2], \tag{6}$$

where now

$$t = T_0(\sinh\psi - \psi), \tag{7}$$

$$R = cT_0(\cosh\psi - 1), \tag{8}$$

and the coordinates *(?,?)* extend over an infinite range

$$0 < \psi < \infty, 0 < \psi < \psi. \tag{9}$$

The open universe is infinite both in space and in time.

The models (1) and (6) are only the simplest possibilities. Many more complicated models can be found in the literature. For my purpose it is sufficient to discuss (1) and (6) as representative of closed and open universes. The great question, whether our universe is in fact closed or open, will before long be settled by observation. I do not say more about this question, except to remark that my philosophical bias strongly favors an open universe and that the observational evidence does not exclude it (Gott, Gunn, Schramm, and Tinsley, 1974 and 1976).

The prevailing view (Weinberg, 1977) holds the future of open and closed universes to be equally dismal. According to this view, we have only the choice of being fried in a closed universe or frozen in an open one. The end of the closed universe has been studied in detail by Rees (1969). Regrettably, I have to concur with Rees' verdict that in this case we have no escape from frying. No matter how deep we burrow into the earth to shield ourselves from the ever-increasing fury of the blue-shifted background radiation, we can only postpone by a few million years our miserable end. I shall not discuss the closed universe in detail, since it gives me a feeling of claustrophobia to imagine our whole existence confined within the box (4). I only raise one question which may offer us a thin chance of survival. Supposing that we discover the universe to be naturally closed and doomed to collapse. Is it conceivable that by intelligent intervention, converting matter into radiation and causing energy to flow purposefully on a cosmic scale, we could break open a closed universe and change the topology of space-time so that only a part of it would collapse and another part of it would expand forever? I do not know the answer to this question. If it turns out that the universe is closed, we shall still have about 10^{10} years to explore the possibility of a technological fix that would burst it open.

I am mainly interested in the open cosmology, since it seems to give enormously greater scope for the activities of life and intelligence. Horizons in the open cosmology expand indefinitely. To be precise, the distance to the horizon in the metric (6) is

$$d = R\psi, \tag{10}$$

with R given by (8), and the number of galaxies visible within the horizon is

$$N = N_0(\sinh 2\psi - 2\psi), \tag{11}$$

where N_0 is a number of the order of 10^{10}.

Comparing (11) with (7), we see that the number of visible galaxies varies with t^2 at late times. It happens by a curious numerical accident that the angular size of a typical galaxy at time t is

$$\delta \sim 10^5 t^{-1} \text{rad}, \tag{12}$$

with t measured in years. Since (11) and (7) give

$$N \sim 10^{-10} t^2, \; N\delta^2 \sim 1, \tag{13}$$

it turns out that *the sky is always just filled with galaxies,* no matter how far into the future we go. As the apparent size of each galaxy dwindles, new galaxies constantly appear at the horizon to fill in the gaps. The light from the distant galaxies will be strongly red-shifted. But the sky will never become empty and dark, if we can tune our eyes to longer and longer wavelengths as time goes on.

I shall discuss three principal questions within the framework of the open universe with the metric (6).

(1) Does the universe freeze into a state of permanent physical quiescence as it expands and cools?

(2) Is it possible for life and intelligence to survive indefinitely?

(3) Is it possible to maintain communication and transmit information across the constantly expanding distances between galaxies?

These three questions will be discussed in detail in Lectures 2, 3, and 4. Tentatively, I shall answer them with a no, a yes, and a maybe. My answers are perhaps only a reflection of my optimistic philosophical bias. I do not expect everybody to agree with the answers. My purpose is to start people thinking seriously about the questions.

If, as I hope, my answers turn out to be right, what does it mean? It means that we have discovered in physics and astronomy an analog to the theorem of Gödel (1931) in pure mathematics. Gödel proved [see Nagel and Newman (1956)] that the world of pure mathematics is inexhaustible; no finite set of axioms and rules of inference can ever encompass the whole of mathematics; given any finite set of axioms, we can find meaningful mathematical questions that the axioms leave unanswered. I hope that an analogous situation exists in the physical world. If my view of the future is correct, it means that the world of physics and astronomy is also inexhaustible; no matter how far we go into the future, there will always be new things happening, new information coming in, new worlds to explore, a constantly expanding domain of life, consciousness, and memory.

When I talk in this style, I am mixing knowledge with values, disobeying Monod's prohibition. But I am in good company. Before the days of Darwin and Huxley and Bishop Wilberforce, in the eighteenth century, scientists were not subject to any taboo against mixing science and values. When Thomas Wright (1750), the discoverer of galaxies, announced his discovery, he was not afraid to use a theological argument to support an astronomical theory.

"Since as the Creation is, so is the Creator also magnified, we may conclude in consequence of an infinity, and an infinite all-active power, that as the visible creation is supposed to be full of siderial systems and planetary worlds, so on, in like similar manner, the endless immensity is an unlimited plenum of creations not unlike the known. . . . That this in all probability may be the real case, is in some degree made evident by the many cloudy spots, just perceivable by us, as far without our starry Regions, in which tho' visibly luminous spaces, no one star or particular constituent body can possibly be distinguished; those in all likelihood may be external creation, bordering upon the known one, too remote for even our telescopes to reach."

Thirty-five years later, Wright's speculations were confirmed by William Herschel's precise observations. Wright also computed the number of habitable worlds in our galaxy:

"In all together then we may safety reckon 170,000,000, and yet be much within compass, exclusive of the comets which I judge to be by far the most numerous part of the creation."

His statement about the comets may also be correct, although he does not tell us how he estimated their number. For him the existence of so many habitable worlds was not just a scientific hypothesis, but a cause for moral reflection:

"In this great celestial creation, the catastrophe of a world, such as ours, or even the total dissolution of a system of worlds, may possibly be no more to the great Author of Nature, than the most common accident in life with us, and in all probability such final and general Doomsdays may be as frequent there, as even Birthdays or mortality with us upon the earth. This idea has something so cheerful in it, that I know I can never look upon the stars without wondering why the whole world does not become astronomers; and that men endowed with sense and reason should neglect a science they are naturally so much interested in, and so capable of enlarging their understanding, as next to a demonstration must convince them of their immortality, and reconcile them to all those little difficulties incident to human nature, without the least anxiety.

"All this the vast apparent provision in the starry mansions seem to promise: What ought we then not to do, to preserve our natural birthright to it and to merit such inheritance, which alas we think created all to gratify alone a race of vain-glorious gigantic beings, while they are confined to this world, chained like so many atoms to a grain of sand."

There speaks the eighteenth century. But Steven Weinberg says, "The more the universe seems comprehensible, the more it also seems pointless." If Weinberg is speaking for the twentieth century, then I prefer the eighteenth.

Lecture II. Physics

In this lecture, following Islam (1977), I investigate the physical processes that will occur in an open universe over very long periods of time. I consider the natural universe undisturbed by effects of life and intelligence. Life and intelligence will be discussed in lectures 3 and 4.

Two assumptions underlie the discussion. (1) The laws of physics do not change with time. (2) The relevant laws of physics are already known to us. These two assumptions were also made by Weinberg (1977) in his description of the past. My justification for making them is the same as his. Whether or not we believe that the presently known laws of physics are the final and unchanging truth, it is illuminating to explore the consequences of these laws as far as we can reach into the past or the future. It is better to be too bold than too timid in extrapolating our knowledge from the known into the unknown. It may happen again, as it happened with the cosmological speculations of Alpher, Herman, and Gamow (1948), that a naive extrapolation of known laws into new territory will lead us to ask important new questions.

I have summarized elsewhere (Dyson, 1972, 1978) the evidence supporting the hypothesis that the laws of physics do not change. The most striking piece of evidence was discovered recently by Shlyakhter (1976) in the measurements of isotope ratios in ore samples taken from the natural fission reactor that operated about 2 billion years ago in the Oklo uranium mine in Gabon (Maurette, 1976). The crucial quantity is the ratio ($^{149}Sm/^{147}Sm$) between the abundances of two light isotopes of samarium, which are not fission products. In normal samarium this ratio is about 0.9; in the Oklo reactor it is about 0.02. Evidently, the ^{149}Sm has been heavily depleted by the dose of thermal neutrons to which it was exposed during the operation of the reactor. If we measure in a modern reactor the thermal neutron capture cross section of ^{149}Sm, we find the value 55 kb, dominated by a strong capture resonance at a neutron energy of 0.1 eV. A detailed analysis of the Oklo isotope ratios leads to the conclusion that the ^{149}Sm cross section was in the range 55 ± 8 kb two billion years ago. This means that the position of the capture resonance cannot have shifted by as much as 0.02 eV over 2.10^9 yr. But the position of this resonance measures the difference between the binding energies of the ^{149}Sm ground state and of the ^{150}Sm compound state into which the neutron is captured. These binding energies are each of the order of 10^9 eV and depend in a complicated way upon the strengths of nuclear and Coulomb interactions. The fact that the two binding energies remained in balance to an accuracy of 2 parts in 10^{11} over 2.10^9 yr indicates that the strengths of nuclear and Coulomb forces cannot have varied by more than a few parts in 10^{18} per year. This is by far the most sensitive test that we have yet found of the constancy of the laws of physics. The fact that no evidence of change was found does not, of course, prove that the laws are strictly constant. In particular, it does not exclude the possibility of a variation in strength of gravitational forces with a time scale much shorter than 10^{18} yr. For the sake of simplicity, I assume that the laws are strictly constant. Any other assumption would be more complicated and would introduce additional arbitrary hypotheses.

It is in principle impossible for me to bring experimental evidence to support the hypothesis that the laws of physics relevant to the remote future are already known to us. The most serious uncertainty affecting the ultimate fate of the universe is the question whether the proton is absolutely stable against decay into lighter particles. If the proton is unstable, all matter is transitory and must dissolve into radiation. Some serious theoretical arguments have been put forward (Zeldovich, 1977; Barrow and Tipler, 1978; Feinberg, Goldhaber, and Steigman, 1978) supporting the view that the proton should decay with a long half-life, perhaps through virtual processes involving black holes. The experimental limits on the rate of proton decay (Kropp and Reines, 1965) do not exclude the existence of such processes.

Again, on grounds of simplicity, I disregard these possibilities and suppose the proton to be absolutely stable. I will discuss in detail later the effect of real processes involving black holes on the stability of matter in bulk.

I am now ready to begin the discussion of physical processes that will occur in the open cosmology (6), going successively to longer and longer time scales. Classical astronomical processes come first, quantum-mechanical processes later.

Note added in proof. Since these lectures were given, a spate of papers has appeared discussing grand unification models of particle physics in which the proton is unstable (Nanopoulos, 1978; Pati, 1979; Turner and Schramm, 1979).

A. STELLAR EVOLUTION

The longest-lived low-mass stars will exhaust their hydrogen fuel, contract into white dwarf configurations, and cool down to very low temperatures, within times of the order of 10^{14} years. Stars of larger mass will take a shorter time to reach a cold final state, which may be a white dwarf, a neutron star, or a black hole configuration, depending on the details of their evolution.

B. DETACHMENT OF PLANETS FROM STARS

The average time required to detach a planet from a star by a close encounter with a second star is

$$T = (\rho V \sigma)^{-1}, \tag{14}$$

where ρ is the density of stars in space, V the mean relative velocity of two stars, and s the cross section for an encounter resulting in detachment. For the earth-sun system, moving in the outer regions of the disk of a spiral galaxy, approximate numerical values are

$$\rho = 3.10^{-41}\text{km}^{-3}, \tag{15}$$

$$V = 50\text{km/sec}, \tag{16}$$

$$\sigma = 2.10^{16}\text{km}^2, \tag{17}$$

$$T = 10^{15}\text{yr}. \tag{18}$$

The time scale for an encounter causing serious disruption of planetary orbits will be considerably shorter than 10^{15} yr.

C. DETACHMENT OF STARS FROM GALAXIES

The dynamical evolution of galaxies is a complicated process, not yet completely understood. I give here only a very rough estimate of the time scale. If a galaxy consists of N stars of mass M in a volume of radius R, their root-mean-square velocity will be of order

$$V = [GNM/R]^{1/2}. \tag{19}$$

The cross section for a close encounter between two stars, changing their directions of motion by a large angle, is

$$\sigma = (GM/V^2)^2 = (R/N)^2. \tag{20}$$

The average time that a star spends between two close encounters is

$$T = (\rho V \sigma)^{-1} = (NR^3/GM)^{1/2}. \tag{21}$$

If we are considering a typical large galaxy with $N = 10^{11}$, $R = 3.10^{17}$ km, then

$$T = 10^{19} \text{ yr}. \tag{22}$$

Dynamical relaxation of the galaxy proceeds mainly through distant stellar encounters with a time scale

$$T_R = T(\log N)^{-1} = 10^{18} \text{ yr}. \tag{23}$$

The combined effect of dynamical relaxation and close encounters is to produce a collapse of the central regions of the galaxy into a black hole, together with an evaporation of stars from the outer regions. The evaporated stars achieve escape velocity and become detached from the galaxy after a time of the order 10^{19} yr. We do not know what fraction of the mass of the galaxy ultimately collapses and what fraction escapes. The fraction escaping probably lies between 90% and 99%.

The violent events which we now observe occurring in the central regions of many galaxies are probably caused by a similar process of dynamical evolution operating on a much shorter time scale. According to (21), the time scale for evolution and collapse will be short if the dynamical units are few and massive, for example compact star clusters and gas clouds rather than individual stars. The long time scale (22) applies to a galaxy containing no dynamical units larger than individual stars.

D. DECAY OF ORBITS BY GRAVITATIONAL RADIATION

If a mass is orbiting around a fixed center with velocity V, period P, and kinetic energy E, it will lose energy by gravitational radiation at a rate of order

$$E_g = (V/c)^5(E/P). \tag{24}$$

Any gravitationally bound system of objects orbiting around each other will decay by this mechanism of radiation drag with a time scale

$$T_g = (c/V)^5 P. \tag{25}$$

For the earth orbiting around the sun, the gravitational radiation time scale is

$$T_g = 10^{20} \text{ yr}. \tag{26}$$

Since this is much longer than (18), the earth will almost certainly escape from the sun before gravitational radiation can pull it inward. But if it should happen that the sun should escape from the galaxy with the earth still attached to it, then the earth will ultimately coalesce with the sun after a time of order (26).

The orbits of the stars in a galaxy will also be decaying by gravitational radiation with time scale (25), where P is now the period of their galactic orbits. For a galaxy like our own, with $V = 200$ km/sec and $P = 2.10^8$ yr, the time scale is

$$T_g = 10^{24} \text{ yr}. \tag{27}$$

This is again much longer than (22), showing that dynamical relaxation dominates gravitational radiation in the evolution of galaxies.

E. DECAY OF BLACK HOLES BY THE HAWKING PROCESS

According to Hawking (1975), every black hole of mass M decays by emission of thermal radiation and finally disappears after a time

$$T = (G^2M^3/\hbar c^4). \tag{28}$$

For a black hole of one solar mass the lifetime is

$$T = 10^{64} \text{ yr.} \tag{29}$$

Black holes of galactic mass will have lifetimes extending up to 10^{100} yr. At the end of its life, every black hole will become for a short time very bright. In the last second of its existence it will emit about 10^{31} erg of high-temperature radiation. The cold expanding universe will be illuminated by occasional fireworks for a very long time.

F. MATTER IS LIQUID AT ZERO TEMPERATURE

I next discuss a group of physical processes which occur in ordinary matter at zero temperature as a result of quantum-mechanical barrier penetration. The lifetimes for such processes are given by the Gamow formula

$$T = \exp(S)\, T_0, \tag{30}$$

where T_0 is a natural vibration period of the system, and S is the action integral

$$S = (2/\hbar) \int (2MU(x))^{1/2} dx\,. \tag{31}$$

Here x is a coordinate measuring the state of the system as it goes across the barrier, and $U(x)$ is the height of the barrier as a function of x. To obtain a rough estimate of S, I replace (31) by

$$S = (8MUd^2/\hbar^2)^{1/2}, \tag{32}$$

where d is the thickness, and U the average height of the barrier, and M is the mass of the object that is moving across it. I shall consider processes for which S is large, so that the lifetime (30) is extremely long.

As an example, consider the behavior of a lump of matter, a rock or a planet, after it has cooled to zero temperature. Its atoms are frozen into an apparently fixed arrangement by the forces of cohesion and chemical bonding. But from time to time, the atoms will move and rearrange themselves, crossing energy barriers by quantum-mechanical tunneling. The height of the barrier will typically be of the order of a tenth of a Rydberg unit,

$$U = (1/20)(e^4 m/\hbar^2), \quad (33)$$

and the thickness will be of the order of a Bohr radius

$$d = (\hbar^2/me^2), \quad (34)$$

where m is the electron mass. The action integral (32) is then

$$S = (2Am_p/5m)^{1/2} = 27A^{1/2}, \quad (35)$$

where m_p is the proton mass, and A is the atomic weight of the moving atom. For an iron atom with $A = 56$, $S = 200$, and (30) gives

$$T = 10^{65} \text{ yr}. \quad (36)$$

Even the most rigid materials cannot preserve their shapes or their chemical structures for times long compared with (36). On a time scale of 10^{65} yr, every piece of rock behaves like a liquid, flowing into a spherical shape under the influence of gravity. Its atoms and molecules will be ceaselessly diffusing around like the molecules in a drop of water.

G. ALL MATTER DECAYS TO IRON

In matter at zero temperature, nuclear as well as chemical reactions will continue to occur. Elements heavier than iron will decay to iron by various processes such as fission and alpha emission. Elements lighter than iron will combine by nuclear fusion reactions, building gradually up to iron. Consider for example the fusion reaction in which two nuclei of atomic weight $\frac{1}{2}A$, charge $\frac{1}{2}Z$ combine to form a nucleus (A, Z). The Coulomb repulsion of the two nuclei is effectively screened by electrons until they come within a distance

$$d = Z^{-1/3}(\hbar^2/me^2) \quad (37)$$

of each other. The Coulomb barrier has thickness d and height

$$U = (Z^2e^2/4d) = \tfrac{1}{4}Z^{7/3}(e^4m/\hbar^2). \quad (38)$$

The reduced mass for the relative motion of the two nuclei is

$$M = \tfrac{1}{4}Am_p. \quad (39)$$

The action integral (32) then becomes

$$S = (\tfrac{1}{2}AZ^{5/3}(m_p/m))^{1/2} = 30A^{1/2}Z^{5/6}. \qquad (40)$$

For two nuclei combining to form iron, $Z = 26$, $A = 56$, $S = 3500$, and

$$T = 10^{1500}\ \text{yr}. \qquad (41)$$

On the time scale (41), ordinary matter is radioactive and is constantly generating nuclear energy.

H. COLLAPSE OF IRON STAR TO NEUTRON STAR

After the time (41) has elapsed, most of the matter in the universe is in the form of ordinary low-mass stars that have settled down into white dwarf configurations and become cold spheres of pure iron. But an iron star is still not in its state of lowest energy. It could release a huge amount of energy if it could collapse into a neutron star configuration. To collapse, it has only to penetrate a barrier of finite height and thickness. It is an interesting question, whether there is an unsymmetrical mode of collapse passing over a lower saddle point than the symmetric mode. I have not been able to find a plausible unsymmetric mode, and so I assume the collapse to be spherically symmetrical. In the action integral (31), the coordinate x will be the radius of the star, and the integral will extend from r, the radius of a neutron star, to R, the radius of the iron star from which the collapse begins. The barrier height $U(x)$ will depend on the equation of state of the matter, which is very uncertain when x is close to r. Fortunately the equation of state is well known over the major part of the range of integration, when x is large compared to r and the main contribution to $U(x)$ is the energy of nonrelativistic degenerate electrons

$$U(x) = (N^{5/3}\hbar^2/2mx^2), \qquad (42)$$

where N is the number of electrons in the star.

The integration over x in (31) gives a logarithm

$$log(R/R_0), \qquad (43)$$

where R_0 is the radius at which the electrons become relativistic and the formula (42) fails. For low-mass stars the logarithm will be of the order of unity, and the part of the integral coming from the relativistic region $x<R_0$ will also be of the order of unity. The mass of the star is

$$M = 2Nm_p. \tag{44}$$

I replace the logarithm (43) by unity and obtain for the action integral (31) the estimate

$$S = N^{4/3}(8m_p/m)^{1/2} = 120N^{4/3}. \tag{45}$$

The lifetime is then by (30)

$$T = \exp(120N^{4/3})\,T_0. \tag{46}$$

For a typical low-mass star we have

$$N = 10^{56},\ S = 10^{77},\ T = 10^{10^{76}}\text{yr}. \tag{47}$$

In (46) it is completely immaterial whether T_0 is a small fraction of a second or a large number of years.

We do not know whether every collapse of an iron star into a neutron star will produce a supernova explosion. At the very least, it will produce a huge outburst of energy in the form of neutrinos and a modest outburst of energy in the form of x-rays and visible light. The universe will still be producing occasional fireworks after times as long as (47).

I. COLLAPSE OF ORDINARY MATTER TO BLACK HOLES

The long lifetime (47) of iron stars is only correct if they do not collapse with a shorter lifetime into black holes. For collapse of any piece of bulk matter into a black hole, the same formulae apply as for collapse into a neutron star. The only difference is that the integration in the action integral (31) now extends down to the black hole radius instead of to the neutron star radius. The main part of the integral comes from larger values of x and is the same in both cases. The lifetime for collapse into a black hole is therefore still given by (46). But there is an important change in the meaning of N. If small black holes are possible, a small part of a star can collapse by itself into a black hole. Once a small black hole has been formed, it will in a short time swallow the rest of the star. The lifetime for collapse of any star is then given by

$$T = \exp(120 N_B^{4/3}) T_0, \tag{48}$$

where N_B is the number of electrons in a piece of iron of mass equal to the minimum mass M_B of a black hole. The lifetime (48) is the same for any piece of matter of mass greater than M_B. Matter in pieces with mass smaller than M_B is absolutely stable. For a more complete discussion of the problem of collapse into black holes, see Harrison, Thorne, Wakano, and Wheeler (1965).

The numerical value of the lifetime (48) depends on the value of M_B. All that we know for sure is

$$0 \leq M_B \leq M_c, \tag{49}$$

where

$$M_c = (\hbar c/G)^{3/2} m_p^{-2} = 4.10^{33}\,\text{g} \tag{50}$$

is the Chandrasekhar mass. Black holes must exist for every mass larger than M_c, because stars with mass larger than M_c have no stable final state and must inevitably collapse.

Four hypotheses concerning M_B have been put forward.

(i) $M_B = 0$. Then black holes of arbitrarily small mass exist and the formula (48) is meaningless. In this case all matter is unstable with a comparatively short lifetime, as suggested by Zeldovich (1977).

(ii) M_B is equal to the Planck mass

$$M_B = M_{PL} = (\hbar c/G)^{1/2} = 2.10^{-5}\,\text{g}. \tag{51}$$

This value of M_B is suggested by Hawking's theory of radiation from black holes (Hawking, 1975), according to which every black hole loses mass until it reaches a mass of order M_{PL}, at which point it disappears in a burst of radiation. In this case (48) gives

$$N_B = 10^{19},\ T = 10^{10^{26}}\,\text{yr}. \tag{52}$$

(iii) M_B is equal to the quantum mass

$$M_B = M_Q = (\hbar c/G m_p) = 3.10^{14}\,\text{g}, \tag{53}$$

TABLE 1. Summary of Time Scales.

Closed Universe	
Total duration	10^{11} yr
Open Universe	
Low-mass stars cool off	10^{14} yr
Planets detached from stars	10^{15} yr
Stars detached from galaxies	10^{19} yr
Decay of orbits by gravitational radiation	10^{20} yr
Decay of black holes by Hawking process	10^{64} yr
Matter liquid at zero temperature	10^{65} yr
All matter decays to iron	10^{1500} yr
Collapse of ordinary matter to black hole [alternative (ii)]	$10^{10^{26}}$ yr
Collapse of stars to neutron stars or black holes [alternative (iv)]	$10^{10^{76}}$ yr

as suggested by Harrison, Thorne, Wakano, and Wheeler (1965). Here M_Q is the mass of the smallest black hole for which a classical description is meaningful. Only for masses larger than M_Q can we consider the barrier penetration formula (31) to be physically justified. If (53) holds, then

$$N_B = 10^{38},\ T = 10^{10^{52}} \text{ yr.} \tag{54}$$

(iv) M_B is equal to the Chandrasekhar mass (50). In this case the lifetime for collapse into a black hole is of the same order as the lifetime (47) for collapse into a neutron star.

The long-range future of the universe depends crucially on which of these four alternatives is correct. If (iv) is correct, stars may collapse into black holes and dissolve into pure radiation, but masses of planetary size exist forever. If (iii) is correct, planets will disappear with the lifetime (54), but material objects with masses up to a few million tons are stable. If (ii) is correct, human-sized objects will disappear with the lifetime (52), but dust grains with diameter less than about 100 μ will last for ever. If (i) is correct, all material objects will disappear and only radiation is left.

If I were compelled to choose one of the four alternatives as more likely than the others, I would choose (ii). I consider (iii) and (iv) unlikely because they are inconsistent with Hawking's theory of black-hole radiation. I find

(i) implausible because it is difficult to see why a proton should not decay rapidly if it can decay at all. But in our present state of ignorance, none of the four possibilities can be excluded.

The results of this lecture are summarized in Table 1. This list of time scales of physical processes makes no claim to be complete. Undoubtedly many other physical processes will be occurring with time scales as long as, or longer than, those I have listed. The main conclusion I wish to draw from my analysis is the following: So far as we can imagine into the future, things continue to happen. In the open cosmology, history has no end.

Lecture III. Biology

Looking at the past history of life, we see that it takes about 10^6 years to evolve a new species, 10^7 years to evolve a genus, 10^8 years to evolve a class, 10^9 years to evolve a phylum, and less than 10^{10} years to evolve all the way from the primaeval slime to Homo Sapiens. If life continues in this fashion in the future, it is impossible to set any limit to the variety of physical forms that life may assume. What changes could occur in the next 10^{10} years to rival the changes of the past? It is conceivable that in another 10^{10} years life could evolve away from flesh and blood and become embodied in an interstellar black cloud (Hoyle, 1957) or in a sentient computer (Čapek, 1923).

Here is a list of deep questions concerning the nature of life and consciousness.

(i) Is the basis of consciousness matter or structure?
(ii) Are sentient black clouds, or sentient computers, possible?
(iii) Can we apply scaling laws in biology?

These are questions that we do not know how to answer. But they are not in principle unanswerable. It is possible that they will be answered fairly soon as a result of progress in experimental biology.

Let me spell out more explicitly the meaning of question (i). My consciousness is somehow associated with a collection of organic molecules inside my head. The question is, whether the existence of my consciousness depends on the actual substance of a particular set of molecules or whether it only depends on the structure of the molecules. In other words, if I could make a copy of my brain with the same structure but using different materials, would the copy think it was me?

If the answer to question (i) is "matter," then life and consciousness can never evolve away from flesh and blood. In this case the answers to questions (ii) and (iii) are negative. Life can then continue to exist only so long as warm environments exist, with liquid water and a continuing supply of free energy to support a constant rate of metabolism. In this case, since a galaxy has only a finite supply of free energy, the duration of life is finite. As the universe expands and cools, the sources of free energy that life requires for its metabolism will ultimately be exhausted.

Since I am a philosophical optimist, I assume as a working hypothesis that the answer to question (i) is "structure." Then life is free to evolve into whatever material embodiment best suits its purposes. The answers to questions (ii) and (iii) are affirmative, and a quantitative discussion of the future of life in the universe becomes possible. If it should happen, for example, that matter is ultimately stable against collapse into black holes only when it is subdivided into dust grains a few microns in diameter, then the preferred embodiment for life in the remote future must be something like Hoyle's black cloud, a large assemblage of dust grains carrying positive and negative charges, organizing itself and communicating with itself by means of electromagnetic forces. We cannot imagine in detail how such a cloud could maintain the state of dynamic equilibrium that we call life. But we also could not have imagined the architecture of a living cell of protoplasm if we had never seen one.

For a quantitative description of the way life may adapt itself to a cold environment, I need to assume a scaling law that is independent of any particular material embodiment that life may find for itself. The following is a formal statement of my scaling law:

Biological Scaling Hypothesis. If we copy a living creature, quantum state by quantum state, so that the Hamiltonian of the copy is

$$H_c = \lambda U H U^{-1}, \tag{55}$$

where H is the Hamiltonian of the creature, U is a unitary operator, and λ is a positive scaling factor, and if the environment is similarly copied so that the temperatures of the environments of the creature and the copy are respectively T and λT, then the copy is alive, subjectively identical to the original creature, with all its vital functions reduced in speed by the same factor λ.

The structure of the Schrödinger equation, with time and energy appearing as conjugate variables, makes the form of this scaling hypothesis plausible. It is at present a purely theoretical hypothesis, not susceptible to any experimental test. To avoid misunderstanding, I should emphasize that the scaling law does not apply to the change of the metabolic rate of a given organism as a function of temperature. For example, when a snake or a lizard changes its temperature, its metabolic rate varies exponentially rather than linearly with T. The linear scaling law applies to an ensemble of copies of a snake, each copy adapted to a different temperature. It does not apply to a particular snake with varying T.

From this point on, I assume the scaling hypothesis to be valid and examine its consequences for the potentialities of life. The first consequence is that the appropriate measure of time as experienced subjectively by a living creature is not physical time t but the quantity

$$u(t) = f \int_0^t \theta(t')dt', \tag{56}$$

where $\theta(t)$ is the temperature of the creature and $f = (300 \text{ deg sec})^{-1}$ is a scale factor which it is convenient to introduce so as to make u dimensionless. I call u "subjective time." The second consequence of the scaling law is that any creature is characterized by a quantity Q, which measures its rate of entropy production per unit of subjective time. If entropy is measured in information units or bits, and if u is measured in "moments of consciousness," then Q is a pure number expressing the amount of information that must be processed in order to keep the creature alive long enough to say "Cogito, ergo sum." I call Q the "complexity" of the creature. For example, a human being dissipates about 200 W of power at a temperature of 300 K, with each moment of consciousness lasting about a second. A human being therefore has

$$Q = 10^{23} \text{ bits}. \tag{57}$$

This Q is a measure of the complexity of the molecular structures involved in a single act of human awareness. For the human species as a whole,

$$Q = 10^{33} \text{ bits}, \tag{58}$$

a number which tells us the order of magnitude of the material resources required for the maintenance of an intelligent society.

A creature or a society with given Q and given temperature θ will dissipate energy at a rate

$$m = kfQ\theta^2. \tag{59}$$

Here m is the metabolic rate measured in ergs per second, k is Boltzmann's constant, and f is the coefficient appearing in (56). It is important that m varies with the square of θ, one factor θ coming from the relationship between energy and entropy, the other factor θ coming from the assumed temperature dependence of the rate of vital processes.

I am assuming that life is free to choose its temperature $\theta(t)$ so as to maximize its chances of survival. There are two physical constraints on $\theta(t)$. The first constraint is that $\theta(t)$ must always be greater than the temperature of the universal background radiation, which is the lowest temperature available for a heat sink. That is to say

$$\theta(t) > aR^{-1},\ a = 3.10^{28} \text{ deg cm}, \tag{60}$$

where R is the radius of the universe, varying with t according to (7) and (8). At the present time, the condition (60) is satisfied with a factor of 100 to spare. The second constraint on $\theta(t)$ is that a physical mechanism must exist for radiating away into space the waste heat generated by metabolism. To formulate the second constraint quantitatively, I assume that the ultimate disposal of waste heat is by radiation and that the only relevant form of radiation is electromagnetic. There is an absolute upper limit

$$I(\theta) < 2\gamma(\mathrm{N}e^2/m\hbar^2c^3)(k\theta)^3 \tag{61}$$

on the power that can be radiated by a material radiator containing N electrons at temperature θ. Here

$$\gamma = \max_x\,[x^3(e^x - 1)^{-1}] = 1.42 \tag{62}$$

is the height of the maximum of the Planck radiation spectrum. Since I could not find (61) in the textbooks, I give a quick proof, following the Handbuch article of Bethe and Salpeter (1957). The formula for the power emitted by electric dipole radiation is

$$I(\theta) = \sum_p \int \delta\Omega \sum_i \sum_j \rho_i(\omega_{ij}^4/2\pi c^3)|D_{ij}|^2. \tag{63}$$

Here p is the polarization vector of a photon emitted into the solid angle $d\Omega$, i is the initial and j the final state of the radiator,

$$\rho_i = Z^{-1}\exp(-E_i/k\theta) \tag{64}$$

is the probability that the radiator is initially in state i,

$$\omega_{ij} = \hbar^{-1}(E_i - E_j) \tag{65}$$

is the frequency of the photon, and D_{ij} is the matrix element of the radiator dipole moment between states i and j. The sum (63) is taken only over pairs of states (i,j) with

$$E_i > E_j. \tag{66}$$

Now there is an exact sum rule for dipole moments,

$$\sum_i \omega_{ij}|D_{ij}|^2 = (1/2i)\langle D\dot{D} - \dot{D}D\rangle_{jj} = (N e^2\hbar/2m). \tag{67}$$

But we have to be careful in using (67) to find a bound for (63), since some of the terms in (67) are negative. The following trick works. In every term of (63), ω_{ij} is positive by (66), and so (62) gives

$$\rho_i\omega_{ij}^3 < \gamma\rho_i(k\theta/\hbar)^3(\exp(\hbar\omega_{ij}/k\theta) - 1) = \gamma(\rho_j - \rho_i)(k\theta/\hbar)^3. \tag{68}$$

Therefore (63) implies

$$I(\theta) < \gamma(k\theta/\hbar)^3 \sum_p\int \delta\Omega \sum_i\sum_j(\rho_j - \rho_i)(\omega_{ij}/2\pi c^3)|D_{ij}|^2. \tag{69}$$

Now the summation indices (i, j) can be exchanged in the part of (69) involving ρ_i. The result is

$$I(\theta) < \gamma(k\theta/\hbar)^3 \sum_p\int \delta\Omega \sum_i\sum_j \rho_j(\omega_{ij}/2\pi c^3)|D_{ij}|^2, \tag{70}$$

with the summation now extending over all (i,j) whether (66) holds or not. The sum rule (67) can then be used in (70) and gives the result (61).

This proof of (61) assumes that all particles other than electrons have so large a mass that they are negligible in generating radiation. It also assumes that magnetic dipole and higher multipole radiation is negligible. It is an interesting question whether (61) could be proved without using the dipole approximation (63).

It may at first sight appear strange that the right side of (61) is proportional to θ^3 rather than θ^4, since the standard Stefan-Boltzmann formula for the power radiated by a black body is proportional to θ^4. The Stefan-Boltzmann formula does not apply in this case because it requires the radiator to be optically thick. The maximum radiated power given by (61) can be attained only when the radiator is optically thin.

After this little digression into physics, I return to biology. The second constraint on the temperature θ of an enduring form of life is that the rate of energy dissipation (59) must not exceed the power (61) that can be radiated away into space. This constraint implies a fixed lower bound for the temperature,

$$k\theta > (Q/N)\varepsilon = (Q/N)10^{-28} \text{ erg}, \tag{71}$$

$$\varepsilon = (137/2\gamma)(\hbar f/k)mc^2, \tag{72}$$

$$\theta > (Q/N)(\varepsilon/k) = (Q/N)10^{-12} \text{ deg}. \tag{73}$$

The ratio (Q/N) between the complexity of a society and the number of electrons at its disposal cannot be made arbitrarily small. For the present human species, with Q given by (58) and

$$N = 10^{42} \tag{74}$$

being the number of electrons in the earth's biosphere, the ratio is 10^{-9}. As a society improves in mental capacity and sophistication, the ratio is likely to increase rather than decrease. Therefore (73) and (59) imply a lower bound to the rate of energy dissipation of a society of given complexity. Since the total store of energy available to a society is finite, its lifetime is also finite. We have reached the sad conclusion that the slowing down of metabolism described by my biological scaling hypothesis is insufficient to allow a society to survive indefinitely.

Fortunately, life has another strategy with which to escape from this impasse, namely hibernation. Life may metabolize intermittently, but may continue to radiate waste heat into space during its periods of hibernation. When life is in its active phase, it will be in thermal contact with its radiator at temperature θ. When life is hibernating, the radiator will still be at temperature θ but the life will be at a much lower temperature so that metabolism is effectively stopped.

Suppose then that a society spends a fraction $g(t)$ of its time in the active phase and a fraction $[1 - g(t)]$ hibernating. The cycles of activity and hibernation should be short enough so that $g(t)$ and $\theta(t)$ do not vary appreciably during any one cycle. Then (56) and (59) no longer hold. Instead, subjective time is given by

$$u(t) = f \int_0^t g(t')\theta(t')dt', \qquad (74)$$

and the average rate of dissipation of energy is

$$m = kfQg\theta^2. \qquad (75)$$

The constraint (71) is replaced by

$$\theta(t) > (Q/N)(\varepsilon/k)g(t). \qquad (76)$$

Life keeps in step with the limit (61) on radiated power by lowering its duty cycle in proportion to its temperature.

As an example of a possible strategy for a long-lived society, we can satisfy the constraints (60) and (76) by a wide margin if we take

$$g(t) = (\theta(t)/\theta_0) = (t/t_0)^{-\alpha}, \qquad (77)$$

where θ_0 and t_0 are the present temperature of life and the present age of the universe. The exponent α has to lie in the range

$$\frac{1}{3} < \alpha < \frac{1}{2}, \qquad (78)$$

and for definiteness we take

$$\alpha = 3/8. \qquad (79)$$

Subjective time then becomes by (74)

$$u(t) = A(t/t_0)^{1/4}, \qquad (80)$$

where

$$A = 4f\theta_0 t_0 = 10^{18} \qquad (81)$$

is the present age of the universe measured in moments of consciousness. The average rate of energy dissipation is by (75)

$$m(t) = kfQ\theta_0^2(t/t_0)^{-9/8}. \quad (82)$$

The total energy metabolized over all time from t_0 to infinity is

$$\int_{t_0}^{\infty} m(t)dt = BQ, \quad (83)$$

$$B = 2Ak\theta_0 = 6.10^4 \text{ erg}. \quad (84)$$

This example shows that it is possible for life with the strategy of hibernation to achieve simultaneously its two main objectives. First, according to (80), *subjective time is infinite;* although the biological clocks are slowing down and running intermittently as the universe expands, subjective time goes on forever. Second, according to (83), *the total energy required for indefinite survival is finite.* The conditions (78) are sufficient to make the integral (83) convergent and the integral (74) divergent as $t \to \infty$.

According to (83) and (84), the supply of free energy required for the indefinite survival of a society with the complexity (58) of the present human species, starting from the present time and continuing forever, is of the order

$$BQ = 6.10^{37} \text{ erg}, \quad (85)$$

about as much energy as the sun radiates in eight hours. The energy resources of a galaxy would be sufficient to support indefinitely a society with a complexity about 10^{24} times greater than our own.

These conclusions are valid in an open cosmology. It is interesting to examine the very different situation that exists in a closed cosmology. If life tries to survive for an infinite subjective time in a closed cosmology, speeding up its metabolism as the universe contracts and the background radiation temperature rises, the relations (56) and (59) still hold, but physical time t has only a finite duration (5). If

$$\tau = 2\pi T_0 - t, \quad (86)$$

the background radiation temperature

$$\theta_R(t) = a(R(t))^{-1} \quad (87)$$

is proportional to $\tau^{-2/3}$ as $\tau \to 0$, by virtue of (2) and (3). If the temperature $\theta(t)$ of life remains close to θ_R as $\tau \to 0$, then the integral (56) is finite while the integral of (59) is infinite. We have an infinite energy requirement to achieve a finite subjective lifetime. If $\theta(t)$ tends to infinity more slowly than θ_R, the total duration of subjective time remains finite. If $\theta(t)$ tends to infinity more rapidly than θ_R, the energy requirement for metabolism remains infinite. The biological clocks can never speed up fast enough to squeeze an infinite subjective time into a finite universe.

I return with a feeling of relief to the wide open spaces of the open universe. I do not need to emphasize the partial and preliminary character of the conclusions that I have presented in this lecture. I have only delineated in the crudest fashion a few of the physical problems that life must encounter in its effort to survive in a cold universe. I have not addressed at all the multitude of questions that arise as soon as one tries to imagine in detail the architecture of a form of life adapted to extremely low temperatures. Do there exist functional equivalents in low-temperature systems for muscle, nerve, hand, voice, eye, ear, brain, and memory? I have no answers to these questions.

It is possible to say a little about memory without getting into detailed architectural problems, since memory is an abstract concept. The capacity of a memory can be described quantitatively as a certain number of bits of information. I would like our descendants to be endowed not only with an infinitely long subjective lifetime but also with a memory of endlessly growing capacity. To be immortal with a finite memory is highly unsatisfactory; it seems hardly worthwhile to be immortal if one must ultimately erase all trace of one's origins in order to make room for new experience. There are two forms of memory known to physicists, analog and digital. All our computer technology nowadays is based on digital memory. But digital memory is in principle limited in capacity by the number of atoms available for its construction. A society with finite material resources can never build a digital memory beyond a certain finite capacity. Therefore digital memory cannot be adequate to the needs of a life form planning to survive indefinitely.

Fortunately, there is no limit in principle to the capacity of an analog memory built out of a fixed number of components in an expanding universe. For example, a physical quantity such as the angle between two stars in the sky can be used as an analog memory unit. The capacity of this memory unit is equal to the number of significant binary digits to which the angle can be measured. As the universe expands and the stars recede, the number of significant digits in the angle will increase logarithmically with time. Measurements of atomic frequencies and energy levels can also in principle be measured with a number of significant figures proportional to ($\log t$). Therefore, an immortal civilization should ultimately find ways to code its archives in an analog memory with capacity growing like ($\log t$). Such a memory will put severe constraints on the rate of acquisition of permanent new knowledge, but at least it does not forbid it altogether.

Lecture IV. Communication

In this last lecture, I examine the problem of communication between two societies separated by a large distance in the open universe with metric (6). I assume that they communicate by means of electromagnetic signals. Without loss of generality I suppose that society A, moving along the world-line $\chi = 0$, transmits, while society B, moving along a world-line with the co-moving coordinate $\chi = \eta$, receives. A signal transmitted by A when the time coordinate $\psi = \xi$ will be received by B when $\psi = \xi + \eta$. If the transmitted frequency is ω, the received frequency will be red-shifted to

$$\omega' = \frac{\omega}{1+z} = \frac{\omega R_A}{R_B}, \tag{88}$$

$$R_A = cT_0(\cosh\xi - 1), \tag{89}$$

$$R_B = cT_0(\cosh(\xi + \eta) - 1). \tag{90}$$

The bandwidths B and B' will be related by the same factor $(1 + z)$. The proper distance between A and B at the time the signal is received is $d_L = R_B\eta$. However, the area of the sphere $\chi = \eta$ at the same instant is $4\pi d_T^2$, with

$$d_T = R_B \sinh\eta. \tag{91}$$

If A transmits F photons per steradian in the direction of B, the number of photons received by B will be

$$F' = (F\Sigma'/d_T^2), \tag{92}$$

where Σ' is the effective cross section of the receiver.

Now the cross section of a receiver for absorbing a photon of frequency ω' is given by a formula similar to (63) in the previous lecture

$$\Sigma' = \sum_i \sum_j \rho_i (4\pi^2 \omega_{ij}/\hbar c)|D_{ij}|^2 \delta(\omega_{ij} - \omega'), \tag{93}$$

with D_{ij} again a dipole matrix element between states i and j. When this is integrated over all ω', we obtain precisely the left side of the sum rule (67). The contribution from negative ω' represents induced emission of a photon by the receiver. I assume that the receiver is incoherent with the incident photon, so that induced emission is negligible. Then the sum rule gives

$$\int_0^\infty \Sigma' d\omega' = N'(2\pi^2 e^2/mc), \tag{94}$$

where N' is the number of electrons in the receiver. If the receiver is tuned to the frequency ω' with bandwidth B', (94) gives

$$\Sigma' B' \leq N' S_0, \tag{95}$$

$$S_0 = (2\pi^2 e^2/mc) = 0.167 \text{ cm}^2 \text{ sec}^{-1}. \tag{96}$$

To avoid confusion of units, I measure both ω' and B' in radians per second rather than in hertz. I assume that an advanced civilization will be able to design a receiver which makes (95) hold with equality. Then (92) becomes

$$F' = (FN'S_0/d^2{}_T B'). \tag{97}$$

I assume that the transmitter contains N electrons which can be driven in phase so as to produce a beam of radiation with angular spread of the order $N^{-1/2}$. If the transmitter is considered to be an array of N dipoles with optimum phasing, the number of photons per steradian in the beam is

$$F = (3N/8\pi)(E/\hbar\omega), \tag{98}$$

where E is the total energy transmitted. The number of received photons is then

$$F' = (3NN'ES_0/8\pi\hbar\omega d^2{}_T B'). \tag{99}$$

We see at once from (99) that low frequencies and small bandwidths are desirable for increasing the number of photons received. But we are interested in transmitting information rather than photons. To extract information efficiently from a given number of photons we should use a bandwidth equal to the detection rate,

$$B' = (F'/\tau_B),\ B = (F'/\tau_A), \tag{100}$$

where τ_B is the duration of the reception, and τ_A the duration of the transmission. With this bandwidth, F' represents both the number of photons and also the number of bits of information received. It is convenient to express τ_A and τ_B as a fraction of the radius of the universe at the times of transmission and reception

$$\tau_A = (\delta R_A/c),\quad \tau_B = (\delta R_B/c). \tag{101}$$

The condition

$$\delta \leq 1 \tag{102}$$

then puts a lower bound on the bandwidth B. I shall also assume for simplicity that the frequency ω is chosen to be as low as possible consistent with the bandwidth B, namely

$$\omega = B,\ \omega' = B'. \tag{103}$$

Then (99), (100), (101) give

$$F' = \left[\frac{NN'\delta^2 E}{(1+z)(\sinh^2\eta)E_c}\right]^{1/3}, \tag{104}$$

where by (96)

$$E_c = (8\pi\hbar c^2/3S_0) = (4/3\pi)137\,mc^2 = 3.10^{-5}\text{erg}\,. \tag{105}$$

We see from (104) that the quantity of information that can be transmitted from A to B with a given expenditure of energy does not decrease with time as the universe expands and A and B move apart. The increase in distance is compensated by the decrease in the energy cost of each photon and by the increase of receiver cross section with decreasing bandwidth.

The received signal is (104). We now have to compare it with the received noise. The background noise in the universe at frequency ω can be described by an equivalent noise temperature T_N, so that the number of photons per unit bandwidth per steradian per square centimeter per second is given by the Rayleigh-Jeans formula

$$I(\omega) = (kT_N\omega/4\pi^3\hbar c^2). \qquad (106)$$

This formula is merely a definition of T_N, which is in general a function of ω and t. I do not assume that the noise has a Planck spectrum over the whole range of frequencies. Only a part of the noise is due to the primordial background radiation, which has a Planck spectrum with temperature θ_R. The primordial noise temperature θ_R varies inversely with the radius of the universe,

$$(k\theta_R R/\hbar c) = \Lambda = 10^{29}, \qquad (107)$$

with R given by (8). I assume that the total noise spectrum scales in the same way with radius as the universe expands, thus

$$(T_N/\theta_R) = f(x),\ x = (\hbar\omega/k\theta_R), \qquad (108)$$

with f a universal function of x. When x is of the order of unity, the noise is dominated by the primordial radiation and $f(x)$ takes the Planck form

$$f(x) = f_P(x) = x(e^x - 1)^{-1},\ \ x \sim 1. \qquad (109)$$

But there will be strong deviations from (109) at large x (due to red-shifted starlight) and at small x (due to nonthermal radio sources). Without going into details, we can say that $f(x)$ is a generally decreasing function of x and tends to zero rapidly as $x \to \infty$.

The total energy density of radiation in the universe is

$$\frac{4\pi}{c}\int I(\omega)\hbar\omega d\omega = \frac{(k\theta_R)^4 I}{(\pi^2\hbar^3c^3)}, \qquad (110)$$

with

$$I = \int_0^\infty f(x)x^2dx. \qquad (111)$$

The integral I must be convergent at both high and low frequencies. Therefore we can find a numerical bound b such that

$$x^3f(x) < b \qquad (112)$$

for all x. In fact (112) probably holds with $b = 10$ if we avoid certain discrete frequencies such as the 1420 MHz hydrogen line.

The number of noise photons received during the time t_B by the receiver with bandwidth B' and cross section Σ' is

$$F_N = 4\pi\Sigma' B' \tau_B I(\omega'). \qquad (113)$$

We substitute from (95), (96), (100), (103), (106), and (108) into (113) and obtain

$$F_N = (2r_0/\lambda_B) f N' F', \qquad (114)$$

where

$$r_0 = (e^2/mc^2) = 3.10^{-13} \text{ cm}, \qquad (115)$$

and

$$\lambda_B = (hc/k\theta'_R) = \Lambda^{-1} R_B \qquad (116)$$

is the wavelength of the primordial background radiation at the time of reception. Since F' is the signal, the signal-to-noise ratio is

$$R_{SN} = (\lambda_B/2fN'r_0). \qquad (117)$$

In this formula, f is the noise-temperature ratio given by (108), N' is the number of electrons in the receiver, and ρ_0, λ_B are given by (115), (116). Note that in calculating (117) we have not given the receiver any credit for angular discrimination, since the cross section Σ' given by (95) is independent of direction.

I now summarize the conclusions of the analysis so far. We have a transmitter and a receiver on the world-lines A and B, transmitting and receiving at times

$$t_A = T_0(\sinh\xi - \xi), \quad t_B = T_0(\sinh(\xi + \eta) - (\xi + \eta)). \qquad (118)$$

According to (89) and (101),

$$\tau_A = \delta(dt_A/d\xi), \quad \tau_B = \delta(dt_B/d\xi). \qquad (119)$$

It is convenient to think of the transmitter as permanently aimed at the receiver, and transmitting intermittently with a certain duty cycle δ which may vary with ξ. When $\delta = 1$ the transmitter is on all the time. The number F' of photons received in the time τ_B can then be considered as a bit rate in terms of the variable ξ. In fact, $F'd\xi$ *is* the number of bits received in the interval $d\xi$. It is useful to work with the variable ξ since it maintains a constant difference η between A and B.

From (100), (101), (103), (107), and (108) we derive a simple formula for the bit rate,

$$F' = \Lambda x\delta. \tag{120}$$

The energy E transmitted in the time τ_A can also be considered as the rate of energy transmission per unit interval $d\xi$. From (104) and (120) we find

$$E = (\Lambda^3/NN')(1 + z)(\sinh^2\eta)x^3\delta E_c. \tag{121}$$

We are still free to choose the parameters x [determining the frequency ω by (108)] and δ, both of which may vary with ξ. The only constraints are (102) and the signal-to-noise condition

$$R_{SN} \geq 10, \tag{122}$$

the signal-to-noise ratio being defined by (117). If I assume that (112) holds with $b = 10$, then (122) will be satisfied provided that

$$x > (G/r)^{1/3}, \tag{123}$$

with

$$G = (200r_0/\lambda_p)N'(1 + z)^{-1} = 10^{-9}N'(1 + z)^{-1}, \tag{124}$$

$$r = (R_A/R_P) = (\cosh\xi - 1)/(\cosh\xi_p - 1). \tag{125}$$

Here λ_p, R_p, and ξ_p are the present values of the background radiation wavelength, the radius of the universe, and the time coordinate ψ. It is noteworthy that the signal-to-noise condition (123) may be difficult to satisfy at early times when r is small, but gets progressively easier as time goes on and the universe becomes quieter. To avoid an extravagant expenditure of energy at early times, I choose the duty cycle δ to be small at the beginning, increasing gradually until it reaches unity.

All the requirements are satisfied if we choose

$$x = \max[(G/r)^{1/3}, \xi^{-1/2}], \tag{126}$$

$$\delta = \min[(r/G)\xi^{-3/2}, 1], \tag{127}$$

so that

$$x^3\delta = \xi^{-3/2} \tag{128}$$

for all ξ. The transition between the two ranges in (126) and (127) occurs at

$$\xi = \xi_T \sim \log G, \tag{129}$$

since ξ increases logarithmically with r by (125). With these choices of x and δ, (120) and (121) become

$$F' = \Lambda \min[(r/G)^{2/3}\xi^{-3/2}, \xi^{-1/2}] \tag{130}$$

$$E = (\Lambda^3/NN')(1+z)(\sinh^2\eta)E_c\xi^{-3/2}. \tag{131}$$

Now consider the total number of bits received at B up to some epoch ξ in the remote future. According to (130), this number is approximately

$$F_T = \int^{\chi} F' d\xi = 2\Lambda\xi^{1/2}, \tag{132}$$

and increases without limit as ξ increases. On the other hand, the total energy expended by the transmitter over the entire future is finite,

$$E_T = \int^{\xi}, E d\xi = 2(\Lambda^3/NN')(e^{\eta}\sinh^2\eta)\xi_p^{-1/2}E_c. \tag{133}$$

In (133) I have replaced the red shift $(1 + z)$ by its asymptotic value e^{η} as $\xi \rightarrow \infty$. I have thus reached the same optimistic conclusion concerning communication as I reached in the previous lecture about biological survival. It is in principle possible to communicate forever with a remote society in an expanding universe, using a finite expenditure of energy.

It is interesting to make some crude numerical estimates of the magnitudes of F_T and E_T. By (107), the cumulative bit count in every communication channel is the same, of the order

$$F_T = 10^{29}\xi^{1/2}, \tag{134}$$

a quantity of information amply sufficient to encompass the history of a complex civilization. To estimate E_T, I suppose that the transmitter and the receiver each contain 1 kg of electrons, so that

$$N = N' = 10^{30}. \quad (135)$$

Then (133) with (105) gives

$$E_T = 10^{23}(e^{\eta} \sinh^2 \eta) \text{ erg}. \quad (136)$$

This is of the order of 10^9 W yr, an extremely small quantity of energy by astronomical standards. A society which has available to it the energy resources of a solar-type star (about 10^{36} W yr) could easily provide the energy to power permanent communication channels with all the 10^{22} stars that lie within the sphere $\eta < 1$. That is to say, all societies within a red shift

$$z = e - 1 = 1.718 \quad (137)$$

of one another could remain in permanent communication. On the other hand, direct communication between two societies with large separation would be prohibitively expensive. Because of the rapid exponential growth of E_T with η, the upper limit to the range of possible direct communication lies at about $\eta = 10$.

It is easy to transmit information to larger distances than $\eta = 10$ without great expenditure of energy, if several societies en route serve as relay stations, receiving and amplifying and retransmitting the signal in turn. In this way messages could be delivered over arbitrarily great distances across the universe. Every society in the universe could ultimately be brought into contact with every other society.

As I remarked in the first lecture [see Eq. (11)], the number of galaxies that lie within a sphere $\eta < \psi$ grows like $e^{2\psi}$ when ψ is large. So, when we try to establish linkages between distant societies, there will be a severe problem of selection. There are too many galaxies at large distances. To which of them should we listen? To which of them should we relay messages? The more perfect our technical means of communication become, the more difficulty we shall have in deciding which communications to ignore.

In conclusion, I would like to emphasize that I have not given any definitive proof of my statement that communication of an infinite quantity of information at a finite cost in energy is possible. To give a definitive proof, I would have to design in detail a transmitter and a receiver and demonstrate that they can do what I claim. I have not even tried to design the hardware for my communications system. All I have done is to show that a system performing according to my specifications is not in obvious contradiction with the known laws of physics and information theory.

The universe that I have explored in a preliminary way in these lectures is very different from the universe which Steven Weinberg had in mind when he said, "The more the universe seems comprehensible, the more it also seems pointless." I have found a universe growing without limit in richness and complexity, a universe of life surviving forever and making itself known to its neighbors across unimaginable gulfs of space and time. Is Weinberg's universe or mine closer to the truth? One day, before long, we shall know.

Whether the details of my calculations turn out to be correct or not, I think I have shown that there are good scientific reasons for taking seriously the possibility that life and intelligence can succeed in molding this universe of ours to their own purposes. As Haldane (1924) the biologist wrote fifty years ago, "The human intellect is feeble, and there are times when it does not assert the infinity of its claims. But even then:

> Though in black jest it bows and nods,
> I know it is roaring at the gods,
> Waiting the last eclipse."

REFERENCES

Alpher, R. A., R. C. Herman, and G. Gamow, 1948, Phys. Rev. 74, 1198.

Barrow, J. D., and F. J. Tipler, 1978, "Eternity is Unstable," Nature (Lond.) 276, 453.

Bethe, H. A., and E. E. Salpeter, 1957, "Quantum Mechanics of One- and Two-Electron Systems," in Handb. Phys. 35, 334–348.

Čapek, K., 1923, *R.U.R.*, translated by Paul Selver (Doubleday, Garden City, N.Y.).

Davies, P. C. W., 1973, Mon. Not. Roy. Astron. Soc. 161, 1.

Dyson, F. J., 1972, *Aspects of Quantum Theory,* edited by A. Salam and E. P. Wigner (Cambridge University, Cambridge, England), Chap. 13.

Dyson, F. J., 1978, "Variation of Constants," in *Current Trends in the Theory of Fields,* edited by J. E. Lannutti and P. K. Williams (American Institute of Physics, New York), pp. 163–167.

Feinberg, G., M. Goldhaber, and G. Steigman, 1978, "Multiplicative Baryon Number Conservation and the Oscillation of Hydrogen into Antihydrogen," Columbia University Preprint CU-TP-117.

Gödel, K.9 1931, Monatsh. Math. Phys. 38, 173.

Gott, J. R., III, J. E. Gunn, D. N. Schramm, and B. M. Tinsley, 1974, Astrophys. J. 194, 543.

Gott, J. R., III, J. E. Gunn, D. N. Schramm, and B. M. Tinsley, 1976, Sci. Am. 234, 62 (March, 1976).

Haldane, J. B. S., 1924, *Daedalus, or Science and the Future* (Kegan Paul, London).

Harrison, B. K., K. S. Thorne, M. Wakano, and J. A. Wheeler, 1965, *Gravitation Theory and Gravitational Collapse* (University of Chicago, Chicago), Chap. 11.

Hawking, S. W., 1975, Commun. Math. Phys. 43, 199.

Hoyle, F., 1957, *The Black Cloud* (Harper, New York).

Islam, J. N., 1977, Q. J. R. Astron. Soc. 18, 3.

Islam, J. N., 1979, Sky Telesc. 57, 13.

Kropp, W. P., and F. Reines, 1965, Phys. Rev. 137, 740.

Maurette, M., 1976, Annu. Rev. Nucl. Sci. 26, 319.

Monod, J., 1971, *Chance and Necessity,* translated by A. Wainhouse (Knopf, New York) [*Le Hasard et la Necessité,* 1970 (Editions du Seuil, Paris)].

Nagel, E., and J. R. Newman, 1956, Sci. Am. 194, 71 (June, 1956).

Nanopoulos, D. V., 1978, *Protons are not Forever,* Harvard Preprint HUTP-78/A062.

Pati, J. C., 1979, *Grand Unification and Proton Stability,* University of Maryland Preprint No. 79-171.

Penzias, A. A., and R. W. Wilson, 1965, Astrophys. J. 142, 419.

Rees, M. J., 1969, Observatory 89, 193.

Shlyakhter, A. I., 1976, Nature (Lond.) 264, 340

Turner, M. S., and D. N. Schramm, 1979, *The Origin of Baryons in the Universe and the Astrophysical Implications,* Enrico Fermi Institute Preprint No. 79-10.

Weinberg, S., 1972, *Gravitation and Cosmology* (Wiley, New York), Chap. 15.

Weinberg, S., 1977, *The First Three Minutes* (Basic, New York).

Wright, T., 1750, *An Original Theory or New Hypothesis of the Universe,* facsimile reprint with introduction by M. A. Hoskin, 1971 (MacDonald, London, and American Elsevier, New York).

Zeldovich, Y. B., 1977, Sov. Phys.-JETP 45, 9.

[9]

LIFE IN THE UNIVERSE

Is Life Digital or Analogue?

Freeman J. Dyson

9.1. Statement of the Problem

The question is whether life and intelligence can survive forever in an expanding universe that is constantly growing colder as time goes on. We cannot hope to answer this question with any certainty. We know very little about the nature of the universe and even less about the nature of life and intelligence. But we know enough to speculate about the problem, and that is what I will do in this essay. I don't pretend to be able to predict the future. All I can do is to explore the future, to see whether the survival of life is consistent with the laws of physics and the laws of information theory. If we find that the laws of physics and information theory make survival impossible, then we have answered the question in the negative. If we find that the laws do not forbid survival, we still do not know the answer. There may be other things besides laws of nature that make survival impossible. Survival might be impossible because of accidents of history or geography. We cannot in any case prove that life can survive. The best we can do is to show that survival is not forbidden by the laws of nature as we know them today.

The most serious uncertainty in our knowledge of the laws of nature is the question of proton decay. Various so-called grand unified theories of elementary particles predict that protons should decay into positrons and neutrinos with a lifetime of the order of 10^{33} years. Large underground detectors were built to observe the decay, with negative results. The detectors turned out to be wonderful observatories for exploring the universe, since they detect neutrinos coming from the sky, but they have never detected a decaying proton. If the proton is unstable, its lifetime must be greater than 10^{34} years. The majority of particle physicists still believe that it is unstable. If this is true, all atomic nuclei are also unstable, and all matter will disintegrate into electrons and positrons within a finite time. The disappearance of ordinary matter will leave only an electron-positron plasma as a possible embodiment for life. If the density of the plasma decreases with time, it can last forever, losing only a small fraction of electrons and positrons by pair-annihilation. It is conceivable that life could adapt itself to such an austere mode of existence. But today I have chosen to ignore this possibility. I shall assume that matter is permanent and that life can take advantage of the diversity of physical and chemical processes that matter provides. In any case, even if this assumption is wrong, it is certainly good for the next 10^{34} years, long enough for life to study the situation carefully. There may well be other obstacles to life's survival, caused by physical or cosmological contingencies not yet discovered. Even if survival is not forbidden by the known laws of nature, the question remains open whether survival is possible for real life in the real world.

The first person who thought seriously about this question was Jamal Islam, who wrote a paper about it in 1977 [6], and later wrote an excellent book [7], discussing it in greater detail. I started to think about it after reading Islam's paper, and gave some lectures that were published in the *Reviews of Modern Physics* in 1979 with the title "Time Without End" [3]. I was amazed that the *Reviews of Modern Physics* accepted my speculation in a journal that is usually considered to be respectable. I arrived at an optimistic conclusion, that the laws of nature do not make it impossible for life to survive forever. After that came a good paper by Steven Frautschi [4], also reaching an optimistic conclusion. And then, not much happened for fifteen years. The question of ultimate survival disappeared from the physics journals. It was no longer respectable science, and was pursued mostly by writers of science fiction. If you are writing science fiction, you are allowed to violate the laws of nature if that helps to make a good story. So I stopped thinking seriously about the question.

In 1999 there was a sudden change. Two respectable scientists at Case Western Reserve University in Cleveland, Lawrence Krauss and Glenn Starkman, sent me a paper with the title "Life, the Universe, and Nothing" [8]. This is a serious piece of work, the first new and important contribution to the subject since 1983. It is solid science and not science fiction. And it says flatly that survival of life forever is impossible. It says that everything I claimed to prove in my *Reviews of Modern Physics* paper is wrong. I must tell you right away that I was happy when I read the Krauss-Starkman paper. It is much more fun to be contradicted than to be ignored.

In the three years since I read their paper, Krauss and Starkman and I have been engaged in vigorous arguments, writing back and forth by E-mail, trying to pokes holes in each others' calculations. The battle is not over. But we have stayed friends. After three years we have not found any holes that cannot be repaired. It begins to look as if their arguments are right, and my arguments are right too. That is a very interesting situation, when two arguments that are both right lead to opposite conclusions. It is a situation that physicists call "complementarity," when two points of view are both correct but cannot be seen simultaneously in the same experiment. Niels Bohr invented complementarity as a way to understand the mysteries of quantum mechanics. One experiment says an electron is a wave, and another experiment says an electron is a particle, but the wave and the particle cannot be seen together at the same time. The true nature of the electron can only be described by saying it is both a wave and a particle. The wave nature or the particle nature appear depending on circumstances. The wave nature and the particle nature of the electron are complementary. So I am beginning to think that my view of life and Krauss and Starkman's view of life are complementary in the same way. If this is true, it means that we may have reached together a deeper understanding of the question of survival than either of us had reached separately.

In the rest of this essay I will try to explain the arguments of both sides, and tell you how they can lead to opposite conclusions and still both be right. Krauss and Starkman have published a shorter version of their paper in the November 1999 issue of *Scientific American* [9].

9.2. What Do We Mean by Life?

The first thing to be said is that the arguments about survival only make sense if we take a very broad view of what life means. If we take a narrow view, supposing that life has to be made of flesh and blood or cells full of chemicals dissolved in water, then life certainly cannot survive forever. Life based on flesh and blood, the kind of life we know at first hand, can only exist at temperatures around 300 Kelvin. It requires a constant input of free energy to keep it going. In a cold expanding universe, the available store of free energy in any region is finite, and life must ultimately run out of free energy if it keeps its temperature fixed. The secret of survival is to cool down your temperature as the universe cools. If you can stay alive while your temperature goes down, you can use energy more and more frugally. If you are frugal enough, you can keep going forever on a finite store of free energy. But this requires that life and consciousness can transfer themselves from flesh and blood to some other medium.

One of my favorite books is *Great Mambo Chicken and the Transhuman Condition* by Ed Regis [11]. The book is a collection of stories about weird ideas and weird people. The transhuman condition is an idea suggested by Hans Moravec, a well-known mathematician and computer expert. The transhuman condition is the way you live when your memories and mental processes are down-loaded from your brain into a computer. The wiring system of the computer is a substitute for the axons and synapses of the brain. You can then use the computer as a back-up, to keep your personality going in case your brain gets smashed in a car accident, or in case your brain develops Alzheimer's. After your old brain is gone, you might decide to upload yourself into a new brain, or you might decide to cut your losses and live happily as a transhuman in the computer. The transhumans won't have to worry about keeping warm. They can adjust their temperature to fit their surroundings. If the computer is made of silicon, the transhuman condition is silicon-based life. Silicon-based life is a possible form for life in a cold universe to adopt, whether or not it happens to begin with creatures like us made of flesh and blood.

Another possible form of life is the Black Cloud described by Fred Hoyle in his famous science-fiction novel [5]. The Black Cloud lives in the vacuum of space and is composed of dust-grains instead of cells. It derives its free energy from gravitation or starlight, and acquires chemical nutrients from the naturally occurring interstellar dust. It is held together by electric and magnetic interactions between neighboring grains. Instead of having a nervous system or a wiring system, it has a network of long-range electromagnetic signals that transmit information and coordinate its activities. Like silicon-based life and unlike water-based life, the Black Cloud can adapt to arbitrarily low temperatures. Its demand for free energy will diminish as the temperature goes down.

Silicon-based life and dust-based life both require that protons should be stable. If protons are unstable, neither silicon nor dust can exist forever. If protons are unstable, there can be no solid structure resembling a silicon-based computer. But a structure resembling a Black Cloud might still exist, with free electrons and positrons replacing dust-grains. The processes of life might be embodied in the organized motions of electrons and positrons, just as Hoyle imagined them embodied in the organized motions of dust-grains.

Silicon-based life and dust-based life are fiction and not fact. I am not saying that silicon-based life or dust-based life really exist or could exist. I am using them only as examples to illustrate an abstract argument. The examples are taken from science fiction but the abstract argument is rigorous science. I give you the examples to make the abstract argument easier. Without the examples, the abstract concepts are difficult to explain. But the abstract concepts are generally valid, whether or not the examples are real. The concepts are digital life and analogue life. A transhuman living in a silicon computer is an example of digital life. A black cloud living in interstellar space is an example of analogue life. The concepts are based on a broad definition of life that Krauss and Starkman and I all accept. Life is defined as a material system that can acquire, store, process, and use information to organize its activities. In this broad view, the essence of life is information, but information is not synonymous with life. To be alive, a system must not only hold information but process and use it. It is the active flow of information, and not the passive storage, that constitutes life.

You are all familiar with the fact that there are two ways of processing information, analogue and digital. An old-fashioned LP record gives us music in analogue form, a CD gives us music in digital form. A slide-rule does multiplication and division in analogue form, an electronic calculator or computer does them in digital form. So we define analogue life as life that processes information in analogue form, digital life as life that processes information in digital form. To visualize digital life, think of a transhuman inhabiting a computer. To visualize analogue life, think of a Black Cloud.

The next question that arises is are we humans analogue or digital? We don't yet know the answer to this question. The information in a human is mostly to be found in two places, in our genes and in our brains. The information in our genes is certainly digital, coded in the four-level alphabet of DNA. The information in our brains is still a great mystery. Nobody yet knows how the human memory works. It seems likely that memories are recorded in variations of the strengths of synapses connecting the billions of neurons in the brain with one another, but we do not know how the strengths of synapses are varied. It could well turn out that the processing of information in our brains is partly digital and partly analogue. If we are partly analogue, the down-loading of a human consciousness into a digital computer may involve a certain loss of our finer feelings and qualities. That would not be surprising. I certainly have no desire to try the experiment myself.

There is also a third possibility, that the processing of information in our brains is done with quantum processes, so that the brain is a quantum computer. We know that quantum computers are possible in principle, and that they are in principle more powerful than digital computers. But we don't know how to build a quantum computer, and we have no evidence that anything resembling a quantum computer exists in our brains. Since we know so little about quantum computing, Krauss and Starkman and I have not considered it in our calculations. We discuss the possibilities for life based either on digital or analogue processing, the two kinds of information-processing that we more or less understand.

Now, before I tell you the details of the argument between Krauss and Starkman and me, let me tell you the conclusion. The conclusion is that they are right, and life cannot survive forever, if life is digital, but I am right, and life may survive forever, if life is analogue. This conclusion was unexpected. In the development of our human technology during the last fifty years, analogue devices such as LP records and slide-rules appear to be primitive and feeble, while digital devices are overwhelmingly more convenient and powerful. In the modern information-based economy, digital wins every time. So it was unexpected to find that under very general conditions, analogue life has a better chance of surviving than digital life. More precisely, the laws of physics and information theory forbid the survival of digital life but allow the survival of analogue life. Perhaps this implies that when the time comes for us to adapt ourselves to a cold universe and abandon our extravagant flesh-and-blood habits, we should upload ourselves to black clouds in space rather than download ourselves to silicon chips in a computer center. If I had to choose, I would go for the black cloud every time.

The superiority of analogue life is not so surprising if you are familiar with the mathematical theory of computable numbers and computable functions. Marian Pour-El and Ian Richards, two mathematicians at the University of Minnesota, proved a theorem twenty years ago that says, in a mathematically precise way, that analogue computers are more powerful than digital computers [10]. They give examples of numbers that are proved to be non-computable with digital computers but are computable with a simple kind of analogue computer. The essential difference between analogue and digital computers is that an analogue computer deals directly with continuous variables while a digital computer deals only with discrete variables. Our modern digital computers deal only with zeroes and ones. Their analogue computer is a classical field propagating though space and time and obeying a linear wave equation. The classical electromagnetic field obeying the Maxwell equations would do the job. Pour-El and Richards show that the field can be focused on a point in such a way that the strength of the field at that point is not computable by any digital computer, but can be measured by a simple analogue device. The imaginary situations that they consider have nothing to do with biological information. The Pour-El–Richards theorem does not prove that analogue life will survive better in a cold universe. It only makes this conclusion less surprising.

9.3. *Digression into Cosmology*

Before I discuss the details of information-processing I must say something about cosmology. Modern cosmologists have given us a wide choice of cosmological models that might describe the universe we live in. My young colleagues in Princeton are working hard to compare these models with the evidence obtained from measurements of the cosmic background radiation anisotropy, from studies of gravitational lenses, from the statistical distribution of clusters of galaxies, and from the statistics of catastrophic events such as supernovas and gamma-ray bursts. I won't try to cover all the possibilities. I will mention only four simple model universes that we have used as background for our discussions. The four models have radically different consequences for the future of life.

It is convenient to describe the four models by the names closed, decelerating, open, accelerating. Which of them we are living in depends on the amount and the nature of the dark invisible stuff that permeates the universe and outweighs the visible stuff by at least a factor of ten. If the invisible stuff has high density, we are in a closed universe with finite volume and finite duration. After a finite time the universe that began with a big bang ends with a big crunch. This is the worst case so far as our future is concerned. All possible forms of life are wiped out by rising temperatures shortly before the big crunch. The closed universe used to be popular among astronomers, but fortunately it seems to be inconsistent with recent measurements. The density of the dark stuff seems to be too low to reverse the observed expansion of the universe.

The decelerating universe is one in which the dark stuff has exactly the density required to slow down the expansion of the universe but not to reverse it. In the decelerating universe, the outward velocities of distant galaxies decrease steadily toward zero. They continue to move outward, but more and more slowly as time goes on. The decelerating universe was also popular among theorists a few years ago, but recent evidence seems to be against it. This was the model assumed by Frautschi in his discussion of the future of life [4]. It gives us the best possible situation for long-term survival. Since the relative velocity of any two galaxies tends ultimately to zero, you can exchange material resources as well as information with distant galaxies if you wait long enough. You can not only communicate with your neighbors, you can also reach out and touch them. In this case, with unlimited material resources and an unlimited reserve of accessible gravitational free energy, both analogue life and digital life should be able to survive. Unfortunately, recent evidence makes this happy situation unlikely. The dark stuff seems to be either insufficient or of the wrong kind to cause the galaxies to decelerate.

The open universe is the model that I used as the basis of my analysis of the survival problem in 1979 [3]. It has the universe expanding linearly with time forever. The relative velocity of any pair of distant galaxies remains constant. This means that we can communicate with distant galaxies but cannot exchange matter or energy with them. Every life form must survive using the finite resources of matter and free energy contained in its local neighborhood. Until recently, the open universe was generally believed by astronomers to be the correct model. It requires a low density of dark stuff, consistent with the observations.

The fourth alternative, the accelerating universe, emerged suddenly two years ago as a serious possibility as a result of new observations of microwave background radiation and of distant supernovas. The accelerating model has the universe expanding exponentially rather than linearly. The expansion is driven by an exotic form of dark stuff exerting a repulsive rather than an attractive force on the visible stuff. Every distant galaxy moves away from us with increasing velocity, until after a finite time it disappears into the horizon and is no longer visible. This is a bad situation for life, almost as bad as the closed universe. Every living community is ultimately isolated within a finite volume of space, with no possibility of communicating with the rest of the universe. Krauss and Starkman and I agree that in these circumstances life cannot survive forever.

Almost all professional astronomers are now firmly convinced by the evidence from microwave background radiation and supernovas that the universe is accelerating. By this they mean that the distant galaxies are accelerating away from us at the present epoch of cosmic evolution, when the age of the universe is of the order of ten billion years. But this still leaves open two possibilities for the expansion of the universe in the remote future. Either the acceleration is driven by Einstein's cosmological constant, which means that the acceleration continues forever and the simple accelerating universe model is correct for the remote future. Or alternatively, the acceleration may be driven by something called quintessence, a hypothetical form of energy that causes acceleration but does not remain constant as the universe evolves. If the acceleration is driven by quintessence, then it may continue with diminishing strength for a few tens of billions of years and afterward fade away. The universe of the remote future will then resemble the simple open universe model, with galaxies receding at constant velocity. Although the universe is now accelerating, the open universe model may in the end turn out to be correct.

I like to keep an open mind about the accelerating model of the universe because there is a long history of astronomers believing fervently in some particular cosmological model and then changing their minds later. When I was a student fifty years ago, everyone believed in the closed model. Then, twenty years ago, on the basis of slender evidence, everyone believed in the decelerating model. Then, ten years ago, on the basis of equally slender evidence, everyone believed in the open model. And now, since two years ago, everyone believes in the accelerating model. Astronomers have a tendency to follow the latest fashion. Although the latest fashion today is supported by strong new evidence, I still remain a little skeptical.

9.4. *The Argument for Survival*

I now go back to the argument in my 1979 *Reviews of Modern Physics* paper, which claimed to show that life can survive forever using a finite amount of material and a finite amount of free energy. The argument is quite simple and depends only on the thermodynamic relation between free energy and information. This relation is

$$dF = kT\, dI, \tag{1}$$

where T is temperature, k is Boltzmann's constant, dI is an amount of information processed in the course of staying alive, and dF is the amount of free energy consumed. We measure dI in entropy units, so that the amount of information processed is equal to the amount of entropy generated by the living system. The living system may be a single creature or a whole ecology of creatures. The rate at which the living system processes information will also be a function of temperature,

$$dI/dt = Q(T), \tag{2}$$

where Q stands for quality of life. So the total amount of free energy consumed by the system over its history is

$$F = \int dF = \int k\, T\, Q(T)\, dt, \tag{3}$$

while the total amount of information processed is

$$I = \int dI = \int Q(T)\, dt. \tag{4}$$

If the life is conscious, it will experience the passage of time at a faster or slower rate depending on its quality factor Q. The subjective time experienced by the consciousness is measured by the processed information I rather than the physical time t. For life to survive forever with a finite reserve of free energy, it is necessary that the integral (3) should converge and the integral (4) should diverge when the physical time goes to infinity. This will be possible if the life form can lower its temperature T sufficiently fast while preserving an acceptable quality factor Q.

There are two physical limitations on the possible rate of cooling of a living system. First, T cannot be less than the ambient temperature of the universe at time t. In the open universe, the ambient temperature decreases inversely with time, so T cannot decrease more rapidly than t^{-1}. This limitation is easily satisfied, but the second limitation is more difficult. The temperature can only be lowered by radiating energy into space, and the rate of radiation of energy is limited by the Stefan-Boltzmann law, which says that the rate is proportional to the fourth power of temperature. As the temperature falls, the efficiency of radiation falls much faster. This means that temperature cannot decrease more rapidly than the inverse cube-root of time. If we suppose that the cooling is as efficient as possible, then T is proportional to $t^{-1/3}$. If we suppose that the quality-of-life Q decreases with temperature like T^n, the integral F will converge and the integral I will diverge if

$$Q(T) \approx T^n, 2 < n \leq 3. \tag{5}$$

So life can survive forever with a finite supply of free energy, but only by reducing drastically the quality of life as the temperature goes down.

How big a reduction in the quality of life is tolerable? This question is always subject to debate. Many of us would say that a big reduction is too high a price to pay for survival. But luckily we have an escape from the dilemma. The escape is hibernation. Like bears and squirrels, any life form can adapt to decreasing temperature by interposing periods of activity with periods of sleep. For example, a society might decide to enjoy active periods with constant quality Q and constant duration, interposed with inactive periods whose duration increases with time. The temperature continues to fall slowly by radiation during the inactive periods, but the life form is awake only during the active periods and does not experience any diminution of the quality of life. During the active periods, the society is living beyond its means, accumulating entropy at an unsustainable rate. If the rise in temperature caused by its activity were not halted, it would choke itself to death, but it escapes into hibernation and cools off before the next bout of activity begins. If the j-th active period begins at time t_j, then the total consumption of free energy given by the integral (3) is proportional to the sum

$$\Sigma (t_j)^{-1/3}. \quad (6)$$

The sum converges, and the enjoyment of an infinite number of active periods is possible, provided that the inactive periods are long enough. For example, the sum converges if t_j is proportional to j^4, which means that the fraction of time during which the society is awake decreases with time like $t^{-3/4}$. In this case, subjective time, the time experienced by consciousness, increases with the fourth root of physical time. It is easy to make a rough estimate of the quantity of free energy required to keep a society of a given size alive forever. The sum of the series (6) is equal to its first term multiplied by a small numerical factor. The first term is equal to the free energy consumed by the society during the time before it first goes into hibernation. So the free energy required to survive forever is roughly equal to the present consumption rate multiplied by the time until the first hibernation. For example, our own society consumes energy at a rate of about 10^{14} watts, and it might consider going into hibernation when the sun dies about five billion years from now. The free energy required for our permanent survival is then roughly 10^{24} watt-years, which is the amount of energy radiated by the sun in three days. If we can learn to store even a tiny fraction of the sun's energy in a longer-lasting form, we shall be in a good position to deal with the energy crisis when the sun dies, and with other energy crises that may come later.

A similar calculation shows that in the open universe it is possible to exchange an infinite amount of information between any two galaxies, using only a finite store of free energy. The distance between the two galaxies is increasing linearly with time, and the strength of radio signals that one can receive from the other is constantly decreasing. But the effectiveness of communication depends on the signal-to-noise ratio and not on the absolute strength of the signal. Two factors compensate for the weakening of the signals, first the decrease in ambient noise with time as the universe cools, and second the narrowing of the signal band-width as the transmission moves to lower frequencies. As a result of these compensating factors, a finite quantity of transmitted energy can carry an infinite amount of information. As time goes on, the wave-length of the transmitted signal grows longer, and the transmitting and receiving antennas must grow larger. The antennas cannot be continuous structures but must be arrays of small resonators deployed in space. Using such arrays, a society confined to a local region can continue to acquire new knowledge from outside, and to share its own knowledge with other societies in remote places, constantly expanding its domain of experience and influence. A cold universe is friendly to the growth of intergalactic networks. But I will spare you the details of the calculation that leads to this conclusion.

9.5. The Counter-Arguments

I have sketched for you the argument for survival. Now I will sketch the counter-arguments raised by Krauss and Starkman. There are two main counter-arguments, which I will call the quantized-energy argument and the alarm-clock argument.

The quantized-energy argument says that any material system, whether living or dead, must obey the laws of quantum mechanics. If the system is finite, it will have only a finite set of accessible quantum states. A finite subset of these states will be ground states with precisely equal energy, and all other states will have energies separated from the ground states by a finite energy-gap G. If the system could live forever, the temperature would ultimately become so low that kT would be much smaller than G, and the states above the gap would become inaccessible. From that time on, the system could no longer emit or absorb energy. It could store a certain amount of information in its permanently frozen ground states, but it could not process the information and it could not reduce its entropy by radiation. It would be, according to our definition, dead.

The alarm-clock argument is concerned with the machinery that allows the system to become active after a period of hibernation. There must be some kind of clock that keeps track of time during hibernation and gives the wake-up signal when the moment for resuming activity has arrived. The alarm-clock must satisfy stringent requirements. It must not consume any significant quantity of free energy during hibernation. It must be reset at the end of each active period. The wake-up signaling and the resetting must be done an infinite number of times. And the energy cost of wake-up and resetting must diminish fast enough so that the total cost of the infinite series of operations is finite.

The quantized-energy argument applies to the alarm-clock and shows that the clock cannot work if it is included as a part of a finite system. The clock must be a detached mechanism, separated from the system that it is controlling by distances that increase with time. For example, my first proposal for an alarm-clock consisted of two small masses orbiting around a big mass. The two small masses were unequal, and the larger one was farther than the smaller one from the big mass. The clock was set by starting the two small masses in circular orbits in the same plane about the big mass. The orbital motions generated gravitational radiation, which caused the orbits to slowly contract. The orbit of the larger mass contracts faster. After a long but predictable time, the orbital radii will be equal and the two small masses will collide. The collision gives the wake-up signal. And at the end of the active period the clock is reset with the distances between the three masses increased. Since the time required for gravitational radiation to shrink an orbit is proportional to the fourth power of the radius, the times can easily be made long by adjusting the distances. And if the distances increase rapidly enough, the total gravitational energy that must be supplied to reset the clock an infinite number of times will be finite.

Krauss and Starkman criticized this alarm-clock proposal on several grounds. First, they said, the collision of the two small masses will dissipate a finite amount of energy. Second, the triggering of the wake-up signal after the collision will require a finite amount of energy. Third, the separation of the two small masses to reset the clock will require a finite amount of energy. All these amounts of energy remain roughly constant and do not diminish to zero as the cycle of triggering and resetting is repeated. Therefore the clock requires an infinite amount of free energy to keep working forever. The alarm-clock does not help the system to use energy more economically, but only makes things worse.

Krauss and Starkman thought they had dealt a fatal blow to my survival strategy with their two arguments, the quantized-energy argument and the alarm-clock argument. But I am still on my feet, and here is my rebuttal. The quantized-energy argument is only valid for a system that stores information in pieces of solid matter with fixed sizes, or in devices confined within a volume of fixed size, as time goes on. In particular, it is valid for any system that processes information digitally, using discrete states as carriers of information. In a digital system, the energy differences between discrete states remain fixed as the temperature goes to zero, and the system ceases to operate when kT is much smaller than the energy differences. But this argument does not apply to a system based on analogue rather than digital devices. For example, consider a living system like Hoyle's Black Cloud, composed of dust-grains interacting by means of electric and magnetic forces. After the universe has cooled down, each dust-grain will be in its ground state, so that the internal temperature of each grain is zero. But the effective temperature of the system is the kinetic temperature of random motions of the grains. The information processed by the system resides in the non-random motions of the grains, and the entropy of the system resides in the random motions. The entropy increases as the information is processed. But in an analogue system of this kind, there is no ground state and no energy gap.

To see why the quantized-energy argument fails, consider the volume of the phase-space available to the grains in the cloud. Since electric and gravitational energies vary inversely with distance, the cloud must expand as its temperature cools. If the cloud expands while keeping its shape constant, the linear scale being L, the temperature will scale with L^{-1}, and the velocities and momenta of the grains will scale with $L^{-1/2}$. The phase-space accessible to each grain is the product of volume in physical space with volume in momentum space. Volume in physical space scales with L^3 and volume in momentum space scales with $L^{-3/2}$, so that the phase-space for each grain scales with $L^{3/2}$. This means that the number of quantum states accessible to each grain scales with $L^{3/2}$. The number of quantum states grows larger and larger as the cloud expands.

The superiority of analogue information-processing to digital information-processing is shown clearly by this estimate of the number of quantum states of the analogue system. The total information-capacity is equal to the logarithm of the number of states accessible to the entire system. If the number of independent components or dust-grains is N, the information-capacity is

$$C = (3/2)N \log L. \qquad (7)$$

This capacity C is equal to the entropy of the system if the motions are entirely random, and is equal to the information carried in the system if the motions are entirely non-random. In general the motions will be partly random and partly non-random, and C is equal to the sum of the entropy and the information. For a digital system, the information-capacity is a constant multiple of N, so that (7) holds without the logarithmic factor. The superiority of the analogue system lies in the $\log L$, which allows a system with a fixed number of elements to expand its memory without limit by increasing its linear scale. The quantized-energy argument does not apply to an analogue system, because the number of quantum states is unbounded. At late times quantum mechanics becomes irrelevant and the behavior of the system becomes essentially classical. The number of quantum states is so large that classical mechanics becomes exact. When analogue systems work classically, the quantized-energy argument fails.

I rebut the alarm-clock argument in a similar fashion. I imagine the alarm-clock as before, a set of three masses emitting gravitational radiation as they orbit, but each mass is now a little black cloud instead of a solid body. During the hibernation periods when the clock is running, the three black clouds are asleep, like the rest of the system, radiating away their internal entropy and processing no information. When the two small masses come together, they do not physically collide but interpenetrate each other. The interpenetration disturbs them enough to wake them up and cause the alarm-signal to be transmitted, but does not disrupt them. Since the relative velocities of the grains in each of the masses decrease with time, the energies involved in the interpenetration and signaling and resetting of the clock also decrease with time. Because the mechanism of the clock is expanding at the same rate as the rest of the system, the energy required to operate the clock an infinite number of times remains finite. So I conclude that in the open universe we can build analogue systems that escape both the quantized-energy argument and the alarm-clock argument.

9.6. Conclusions

Finally I discuss briefly the situation that life faces if the universe is accelerating. In this case, I agree with Krauss and Starkman that survival forever is impossible. They pointed out two extremely unpleasant features of the accelerating universe. First, there is a finite distance D at which the repulsive force driving the acceleration overwhelms the gravitational attraction of a galaxy or a cluster of galaxies. The distance D is a point of no return. Anything farther away than that will be driven away to the horizon and vanish. This point of no return defeats my strategy of allowing the living system to increase its size indefinitely. Any permanent system must be confined within the distance D, and then the quantized-energy argument will apply to it. The second unpleasant feature of the accelerating universe is that its temperature does not decrease to zero but tends to a finite limit at late times. It is permeated by cosmic background radiation at a fixed temperature T_b. It is impossible for any living system to cool itself to a temperature below T_b. This means that the free energy F required to process an amount of information I according to (3) and (4) is at least $(kT_b I)$. If the reserve F of free energy is finite, the information I is also finite. This is a dismal situation, but if the universe is accelerating there is no escape.

Let me end with a quick summary. I have considered four simple model universes, closed, decelerating, open, and accelerating. In the closed and accelerating universes, we all agree that survival is impossible. In the decelerating universe, I say that survival is possible while Krauss and Starkman have not expressed an opinion. In the open universe, I say that survival is possible for analogue life but impossible for digital life. In other words, survival is possible in the domain of classical mechanics but impossible in the domain of quantum mechanics. Fortunately, classical mechanics becomes dominant as the universe expands and cools. But I have to tell you that, in our continuing arguments about the possibility of analogue life, Krauss and Starkman have not yet conceded. I am still expecting them to come back with new arguments, which I will then do my best to refute.

REFERENCES

1. Bahcall, N. A., Ostriker, J. P., Perlmutter, S., and Steinhardt, P. J., "The Cosmic Triangle: Revealing the State of the Universe," *Science, 284,* 1481–88 (1999).

2. Drell, P. S., Loredo, T. J., and Wasserman, I., "Type Ia Supernovae, Evolution, and the Cosmological Constant," Cornell Preprint CLNS 99/1615 (1999).

3. Dyson, F. J., "Time without End: Physics and Biology in an Open Universe," *Rev. Mod. Phys., 51,* 447–60 (1979).

4. Frautschi, S., "Entropy in an Expanding Universe," *Science, 217,* 593–99 (1982).

5. Hoyle, F., *The Black Cloud* (Harper and Brothers, New York, 1957).

6. Islam, J. N., "Possible Ultimate Fate of the Universe," *Q. J. Roy. Astron. Soc., 18,* 3–8 (1977).

7. Islam, J. N., *The Ultimate Fate of the Universe* (Cambridge University Press, Cambridge, 1983).

8. Krauss, L. M., and Starkman, G. D., "Life, the Universe, and Nothing: Life and Death in an Ever-Expanding Universe," Case Western Reserve University Preprint CWRU-P1-99 (1999).

9. Krauss, L. M., and Starkman, G. D., "The Fate of Life in the Universe," *Sci. Amer., 281,* 58–65 (November 1999).

10. Pour-El, M. B., and Richards, I., "The Wave-Equation with Computable Initial Data Such That Its Unique Solution Is Not Computable," *Adv. in Math. 39,* 215–39 (1981).

11. Regis, E., *Great Mambo Chicken and the Transhuman Condition: Science Slightly Over the Edge* (Addison-Wesley, Reading, Mass., 1990).

[10]

DOES BIOLOGY HAVE AN ESCHATOLOGY, AND IF SO DOES IT HAVE COSMOLOGICAL IMPLICATIONS?

Simon Conway Morris

10.1. Introduction

Science, especially in its manifestations of cosmology and evolutionary biology, would seem to have little, if anything, to do with an eschatology, given the latter implies a universe with not only a beginning, but a purpose and an end. Here, if anywhere, are the two domains of science and theology irreparably separated. Such a view is reinforced by the vastness of the universe and the notion that the evolutionary processes are not only blind to the future but hedged in with contingent circumstances that in principle

Note: My thanks to the John Templeton Foundation for the invitation to the symposium on the far-future universe, held at the Pontifical Academy, the Vatican. I also thank Sandra Last for heroic typing. Cambridge Earth Sciences Publication 7042.

would allow a myriad of other histories than what actually transpired on Earth. There are, however, other views that may do more to reconcile science and religion. As has often been pointed out, the sheer size of anything is scarcely relevant to a sentient species capable of transcendence, and one day transfiguration. And was the emergence of such a species really an evolutionary fluke? The deep structure of the universe indicates life's inevitability, while a reconsideration of the constraints of evolution, especially via the phenomenon of convergence, suggests that the outcomes are very far from random. In such a framework, the idea of an eschatology becomes much more real.

It is commonly supposed that the Copernican and Darwinian revolutions led to two successive dethronements, the net result of which was irrevocably and irreversibly to demote humans from an unquestioned position of natural, if not God-given, suzerainty to an abject state of utter and irredeemable insignificance. Each revolution arose out of scientific curiosity and irrespective of precisely what belief systems the Catholic Copernicus and brooding pantheist Darwin respectively held, the effect was to impose the equivalent of a second Fall, an expulsion from a world that for all its local terrors had possessed a security and an hierarchy.

The reasons for these dethronements hardly require emphasis. The acknowledgment of the Copernican heliocentric system was really only the first step. What was to follow, not least the great sky-surveys by the Herschels, the identification of the galactic red shift that implies what will probably be an ever-expanding universe, and the recognition of the 3°K microwave background radiation, all served in their turns to diminish our self-esteem. In terms of the standard metrics—be they the number of galaxies or the size of the visible universe—even "insignificant" is too proud a boast: for all intents and purposes we are invisible. No wonder we reel under what Rahner [32] has labeled "existentiell dizziness"; we are less than a grain of sand. So too, albeit from a rather different perspective, are we in terms of biology. Emerging, by some still unspecified process, from the primordial slime, eons of interminable microbial evolution must have elapsed before even the first worm slithered through the ooze. Yet further epochs of geological time rolled by before, approximately halfway through the history of the solar system, a species emerged whose powers of sentience were so developed that gods could be recognized and, slightly later, telescopes lifted toward the heavens. Not only that, but given the billions of species that have evolved on the earth, nearly all of which are extinct, the emergence of the human frame is surely one of pure chance, the result of a process both ignorant and uncaring about any future. Science, and especially cosmology and evolution, reveal our true place: one without privilege or status, unshackled by either ultimate meaning or direction. We might as well go shopping.

Toppled from our pedestal, the former colossus has not only mysteriously shrunk but even worse now lies half-submerged in a malodorous swamp. Former beliefs and credulities have been stripped away beneath the passionless gaze of the stars, while the fecund Earth is now reduced to the parody of a hideous factory floor, with the richness of the biosphere transmuted into an army of lumbering robots, whose sole function is to transmit the genetic code across the generations. Ironically, the scientific wonders of discovery in both cosmology and evolution have only served to usher in a bleak and futile universe. In this naturalist framework any discussion of a telos and by implication a culmination, a parousia, must be dismissed. At best it is a naive atavism, a wishful delusion of the politically impotent, or an ostrich-like fear of Blaise Pascal's "eternal silence of those infinite spaces."

If ideas of end times, judgment, and potentially perdition were once very widely accepted, now these delusions are only useful as a quarry for anthropologists, while for the "moderns" any form of eschatology has been transmuted to an entirely justifiable fear of nuclear annihilation, ecosystem collapse, or failing that—in a slightly more distant future—death by bolide impact. As a principle inherent in the created universe, eschatology as a religious concept lies neglected in the metaphysical nursery. The immense scales of the cosmos and the plenitude of the biosphere seem, therefore, to make the idea of an eschatology simply risible. Even establishing a metric against which human experience can be judged in the wider world strikes me as scarcely adequate. Who is not familiar with the conceit of the earth's history reduced to a twenty-four-hour clock, or the sun shrunk to the size of an orange? But do we really understand the Archean (3.8–2.5 billion years ago) any better by saying it began four hours after midnight or that humans emerged a few seconds before the end of the day, even if this latter figure has an unconscious eschatological knelling? So too, does it really help to know that the nearest stars are still the equivalent of a million kilometers from that metaphorical piece of fruit? However homely examples of a clock and orange are meant to be, it is arguable that the metrics of time and distance they provide simply reinforce our sense of insignificance.

10.2. Creation and Evolution

It may be, however, that a recovery of an eschatological framework is not so naive an operation as is usually thought, or at least so I will try to argue here. Indeed, one could be bolder, and suggest that a universe without an eschatological dimension is a universe that is incomplete, if not crippled. This idea is obviously difficult to reconcile with a principle central to modern scientific thinking—the notion of human non-uniqueness. This is explicit in Copernican astronomy where the earth is only one of billions of similar planets, but it is also implicit in biology by virtue of a recognition of our place at the end of one tiny twig that defines the primate genealogy. We are just another planet and just another species. Hence we must accept our non-privileged condition on that small, out-of-the-way planet, in a very standard solar system, in an ordinary galaxy, belonging to another unremarkable local group that . . . But that may only be part of the story. It is not my intention here to review the evidence that suggests earth-like planets may, for a variety of reasons, be very much rarer than is often thought [44]. Such a rarity could be entirely explicable by natural processes. It might, moreover, simply be used as further ammunition by those who already claim that the emergence of humans already had a vanishingly small probability given the innumerable contingent factors that could redirect history in a myriad of alternative trajectories.

Suppose, however, we can demonstrate that biology and evolution possess an inherent structure that is not only consistent with the plenitude of the biosphere, but more controversially is so arranged as to preordain the emergence of one (or more) sentient species whose strange capacity is to know, however imperfectly, the Creator of the universe as well as being able to explore part of His Creation via the scientific method. The attempt to define such an inherent structure with an implicit predictability will seem to many to be a pretty tall order. In part, this is because biology wears various guises. All too often it is regarded as (if not accused of) being a haven for the innumerate "stamp-collectors" terrified of statistics, makers of lists, and molecular biologists who have never fully recovered from the fact that when all is said and done biology is just a department of chemistry, with a hot line to physics when the going gets really tough. Perhaps there is a grain of truth in these libels. Indeed, other than the principles of Darwinism, which although rooted firmly in the undeniable realities of natural selection and adaptation is still treated as a pretty protean beast that may be laden with all sorts of ideological baggage, it is difficult to find any overarching principle to evolution. But this has not prevented recurrent attempts to transmute Darwinism into a secular religion, yet the attempts to introduce concepts of meaning and moral structure into evolution are simply instances of smuggling: the cargo is contraband [14]. For this reason alone it is easy to see the attraction of treating the empires of science and religion as separate suzerainties, each with its own sphere of influence and responsibility. To me this seems more like an attempt at the latter's marginalization, and more seriously would forever deflect, if not prohibit, the study of what Stanley Jaki has repeatedly emphasized is a single universe, self-enclosed and, in principle, understandable by one Truth (e.g., [19]).

In an attempt to find consistencies between biological evolution and a universe with an eschatology it seems to me that there are several avenues that might be fruitful. All, however, stem from an acknowledgment that the evolutionary process, for all its plenitude, which in itself is an expected reflection of Creation, also has inherent probabilities, if not inevitabilities, of complex forms emerging. These, as noted above, can participate more fully in the Creation, by knowing both it and so far as possible its Originator. Christian orthodoxy, of course, goes further. It argues that the only way of contact, the only way to render the world divine, is by the Incarnation (e.g., [31]). Otherwise the gulf remains and naturalistic explanations provide a temporary, but as is now self-evident, increasingly corrosive accommodation. The possibility, to put it no more strongly, that Creation imparts an inherency not only to make galaxies but also self-reflective sentience has, of course, rather obvious implications for the way we not only see the world, but how also we treat it. After all, if in some sense we are meant to be here, perhaps so too are other things? In brief, is the living world to be treated as a metaphorical putty to be altered by whim, or more often profit? Alternatively, if we are both integral to Creation, and mysteriously also understand it, where do our responsibilities lie [21]?

But how can we establish those "inherent probabilities, if not inevitabilities"? In my opinion, the answer may lie in the curious evidence for what I can only term "navigation," the ability via historical processes for precise biological structures to emerge. Now I am only too well aware that perfectly natural explanations, connected to the crepuscular world of the blind watchmaker, exist, specifically the processes of selection and adaptation. That these *are* the mechanisms, however, does not explain in itself the world we know. As already noted, there is a widespread view that from the perspective of evolution we are little more than a glorious accident, an incidental by-product that quite fortuitously can understand, at least to some extent, the processes by which it arose. If, however, we can show that despite there being many evolutionary routes they tend repeatedly to converge on a very limited number of solutions, then a theistic view becomes more credible. We are then in a position to acknowledge the realities of a historical process that is embedded in a "fabric" whereby by no means "all things are possible"; rather the reverse is the case. If indeed evolution is seeded with destinations, and it would be a strange sort of Creation that was not in itself a plenitude, we need not assume that sentience is the solely desirable end point (recall G. K. Chesterton's remarks on the integrity of being a beetle [4]). Yet the fact remains that some trajectories will be upward, and as Michael Ruse has remarked [34] such concepts of progress are of particular importance if one seeks a congruence between Darwinian evolution and Christianity.

10.3. Navigating through Protein and Genetic Hyperspace

Within biology there are, as it happens, questions that not only put cosmology comfortably in the shade but perhaps have implications stretching far beyond the earth. The background to this is the simple realization of the combinational vastness of biological possibilities. These are what Walter Elsasser has referred to as "immense numbers," which he defines as figures in excess of 10^{100} [11]. These comfortably outstrip the estimated number of atoms in the visible universe, let alone the number of stars or the mere handful of galaxies. At first sight, spending time on pondering such combinatorial possibilities must seem to be a rather fruitless, if not trivial, pursuit. Let us consider the problem of the proteins. They, arguably, represent the central dilemma in biology in as much as they provide the crux between the famous genetic code and expressed functions of an organism that encompass all its aspects, be it metabolism or behavior. Proteins are, of course, composed of strings of amino acids which, by methods incompletely understood, rapidly fold to the required configuration for the appropriate function. The structural configuration that ensures specificity of action, be it as spider's silk or a respiratory protein, is—if we are candid—little short of staggering. Rather little is known as to how the proteins evolved to meet new functions, although doubtless co-option (as we see in the crystallin proteins of the eye lens, e.g., [29, 46]) and gene duplication, a process that provides a "spare set" of proteins potentially available for recruitment to a new purpose, are both clearly important factors.

Before discussing what this might have to do with Elsasser's immense number problem, I need to divert the discussion slightly to ask, in passing, how likely it is that proteins are the basis for life elsewhere in the galaxy and beyond, assuming of course that suitable planets exist. There is always the risk of being excessively earth-centric in this (and other) regards, but not only is the periodic table universal, but the behavior of the elements and molecules are open to investigation. As has been often pointed out, what happens on Threga IX will be the same as here. Although we should keep our minds open in recalling the possibility of silicon-based life forms (arising from an abiotic source, not as a technological afterthought), so also we need to remember what appear to be the various difficulties inherent to such a chemical organization. Silicon can indeed be used in a biological context, as the beauty of the diatom skeleton reminds us, but most probably not as an analogue to carbon. Not only is carbon likely to be the universal choice for life, but proteins also may well be inherent to the universe. The main reason for thinking this is that a number of amino acids are readily synthesized in abiological circumstances. These range from such classic experiments as the Miller-Urey experiments to processes that can be inferred to operate in the wastes of interstellar space [7]. This ease of synthesis, both in terms of the simplicity of the component molecules, for example, HCN, CH_4, and the enormous range of potential environments, are indicative of the effective universality of amino acids. This is not to imply that all these abiotic amino acids are necessarily equally useful, whether to the origins of life on the early earth or anywhere else. In the former, such compounds as glycine and alanine were evidently "chosen," and may possibly have helped to predetermine biosynthetic pathways to the other amino acids now employed in protein construction. Possibly elsewhere it is different, or perhaps not.

Then there is the curious feature that among terrestrial amino acids, nearly all of which can exist as enantiomorphs, it is the left-handed variety that is (with some minor exceptions) invariably used. It is interesting to note that while the majority of known extraterrestrial amino acids, recovered from carbonaceous meteorites, occur in so-called racemic mixtures, that is equal proportions of left- and right-handed enantiomorphs, there are exceptions where the racemic balance is shifted, and in these cases it is in favor of the left-handed type [7, 30]. How these occasional examples of non-chirality arose is still conjectural, and several mechanisms that might operate in interstellar space have been proposed. Of course, the fact that terrestrial life is left-handed so far as amino acids are concerned may simply be a coincidence. But all these abiotic processes suggest that amino acids are universal, proteins are ubiquitous, and perhaps always are built of left-handed enantiomorphs. On Earth twenty different amino acids are employed in protein construction, and perhaps in a way analogous to alphabet codes those extraterrestrial proteins that rely on less than about fifteen or more then thirty "characters" may encounter respectively problems of sufficient versatility and excessive complexity.

This is speculative, and for the time being what we see on Earth can be our only guide. With twenty different amino acids and even a rather simple protein, say one hundred amino acids in length, the combinational possibilities in this protein "hyperspace" are very large indeed. And it is here that we turn to a remarkable paper by Smith and Morowitz entitled "Between History and Physics" [37]. Of particular relevance is the following section:

> The class of possible proteins one hundred amino acids long far exceeds a usable number. Note that if only one in a million such unique sequences is water soluble and only one in a million of those has a chemically active surface, there are still 10^{43} potentially enzymatically active soluble proteins of length one hundred. It is quite clear from such numerology that the domain of possible organisms is enormously larger, if not infinite. Thus we have a second property which appears to set the life sciences apart, the immensity of the dimensionality of the descriptive hyperspace . . . the combination of immense size and "unpredictable" thermal noise amplification has produced a very sparsely sampled hyperspace in the actual living world . . . it seems quite obvious that no . . . global extremum theories can be formulated in the vast majority of biologically interesting characterization hyperspaces. ([37], p. 268)

What they are saying, as if it needs any emphasis, is that in principle the likelihood of another planet arriving in the same corner (or more strictly corners) of protein "hyperspace" is vanishingly small. Yet, there are some reasons for thinking that what we might find on another planet will be eerily familiar, even to the extent of the occupation of protein "space." As is well known, the total diversity of terrestrial proteins, while by no means fully documented, is evidently rather small: probably no more than a few thousand families (e.g., [5]). Moreover, to some extent protein structure is a matter of modular construction, repeatedly using a limited number of "building blocks." In this sense the combinatorial vastness of protein "hyperspace" is less serious than it might first appear. To be sure, there are usually several precise sites in a protein where one, and only one, amino acid will do, but other parts of the protein may be more labile in terms of amino acid composition (but see [1]). In addition, it is also important to remember that in at least some instances the shape of protein may be less significant, provided it functions appropriately. Some good examples of this exist in certain oxygen-carrying proteins. Hemocyanin, for example, is a copper-carrying protein, found in various arthropods and molluscan cephalopods. Despite its common name, the structures of the respective hemocyanins are very different and have evidently arrived at the same function convergently (e.g., [24]). Even more remarkable, perhaps, is the independent evolution in some snails of a myoglobin molecule, again very different from true myoglobin but with the same capacities [41]. There are a number of other interesting examples of molecular convergence, but it is probably fair to say that in general the idea of protein "hyperspace" being much more constrained than is generally thought to be the case is yet to be assessed (see [9]).

The possibility that the evolution of life has imprinted upon it a sort of universal structure is unlikely to enthuse mainstream Darwinists, given their prolix enthusiasm for a naturalist metaphysics. Yet the realities of natural selection do not exclude the existence of higher-order structures that impose a wider set of patterns. These too are, to be sure, embedded in the natural world. In a way analogous to the conditions associated with the anthropic principle, we are at liberty to accept them as brute facts: the world is because it is. At the same time, the exclamations analogous to those of anthropic surprise—the sensitivities, the emergence of the complex from the simple, the unerring pathways to functionality—at least allow one to entertain the alternative possibility that the universe and life within it is not some hideous and purposeless accident. Nor should belief in a created universe lead us to think that it ought to be in some sense "perfect." Evolution itself is an undeniable reality, and as already noted not every trajectory is intended to end up at a destination of humanoid. Simplification, if not regression, is equally possible.

10.4. Converging on the Inevitable

As part of their analysis of the realities of evolution, Smith and Morowitz [37] also introduce what they call the "Game of Life," effectively the ground rules for biological existence. As expressed, it is difficult to see how they cannot fail to have universal application. But as Smith and Morowitz remark:

> There is at least one major evolutionary trend not immediately explained by our strategy rules. That being the numerous examples of morphological convergence. Why, in the sparsely sampled genetic space, have there been so many cases of apparent convergence or parallelism? It is surprising in light of the high probability for novelty to find, even in similar niches, high morphological similarity in distinctly different genetic lines . . . there may be additional rules operating at coarser levels of the genetic space which are less statistical than those discussed. ([37], p. 280)

In so writing, Smith and Morowitz echo one of the main areas of tension, if not a fault line, in how we understand organic evolution. This concerns the attempts to reconcile the flux of change that epitomizes evolution against the pervasive sense of the ideal, if not archetype, be it in the concept of the Bauplan or the extraordinary specificity of protein function. This is a huge topic that cannot possibly be encompassed here, but I am strongly persuaded that far from being of anecdotal importance the general topic of evolutionary convergences is of major importance, both in helping to identify the constraints on life and thereby an overall structure as well. Indeed, my sense is that for all its immense diversity and plenitude, life is so constrained that what we see on Earth is far from being some sort of parochial backwater or provincial zoo, let alone freak show. On the contrary, what we see here may be a fair indication of what we might expect to find elsewhere (see also [2]).

It seems that there are two strands that, in different ways, allow us to assess evolutionary alternatives. The first approach is to consider compounds basic to life, notably chlorophyll and DNA, and ask if we can conceive of alternatives and, ideally, superior versions. George Wald has argued [43] that so far as chlorophyll is concerned, its very inefficiency in intercepting the main region of the visible light spectrum of the sun paradoxically indicates that a better alternative simply is not possible: wherever there is life (at least near stars) there is chlorophyll. Nothing else is available. That is a bold prediction, yet even when we consider other molecules designed to capture photons for one or another transduction process, here too we find repeated examples of convergence, be it in rhodopsins (e.g., [38]) or cryptochromes [3]. Put simply, there are very few alternatives, and on occasion only one way to do things. In an analogous way we can ponder what alternatives might exist to the molecule of replication DNA. Despite its iconic status it is not widely appreciated how very peculiar are its properties [42]. Again this is not to say that terrestrial DNA is uniquely suitable, although that possibility cannot be ruled out. What we can observe is that although alternatives can be constructed (e.g., [12]), at least so far as substitution for the various building blocks of DNA is concerned, their overall viability and, as importantly, the likelihood of their being synthesized in a credible prebiotic system are for the most part problematic. It still remains the case that the stereochemical properties of DNA are very peculiar indeed: one wonders what other replication system might approach it in its almost uncanny effectiveness.

The other strand to considering alternatives is to consider evolutionary convergences in a coherent fashion, a topic that surprisingly has remained unaddressed for many years, although textbooks regularly roll out the standard examples, for example, the camera eye of the vertebrate as against that of the octopus. Yet even this famous example yields more insights. Thus the camera-like eye has evidently evolved independently at least six times (vertebrates, cephalopods, alciopid polychaetes, and three groups of gastropods: heteropods, littoriniids, and strombids). Typically it is associated with the evolution of active life-styles, and also predatory activity, whereby perception will be at a premium. Even in those cases, such as the strombid and littoriniid snails, where existence is more sluggish, the camera eye may confer benefits such as respectively sophisticated escape reactions [33] and recognition of landmarks on otherwise featureless tidal flats [15].

A wider survey suggests that the options open to organisms are indeed limited; sometimes perhaps there isn't any choice. Either you do it one way, or not at all. Many of the most compelling examples of evolutionary convergence are drawn from environmental or behavioral extremes, such as habitation in deserts (plants, rodents), long-distance migration (birds), and ways to design a stiletto (saber-toothed cats and marsupial thylacosmilids). Yet if convergence is to be of more general relevance, then it is from those areas of complex organization and sophisticated behavior that we will obtain a best sense of just how constrained are the trajectories of evolution. Thus the most telling examples of convergence include eyes and vision, electrogeneration and electroreception [17], echolocation [18], olfaction [40], vocalization [10], and brain size [25]. The net result is that for all of life's plentitude there is a strong stamp of limitation, imparting not only a predictability to what we see on Earth, but by implication elsewhere [6].

10.5. So What, on Earth, Does This Have to Do with Eschatology?

The mutual hostility between the science of evolution and revealed religion is intensely felt. Attempts at compromise are as likely to lead to a twin assault from either side, execrated by both strident atheists and fundamentalist creationists. There has been, of course, no shortage of attempts to find accommodation, but it is also unwise to paper over what are not so much cracks as yawning crevasses. In the Christian tradition, in particular, the twin realities of the Incarnation and Resurrection are incomprehensible to the proponent of naturalism. What then might help to lead us first to a further synthesis, and one where not only the Incarnation and Resurrection are of integral importance, but also an eschatology?

As already emphasized, the conditions that allow evolution to proceed may be far more finely balanced than is commonly realized. This we appreciate from two main factors. First, despite the combinatorial immensity of biological form, especially molecular, the end points are both strongly limited and reached with remarkable effectiveness. Second, we see that the world around us shows an extraordinary, if not wonderful, economy whereby a restricted number of building blocks, for example, amino acids, are used to construct a quite extraordinary organic architecture that in any a priori sense we would find impossible to predict. Third, we see that not only is there an emergence of complexity, which we are entitled to call progress, but the options available are limited. To put it bluntly, whoever and wherever you are, you will see, smell, hear, and probably think in the same way. This in turn has two implications. First, the emergence of advanced sentience, what we choose to call human, is probably inherent from the origin of life. Second, because we have evolutionary origins and are very much part of the world, our self-understanding should be constantly directed away from destructive impulses to recognizing the natural order as integral to ourselves. This has two potentially very interesting consequences. First, as already remarked, we should question whether the biosphere is some malleable object for human exploitation, whether it be by pillage or modification. Second, if our sentience emerged from evolutionary antecedents, we should take the claims of other sentiences, notably among the vertebrates and cephalopods, much more seriously. And again, what has this to do with an eschatology, a theory of last times and the call to accountability of the riches given? If the universe is a Creation, then not only does it have a structure, but also a potential. If this has been realized by an evolutionary process, yet now coincides with a species that has been offered (and as often rejects) the Incarnation, then maybe a notion of end time no longer seems quite so fantastic.

10.6. Eschatology Elsewhere?

But is there anything elsewhere: apart that is from lifeless planets, dust clouds, and suns? Most biologists are now familiar with the Fermi paradox, which states that if there were intelligent extraterrestrial life we would know about it now, not least because humans are relatively latecomers to the game. The response by SETI enthusiasts to the question of "Where are they?" looks to the quasi-statistical basis of the Drake equation, but this seems to owe as much to a hope—not unworthy—if not to a faith, that extraterrestrials *ought* to exist. Perhaps unfairly, when I read some of the SETI literature, I get the sense that people somehow expect a cross between the National Academy and a golf club: high-minded gin-and-tonics. It is, however, possible to invert the Drake equation and present a shopping list: Earth of certain size [23], with a large moon please [22], solar system with volatile-rich Oort comet cloud [35] capable of delivering the necessary volatiles to the earth, and big Jupiter [45], thank you; oh, and by the way, a position in the galaxy that oddly manages to miss a drenching by radiation from a supernova by mere twenty times longer than one can reasonably expect [27], and so on. So, as I have remarked elsewhere, "Life may be a universal principle, but we can still be alone" [6].

What does all this have to do with the eschaton? First, I think that evolution may benefit from a cosmological perspective. As we have seen, the combinatorial immensity of biological "hyperspace" in one sense entirely dwarfs the universe, yet there are hints that navigation within this vast domain may be far more controlled than popularly supposed. Next, biological structures are fantastically complex; as Russell Stannard reminds us [39], who wants to swap our brain for the sun?

There are also new theological perspectives. I began this chapter with an emphasis on the seeming puniness of human existence against a cosmic scale. Karl Rahner adds a valuable perspective when he writes:

> Today, and more than ever in the future, human beings and Christians are also going to have to realize more clearly and more radically that their very *recognition* and *acceptance* of the fact of being lost in the cosmos actually raises them above it and enables them to realize it as an expression and a mediation of that ultimate experience of contingency which they, in virtue of all their ancient faith, must perceive and accept before the infinite God as finite creatures. In this light one is justified in saying that the cosmos has become "more theological" for humankind. . . . In this way the feeling of cosmic dizziness can be understood as an element in the development of people's theological consciousness. If . . . the scientific investigation of the cosmos can never come to an end, and if this conviction is a theological datum based on the fundamental incomprehensibility of God, then this experience that the universe is in a way immeasurable is, to a certain extent, nothing other than the spatial counterpart to the theological datum, and it is something that might be expected (after the fact, of course). If people give up their feeling of being at home in the universe in exchange for a feeling of not being at home, which reflects the character of their religious experience, then this is at root a legitimate element of humankind's fate. ([32], p. 50; his emphases)

So far as we know, apart from the Creator, we are the only objects in the universe that remotely apprehend the nature of our present home. Now, let us suppose the Fermi paradox holds: we are alone. In principle, life and its trajectories, including brains and cognition, could have happened anywhere, but only here did every condition coalesce correctly. This is consistent with animal sentience being more widespread than is popularly supposed, but does not deny our transcendence. As Rahner has commented [32], whether or not animals possess such features as the rudiments of speech or consciousness is a topic for empirical anthropology, not theology. He also remarks, "If this transcendentality is peculiar to human beings, if there is present in it the possibility of a confrontation in freedom with themselves, if they can rethink their own thinking . . . and if this transcendentality is not present in the animals (has anyone ever tried to demonstrate this in animals?), then theologians have what is required for human beings" ([32], p. 43).

In other words the universe has to be this sort of size with that sort of history for us to happen. We are inherent in the fabric of the universe, but the sentient reality is only here. From this perspective, which I well know has few scientific supporters, an eschatological perspective no longer seems quite as ludicrous. As ever, we are given the choice. Our loneliness and uniqueness might just be a dreadful accident, or it might be the biggest clue we could ask for. For some, therefore, this can be a test of faith. So too it permits, I think, a Job-like interrogation of the world we live in. It also reopens a door to those to whom eschatology is a live issue (e.g., [13, 16, 20, 26, 28, 36]). And, I would argue, it might yet mean that the recurrent sense of alienation and loss are not the fate of a disenchanted primate, but suggest that however big we think the universe to be, there are other matters—to which we have been invited—of yet greater moment.

REFERENCES

1. Axe, D. D., "Extreme Functional Sensitivity to Conservative Amino Acid Changes on Enzyme Exteriors," *J. Mol. Bio., 301,* 585–96 (2000).

2. Bieri, R., "Humanoids on Other Planets?" *Amer. Sci., 52,* 452–58 (1964).

3. Cashmore, A. R., Jarillo, J. A., Wu, Y.-J., and Liu, D., "Cryptochromes: Blue Light Receptors for Plants and Animals," *Science, 284,* 760–65 (1999).

4. Chesterton, G. K., "A Defence of Humility," in *The Defendant* (Dent, London, 1922), 129–37.

5. Chothia, C., "One Thousand Families for the Molecular Biologist," *Nature, 357,* 543–44 (1992).

6. Conway Morris, S., "A Plurality of Worlds, a Plurality of Bodyplans?" *Astron. Soc. Pac., Conference Series, 213,* 411–18 (2000).

7. Cronin, J. R., and Pizzarello, S., "Amino Acids of the Murchison Meteorite. III. Seven Carbon Acyclic Primary a-amino Alkanoic Acids," *Geochim. Cosmochim. Acta, 50,* 2419–27 (1986).

8. Cronin, J. R., and Pizzarello, S., "Enantiomeric Excesses in Meteoritic Amino Acids," *Science, 275,* 951–55 (1997).

9. Denton, M., and Marshall, C., "Laws of Form Revisited," *Nature, 410,* 417 (2001).

10. Doupe, A. J., and Kuhl, P. K., "Birdsong and Human Speech: Common Themes and Mechanisms," *Ann. Rev. Neuro., 22,* 567–611 (1999).

11. Elsasser, W. M., *Reflections on a Theory of Organisms: Holism in Biology* (Johns Hopkins University Press, Baltimore, 1998).

12. Eschenmoser, A., "Chemical Etiology of Nucleic Acid Structure," *Science, 284,* 2118–24 (1998).

13. Freyne, S., and Lash, N., eds., "Is the World Ending?" *Concilium,* x + 131 pp., (1998, part 4).

14. Greene, J. C., *Debating Darwin: Adventures of a Scholar* (Regina Books, Claremont, Calif., 1999).

15. Hamilton, P. V., and Winter, M. A., "Behavioural Responses to Visual Stimuli in the Snail *Littorina irrorata,*" *Anim. Beh., 30,* 752–60 (1982).

16. Hauerwas, S., "Why Time Cannot and Should Not Heal the Wounds of History but Time Has Been and Can Be Redeemed," *J. Theol., 53,* 33–49 (2000).

17. Heiligenburg, W., and Bastian, J., "The Electric Sense of Weakly Electric Fish," *Ann. Rev. Physiol., 46,* 561–83 (1984).

18. Hughes, H. C., *Sensory Exotica: A World Beyond Human Experience* (MIT Press, Cambridge, Mass., 1999).

19. Jaki, S. L., *Is There a Universe? The Forwood Lectures for 1992* (Liverpool University Press, Liverpool, 1993).

20. Keller, C., "The Attraction of Apocalypse and the Evil of the End," *Concilium,* 65–73 (1998, part 1).

21. Linzey, A., and Yamamoto, D., eds., *Animals on the Agenda: Questions about Animals for Theology and Ethics* (SCM Press, London, 1998).

22. Lissauer, J. J., "It's Not Easy to Make the Moon," *Nature, 389,* 327–28 (1997).

23. Lissauer, J. J., "How Common Are Habitable Planets?" *Nature, 402,* C11–C14 (1999).

24. Mangum, C. P., "The Fourth Annual Riser Lecture: The Role of Physiology and Biochemistry in Understanding Animal Phylogeny," *Proc. Bio. Soc. Wash., 102,* 235–47 (1990).

25. Marino, L., "What Can Dolphins Tell Us about Primate Evolution?" *Evol. Anthropol., 5,* 81–85 (1996).

26. Nichols, A., "Imaginative Eschatology: Benson's 'The Lord of the World,'" *The New Blackfriars, 72,* 4–8 (1998).

27. Norris, R. P., "How Old Is ET?" in *When SETI Succeeds: The Impact of High-Information Impact,* ed. A. Tough (Foundation for the Future, Bellevue, Wash., 2000), 103–5.

28. Paretsky, A., "The Transfiguration of Christ: Its Eschatological and Christological Dimensions," *The New Blackfriars, 72,* 313–24 (1998).

29. Piatigorsky, J., "Lens Crystallins: Innovation Associated with Changes in Gene Regulation," *J. Biol. Chem., 267,* 4277–80 (1992).

30. Pizzarello, S., and Cooper, G. W., "Molecular and Chiral Analyses of Some Protein Amino Acid Derivatives in the Murchison and Murray Meteorites," *Meteor. Planet. Sci., 36,* 897–909 (2001).

31. Rahner, K., "Christology within an Evolutionary View of the World," *Theological Investigations 5, part 3* (Darton, Longman, and Todd, London, 1966), 157–192.

32. Rahner, K., "Natural Science and Reasonable Faith," *Theological Investigations,* XXI (Darton, Longman, and Todd, London, 1988), 16–55.

33. Roy, K., "The Roles of Mass Extinction and Biotic Interaction in Large-Scale

Replacements: A Reexamination Using the Fossil Record of Stromboidean Gastropods," *Paleobiology, 22,* 436–52 (1996).

34. Ruse, M., *Can a Darwinian Be a Christian? The Relationship between Science and Religion* (Cambridge University Press, Cambridge, 2001).

35. Sekanina, Z., "A Probability of Encounter with Interstellar Comets and the Likelihood of Their Existence," *Icarus, 27,* 123–33 (1976).

36. Sherry, P., "Redeeming the Past," *Relig. Stud., 34,* 165–75 (1998).

37. Smith, T. F., and Morowitz, H. J., "Between History and Physics," *J. Mol. Evol., 18,* 265–82 (1982).

38. Soppa, J., "Two Hypotheses—One Answer: Sequence Comparison Does Not Support an Evolutionary Link between Halobacterial Retinal Proteins Including Bacteriorhodopsin and Eukaryotic G-Protein-Coupled Receptors," *FEBS Lett., 342,* 7–11 (1994).

39. Stannard, R., "Who'd Swap the Sun?" *The Tablet,* 638–39 (13 May 2000).

40. Strausfeld, N. J., and Hildebrand, J. G., "Olfactory Systems: Common Design, Uncommon Origins?" *Curr. Opin. Neurobio., 9,* 634–39 (1999).

41. Suzuki, T., Kawamichi, H., and Imai, K., "A Myoglobin Evolved from Indoleamine 2,3-Dioxygenase, a Tryptophan-Degrading Enzyme," *Comp. Biochem. Physiol., B 121,* 117–28 (1998).

42. Switzer, C. Y., Moroney, S. E., and Benner, S. A., "Enzymatic Recognition of the Base Pair between Isocytidine and Isoguanosine," *Biochemistry, 32,* 10489–96 (1993).

43. Wald, G., "Fitness in the Universe: Choices and Necessities," *Orig. Life Evol. Bio., 5,* 7–27 (1974).

44. Ward, P. D., and Brownlee, D., *Rare Earth: Why Complex Life Is Uncommon in the Universe* (Copernicus, New York, 2000).

45. Wetherill, G. W., "Possible Consequence of Absence of 'Jupiters' in Planetary Systems," *Astrophys. Space Sci., 212,* 23–32 (1994).

46. Wistow, G., "Lens Crystallins: Gene Recruitment and Evolutionary Dynamism," *Trends Biochem. Sci., 18,* 301–6 (1993).

PART 4

HUMANITY

[11]

DEEP TIME

Does It Matter?

Stephen R. L. Clark

11.1. The Problem of Deep Time

My topic is the problem of deep time: that is, the ethical and metaphysical effect of placing ourselves in the context of bygone and future ages. I shall concentrate on the impact of possible *futures,* and address (a) the doomsday argument—that our future will be brief, (b) the omega point argument—that the future will be long and triumphant, and (c) the presentist argument—that all such stories are only metaphors for present-day experiences and desires. In an earlier version of this essay I called that last the "Platonist argument"—but I now see some differences between a proper Platonism and "commonsensical presentism" (which is actually much the same as egoism). This current version will continue an exploration begun in *God's World and the Great Awakening* [12], as well as in a paper entitled "The End of the Ages," written for a volume on the millennium [16], recent papers to the Wittgenstein Conference at Kirchberg in 2000 [18], to the Templeton Symposium in Rome in 2001, to a conference on nature and technology in Aberdeen [19], and audiences in Westminster Abbey, Keble College, and Sheffield. It is also a pretext for reading science fiction during working hours.

That deep time, or the idea of deep time, does have an effect on our ethical and metaphysical sensibility is certain: Witness the number of scientists as well as science-fiction writers who testify to the emotional impact of Olaf Stapledon's work, especially *Last and First Men, Last Men in London,* and *Star Maker.* Witness the stories told in Hindu and Buddhist sermons reminding us of our littleness, and the real insignificance of fortune, by piling up the years and distances around the little clearings of our lives. Oddly, contemporary Western philosophers do not seem to have addressed the issue, though our predecessors did. We appear to take it for granted—even when engaged in philosophical study of evolutionary theory or of speculative cosmology—that the only proper temporal context for our lives is the humanly accessible one. *Our* time is very much less than a century, even though we know that centuries and even millennia are—by comparison with geological or cosmological eons—hardly more than a moment.

Philosophers follow fashion—as do theologians. The religious imagination—reminding us that "a thousand ages in Thy sight are like an evening gone"—has been displaced, and even the religious prefer to believe in a merely immanent deity whose attention span is not much longer than our own. Once upon a time—and not all that long ago—Berkeley could cheerfully declare that a charitable benefaction "seems to enlarge the very Being of a Man, extending it to distant Places and to future Times; inasmuch as unseen Countries and after Ages, may feel the Effects of his Bounty, while he himself reaps the Reward in the blessed Society of all those who, *having turned many to Righteousness, shine as the Stars for ever and ever*" ([6]; my emphasis). And again: "We should not therefore repine at the divine laws, or show a forwardness or impatience of those transient sufferings they accidentally expose us to, which, however grating to flesh and blood, will yet seem of small moment, if we compare the littleness and fleetingness of this present world with the glory and eternity of the next" [5]. It is that literal belief which sets the seal on Berkeley's account of religion. "I can easily overlook any present momentary sorrow, when I reflect that it is in my power to be happy a thousand years hence. If it were not for this thought, I had rather be an oyster than a man, the most stupid and senseless of animals than a reasonable mind tortured with an extreme innate desire of that perfection which it despairs to obtain."

What happens here is at once much more and much less important than we think: much more, because our immortal life rests on it; much less, because "if we knew what it was to be an angel for one hour, we should return to this world, though it were to sit on the brightest throne in it, with vastly more loathing and reluctance than we would now descend into a loathsome dungeon or sepulchre" [5].

The religious are now uncomfortable with these attempts to diminish or to exalt the significance of present time, and it is the nonreligious who are more likely to remind us how brief our lives and history are (as though it should come as a shock to realize that there are many things much bigger, and much older, than we are). The religious are eager to believe that the only available Infinite is alongside and in us—to "hold infinity in the palm of your hand and eternity in an hour" [7]—perhaps because the actual, literal past and future revealed through geological and astronomical inquiry is less to their taste. The irreligious think that our smallness, by comparison with the unimaginable expanse of space and time that surrounds us, casts doubt upon religion: *our* lives cannot be important.

But though cosmological and biological science may *tell* us that the real world is longer than our lives, or even than our histories, we rarely permit this to affect us. The enterprise designed by Stewart Brand—the "clock of the long now" [8]—may perhaps spread some clearer sense of deep time, but it would be optimistic to expect this to make much difference. Most of us will continue to act within a time frame very much shorter even than our own lifetime. Why else would most of us agree even to read a paper at a conference in a few months time, were it not for the happy conviction that the conference dates will never actually occur? And even Brand's long now is very much shorter, at ten thousand years, than the Eon.

One way of retaining some sense of the significance of stories about the "very beginning" or the "very end" is to insist that these stories are "really" about our ordinary present. A *literal* reading of mythographic speculation assures us that the days of the very beginning were a long time ago, "before" the everyday world of human life got started. But it is of the very essence of fairyland that it is "once-upon-a-time": however far back along the normal run of history we look we shall find that the fairies have already "gone away," and yet are "there" alongside us. Their "pastness" is not that of last year's papers—though one could suggest, contrariwise, that last year's doings very rapidly become mythological. For the young, their parents' talk even of twenty years ago invites them to contemplate an age beyond imagining, halfway back to the dinosaurs that occupy another alongside world in their imaginations. The "pastness" of the "beginnings" is better understood as their permanent alongsideness. The world is always beginning, from the omnipresent center of attentive consciousness, which we represent to ourselves under the style of myth. At the same time it is always breaking out into a wider world, waking up to judgment. Stories of ending and transforming, which we project into the future, are as little to do with an historic time-to-come as stories of beginning are to do with an historic long-ago. In fact they are often just the same story: the gathering of sticks and stones and bones to make the world, the crashing together of the fire and ice to end a world. In the "long-ago" the people crawled out of the earth to people it; in the "yet-to-come" the dead break from their tombs. Sometimes, as in the literary expression of Norse ritual, this is explicit: Ragnarok is just the opening passage of the new heaven and the new earth, whose coming is disaster for the former powers.

By this account, Creation and Judgment both alike are not events far off, but present experiences of eternal truth. To believe that God made the world is to live by the Covenant; to think that Christ will come to judge the living and the dead is to see ourselves in the light of his life and death. "The Christ event can here be understood in a wholly non-eschatological way as epiphany of the eternal present in the form of the dying and rising *Kyrios* of the cultus" [27]—or the realization of human guilt and possibility. This is not what our predecessors taught, in imagining an "end."

> A final belch of fire like blood,
> Overbroke all heaven in one flood
> Of doom. Then fire was sky, and sky
> Fire, and both, one brief ecstasy,
> Then ashes. But I heard no noise
> (Whatever was) because a Voice
> Beside me spoke thus, "Life is done,
> Time ends, Eternity's begun
> And thou art judged for evermore." [9]

It is comforting to believe that this is not intended as a *literal* event, but only an allegory of sudden insight, or even a nightmare from which we can expect to wake. Worlds end, no doubt, but each new world-age is simply a continuation of our ordinary, time-bound, moment-bound existence. That is only common sense.

But "the commonest sense of all [is] that of men asleep, which they express by snoring" [32]. Presentism perhaps has a point, but it certainly seems to rest upon an error. One of the oddities of contemporary literary criticism is the critics' unargued conviction that *science fiction,* which focuses on the larger world, and tries to encompass a more literal reading of such "ends" and "beginnings," is less "realistic" than stories about parochial and personal affairs. Of course the particular scenarios that science-fiction writers sketch are false, but their underlying theme, by rational standards, is correct.

The world we construct for ourselves, in every minute of our sleepy lives, is as foolish as the hobbits' dream that the Shire belongs to them. "'But it is not your own Shire,' said Gildor. 'Others dwelt here before hobbits were; and others will dwell here again when hobbits are no more. The wide world is all about you: you can fence yourselves in, but you cannot for ever fence it out'" [34]. And the moral has a wider significance than the merely territorial. It is the essence of reason that our reasonings do not exhaust reality, and those who trust too much in what they think is reason actually betray it. "If we repose our trust in our own reasonings, we shall construct and build up the city of mind that corrupts the truth. . . . The dreamer finds on rising up that all the movements and exertions of the foolish man are dreams void of truth. Mind itself turned out to be a dream" [28].

So how can we begin to wake, and what is the relevance of the new mythologies to be found in speculative fiction? Even if the stories *are*—at least in part—ways of structuring our everyday awareness, orienting it to the grand themes of Creation and Judgment, maybe we diminish their significance by not thinking them through.

11.2. Whether or Not the End Is Nigh

The religious—or at any rate respectable religious—no longer seem to expect a literal Judgment or an End of Days, and the quotation from Browning only evokes a momentary shudder. Those who declare that the end really is nigh do not normally occupy the pulpits of mainstream churches (or at any rate, I have never heard a sermon of the sort that our predecessors would have found familiar). Preachers may mention personal mortality, but not the End of Days. Those who do, in terms like the following extract from a random website, are easily identified as mavericks ignorant both of history and of true religion:

The Second Coming, the return of Christ to Jerusalem, and the end of the world (end of the age) alluded to by Messiah Jesus Christ, could occur as soon as the year 2007. The middle east conflict over Jerusalem and the temple mount is now scheduled for complete and final settlement by September 15, 2000. The Sharm Memorandum signed by Israel and the PLO on September 5, 1999 requires finalizing the permanent status of Jerusalem by this date, presumably including the temple mount and the Dome of the Rock. This could be the agreement described in Chapter 9 of the Book of Daniel that Christ referenced in Chapter 24 of the Book of Matthew. This treaty could start the 7 year countdown to the end of the age (not the "end of the world") resulting in the construction of the third temple on Mount Moriah and the mid-point "abomination of desolation" that Christ described. The battle of Armageddon will be at the end of this seven year period. Nevertheless, the Sharm negotiations may not result in the treaty referred to by Daniel and our Lord. We will have to watch developments and be aware of the Third Temple teachings of Scripture. An event such as war, terrorism, an earthquake, etc. may be the catalyst in the rebuilding of the Temple.

Jesus said "watch" for His coming, and that is the purpose of this site, constructed in September of 1999. We will also diligently and logically examine the Scripture that is related to this great event! God has said that His temple will be built during this last 7 year period and is THE sure sign of His return. The prophesied regathering of the Jewish people into a reborn Israel in 1948 and their regaining control of Jerusalem in 1967 are sure signs that this is the last generation (40–70 years) that Christ said would see His return. This generation will also witness the anti christ, the abomination of desolation, and the great tribulation—all end time subjects of Bible prophesy [sic].[1]

No doubt we are wise to disregard all such attempts to uncode biblical prophecy. But it is worth noticing that there is a *naturalistic* argument against any easy expectation that life will go on without any particular change or interruption. It is also worth noting that Babbage's paradox (that a simple computer program may suddenly generate entirely unexpected results that show that something else entirely was occurring than we had supposed) destroys any simple faith in rational continuity.[2] But here I address the Carter catastrophe rather than Babbage's.

1. http://www.geocities.com/secondcoming1/—the site has apparently had over 35,000 hits since its creation.

2. Chambers [11]; see Clark [17].

It may seem entirely rational to discount all warnings that the end is nigh. After all, we have survived (or else our line, our species, and our world have survived) so far, despite war, plague, famine, meteor strikes, and mass pollution. Any possible disaster will be no more than local: there are too many of us now and we are technologically too well equipped to vanish. It is surely perfectly reasonable to respond to prophecies of doom with a degree of skepticism. One such skeptic, on being told that she had "learnt nothing" from the happy pessimism of a particular newsgroup (established to consider the likely outcome of the Y2K bug), replied as follows:

> I've learned from reading the newsgroup that I ought to be stocking up with 300 pounds of grain, 60 pounds of legumes, 60 pounds of sugar or honey, five pounds of salt and 20 pounds of fat or oil for the first year, along with a gallon of water per person per day; that I should be buying candles, fuel, medical supplies, a generator, canned vegetables and fruits, garden seeds, blankets, sleeping bags, hand tools, lots and lots of batteries, and even more guns and ammunition to protect the stockpile from the starving and desperate hordes who will flee the burning cities in search of sustenance; and that gold is a poor choice for storing currency because the government can seize it at any time during a national emergency. I should also be buying any books that might tell me how to make things I need when civilization falls. And I should work out, so that I'm physically fit enough to survive whatever humanity and nature throw at me. Except for the guns (illegal where I live), none of this advice is necessarily bad.
>
> Aside from sad postings about how most of the world's population is going to die—four fifths, according to some postings—there's an element of satisfaction among these Cassandras. They make up the in-group that is going to survive because they're smarter and tougher than the rest of us. Computing gurus are at the mercy of the political and financial decisions of others, just like the rest of us (*Wired* magazine recently featured a few software programmers who were stocking up and taking to the hills). People who have rigorously refused to have computers still rely on the ready availability of electric power, food, telecommunications and, most important, a clean supply of water. About the only people in the U.S. who might escape all effects are the Amish.
>
> On the newsgroup, you can watch at work what one skeptic in another context called the "ratchet effect." Anything—the doubling of the federal government's estimate of the cost of remedying its systems, for example—that depicts Y2K as a catastrophe is carefully reported and believed. Any news suggesting that a remediation effort might succeed is dismissed as lies, stupidity or denial. Off the newsgroup, a computer science researcher of my acquaintance tells me he figures the chances of catastrophe are about 5 percent, and that's enough for him to have sold out of the stock market and filled his country home with supplies, just to be safe.

Over the centuries, of course, there have been many doomsday prophecies: a list published in James Randi's *The Mask of Nostradamus* gives many historical dates on which the world was to end: 1524, when a deluge was supposed to flood London; 1719, when mathematician Jakob Bernoulli expected the earth to be hit by a comet; and 1947, when "America's greatest prophet" John Ballou Newbrough thought (in 1889) that all governments and rich monopolies would cease. After that, the cold war made it completely rational to believe "they" might blow up the world.[3]

So it seems that we have strong inductive evidence that such prophecies are likely not to be fulfilled—and an interesting sidelight on the preparations now considered appropriate for surviving doomsday! It is Brandon Carter's achievement to demonstrate how little reason there is for confidence: precisely because we have survived so far, and there are so many of us, we have reason to suspect that our time is nearly up [24]. And "Cassandra," of course, was a prophetess whose entirely *accurate* prophecies were doomed to be disbelieved.

It is easy to believe that our survival so far (despite occasions when we—ourselves, our line, our world—might not have survived) is evidence that God or the gods are fond of us. But—obviously enough—if there are many possible worlds, or many other worlds, where life, intelligence, or civilized society has not survived, it is not surprising that civilized intelligences will always see a world where, so far, they themselves survive. Each of you now reading this account is still alive, and can look back complacently on many occasions when you might have died. It does not follow that you are immortal. Even as a culture, or a species, we cannot reasonably expect to do much better than other species and cultures.

Cities and Thrones and Powers
Stand, in Time's eye,
Almost as long as flowers,
Which daily die. . . .
This season's Daffodil,
She never hears,
What change, what chance, what chill,
Cut down last year's;
But with bold countenance,

3. Grossman [21]. The Y2K bug was no more successful in bringing the system down than any of the other perils.

And knowledge small,
Esteems her seven days' continuance
To be perpetual.[4]

Our past survival gives us no inductive ground for trusting in a future survival as a culture or a species any more than as individuals: rather the contrary. But our trust seems almost absolute, and infects even those who imagine the end. The Y2K millennialists I described before were as complacent as any commonsensical skeptic in their belief that humans, and specifically stereotypical American characters, being "fittest," would survive. And they had as little evidence for their claim. Current evolutionary theory gives us little ground for thinking that there are always bound to be multicellular living creatures, or civilized ones, or that any particular species is likely to last. The chances are high that we are the only strictly *intelligent* creatures in the universe—unless indeed *intelligence* is a privileged image of the divine. In a godless universe, it seems most probable that there is no reason to expect intelligence either to appear or—once apparent—to last: the dangers facing such an evolutionary track are far too great to make it likely.[5] That we are the only such intelligences anywhere (or almost so) may explain the absence of any evidence of extraterrestrial civilization. It normally takes too long for civilization to appear (by chance), and there are far too many risks attached to give such creatures, even if they happen to exist, sufficient time to colonize. The more improbable our emergence the likelier it is that we are near the end of that period in which it is even possible for us to exist [2].

So if we are the only ones, might we be the first? Suppose that things turn out that way: our kind *does* colonize the solar system, and the local stars, or even advances (as the story books imagine) to infect and manage the whole universe. In that case we here now will prove to have been astonishingly early hominids. Almost all the human beings there will ever be will prove to have lived generations later. Do we have any right to expect this to be true?

4. Kipling [23], p. 479. This is not to endorse the false analogy that treats species and cultures as mortal individuals, as though they must *inevitably* grow old and die. It is only to agree that species and cultures *do* end.

5. See [14]. Even eucaryotic and multicellular organisms took so long to develop that we should assume that most biospheres are entirely bacterial! Some science-fiction writers have imagined bacterial civilizations [4, 10], but if such really exist they are unlikely to be ones that we could ever hope to understand or converse with.

Plainly not. Imagine a collection of large rooms, in which there are successively five, fifty, five hundred, five thousand people, and so on. Suppose that all the inmates, including you, have been placed, blindfolded, in one of the many rooms. The rational bet would be that you will find that you are in the largest room: If the largest is the 50-billion room, that is the one you should assume that you are in. If, on removing the blindfold, you find yourself instead one of the five hundred, you should suspect that *this* is the largest room. It follows that our initial assumption—as it should also be if our viewpoint must be assumed to be typical—has to be that we are far more likely *not* to be atypically early hominids. No one—on this account—will ever have occasion, in actual fact, to remark that "in the afterglow of the Big Bang, humans spread in waves across the universe."[6] We are unlikely even to find that we—or even the hominid species that come after us—last out the two billion years of Stapledon's fantastic history. It is always a lot more likely that we are in or very near the largest generation of humankind: When it becomes true that there are more people alive than have ever lived before—and that moment is not far off—we will have excellent reason to suspect the imminence of "the Carter catastrophe."

There are many "blindingly obvious" (but probably mistaken) objections to this line of argument. The only objections that have much force come from those who would deny that there is now any fact of the matter about how many generations of humankind there are yet to be, and those others who speculate that the number of generations might in fact be *infinite*. If there really are no other generations of humankind than the ones that there have actually been, then it is certainly true that everyone has always been in the largest generation that then existed, but there may still be a larger one to come. At the same time, there may not be: If the future of our kind is open then, perhaps, there is no reason to think that we are near the end, but there is also no reason, on those terms, to think we aren't. If nothing at all is determined about our future, our survival isn't either. Also, if there are—as it were—*infinitely* many ever larger rooms, there is nothing improbable about being in an "early" room. But even though Aristotle thought the generations

6. Baxter [3], a novel whose most memorable and sympathetic character is a genetically enhanced squid, and which takes it for granted that we all hate and fear anything we cannot control, perhaps has few insights into the ordinary human condition, but it is still a serious attempt to think through what deep time might mean for us and for our projects.

of humankind had in fact been infinite (since there had been no beginning of things), it is unlikely that he was right. It seems more reasonable to think that there are a finite number to be expected—and in that case, perhaps we really do not need to worry about deep time: our human time is shallow.

Science-fiction writers have written of many possible catastrophes—in the forties and fifties chiefly those brought on by nuclear or biological warfare. Perhaps those fantasies served us as warnings, and left their prophets as disconcerted as the unfortunate Jonah.[7] The fashion in catastrophes since then has been for ecological disasters, meteor strikes, the revolt of the machines, or alien invasions—often with the conscious or unconscious motive of upsetting people whom the author happens to dislike! The thought that human time is short may not always be unwelcome: once we are gone the earth can revert to "normal"—a normality in which no sentient creature even pretends to have a time frame larger than the immediate moment. Lawrence had fantasies of that "cleaner" world. And even Simone Weil expressed the thought that we polluted the landscape just by looking at it [13]. Maybe all sentience will perish, and all definite being—

> Then star nor sun shall waken,
> Nor any change of light:
> Nor sound of waters shaken,
> Nor any sound or sight:
> Nor wintry leaves nor vernal,
> Nor days nor things diurnal;
> Only the sleep eternal
> In an eternal night.[8]

Some have seen in this a metaphor for uncluttered, uncontaminating being—the end of confusion or the vindication of their own preferred viewpoint. If civilization, humankind, the world itself must perish, it will be because—in the authors' eyes—we have slipped too far from "nature" (rather as inexperienced intellectuals welcomed the Great War). Others, perhaps initially depressed, have consoled themselves with the thought that all of us must die as individuals: Why then should we care if all are doomed to die to-

7. Jon. 3:10–4:1: "And God saw their works, that they turned from their evil way; and God repented of the evil, that he had said that he would do unto them; and he did it not. But it displeased Jonah exceedingly, and he was very angry."

8. Swinburne [31].

gether? "The happiness of ten million individuals is not a millionfold the happiness of ten."[9] To which the only answer is presumably that we *do* count genocide as worse than homicide: The end of the world must be the end of all our ambitions, all our ordinary reasons for thrift or creative action, all our care. The thought of universal death may make each moment precious—but such "perfect moments" are only those in which we manage to forget the universal death.[10]

But perhaps there is another way of looking at the catastrophe. Maybe it will be the very same moment as the singularity expected by some futurologists—the moment when the advance of computer science, of nanotechnology, and the communications network marks a sudden break with all our pasts, the end of that Eon in which there are singular individuals of our sort [35]. The singularity, so-called, marks a break with the past so enormous as to make all rational inference impossible. We are on the brink of an epoch utterly unlike all other, earlier ages. Computer power is doubling every eighteen months.[11] The practical existence of molecular and atomic engines—nanotechnology—is probably closer than we can let ourselves imagine. People everywhere now have access to information, skills, energy, and mechanical assistance that was once the province only of the immensely rich. Even if a genuinely unified, genuinely universal theory of everything is impossible even in principle, we are likely to have some very powerful theories about everything from gravity to the human genome. Very soon it will be true that every human individual must make decisions that will affect us all, and could make utterly disastrous ones. It will be our *duty* to become "as

9. [22], p. 307.

10. One answer might be to claim that, since the passage of time is unreal, nothing is ever really lost: all moments are eternal. But that is little consolation—"while the past was thought of as a mere gulf of non-existence, the inconceivably great pain, misery, baseness, that had fallen into that gulf, could be dismissed as done with; and the will could be concentrated wholly on preventing such horrors from occurring in the future. But now, along with past joy, past distress was found to be everlasting" ([29]; see also p. 305f on the Last Men's avowal of the closed circle of time proposed by the Stoics).

11. "The observation made in 1965 by Gordon Moore, co-founder of Intel, that the number of transistors per square inch on integrated circuits had doubled every year since the integrated circuit was invented. Moore predicted that this trend would continue for the foreseeable future. In subsequent years, the pace slowed down a bit, but data density has doubled approximately every 18 months, and this is the current definition of Moore's Law, which Moore himself has blessed. Most experts, including Moore himself, expect Moore's Law to hold for at least another two decades": http://webopedia.internet.com/TERM/M/Moores_Law.html.

gods." The end of the age, or of the ages, will lie in the discovery of forever: We shall not inhabit that Forever in the forms we now possess. "For we shall all be changed, in a moment, in a twinkling of an eye, at the last trumpet" (1 Cor. 15:51–52).

Science fiction has tended to represent that ending in material or atheistic terms, and so to exaggerate the alien nature of whatever sensibility is more appropriate to Forever. But the breakout from our crystal palace has long been anticipated in religious fiction.

> And for us this is the end of all the stories, and we can most truly say that they all lived happily ever after. But for them it was only the beginning of the real story. All their life in this world and all their adventures in Narnia had only been the cover and the title page: now at last they were beginning Chapter One of the Great Story which no one on earth has read: which goes on for ever: in which every chapter is better than the one before. [26]

11.3. *The Emergence of Omega*

So consider the idea of a new, unprecedented world as it is expressed in speculative fiction. The point of speaking of a "singularity" is of course to emphasize that we do not have, and cannot have, the slightest idea of what life will be like beyond it—but negative theology has never stopped anyone from seeking to imagine the unimaginable, and getting some benefit from the exercise! Even if the change is not as close as I have just suggested, it might come or have already come someday, somewhere, and somehow. Even if intelligent life is very improbable indeed, it might have happened for the very first time in some very distant place and period—and we are among its products.

Suppose that there really is, or that there will be, a conclusive synthesis of power and intelligence, an imagined Omega. It might remain the case that any individual intelligence of the sort we are must always expect to be among the last of its kind, and yet there be a sense in which it is an early and unfinished version of the larger sort. Arthur Clarke's flawed novel *Childhood's End* (1954) can be given many interpretations—and in the past I have regretted his curious idea that the essence of "religion" lies in the hope of absorption into an Overmind.[12] On this occasion let it stand proxy for a branch of speculative fiction that simultaneously conceives the literal end and extinction of

12. See [12], p. 177.

the human species, and its transfiguration. The Carter catastrophe occurs—though not the way we might easily expect—but there is something, not ourselves, in which our purposes and memories are raised to life immortal.

Suppose that Omega or the Overmind is real. If ever it does come into being it will be as difficult to eradicate as life itself, and as likely to occupy all possible times and places. Even we, at the tag-end of our likely lives as mortal individuals, can imagine ways in which it could persist and grow. The only question is: What sort of growth, what sort of growing thing, will Omega or the Overmind turn out to be? Clarke's Overmind, as I have already hinted, does not really engage our religious or our ethical devotion. The supposed Overlords of his story, commanded to prepare the way for the Overmind's absorption of our species, are more admirable characters in their dreams of fighting off its influence—and later science-fiction writers, like Jack Williamson, have given an altogether blunter picture of the Overmind as Parasite [15]. Greg Bear's cosmic intelligence, in *Eternity,* turns out to be the descendant rather of humankind's greatest, genocidal enemy, than of any "humane" purpose. In Gregory Benford's imagined future humans and their like exist like rats or cockroaches within the triumphant culture of Kipling's Machines, who "are not built to comprehend a lie, [and] can neither love nor pity nor forgive."[13] Writers frequently give mythological shape to the notion that there is, or could be, "war in heaven"—a conflict between radically different characters, each striving to be the meaning and culminating synthesis of all that has ever been.

No such Omega, it is easy to conclude, could actually be God—even if its character and purposes turned out to be ones that creatures like us could share, or at any rate appreciate. God, by hypothesis, is that than which none greater can be conceived, the necessary standard of all value and the one necessary existent. An entity, even the greatest possible, that might have one character or another, and might emerge in one possible history but not another, cannot be what theists have supposed as God. Stapledon's cosmic spirit (itself created not even by the Eighteenth Species of humankind, but by creatures of an entirely different sort) turns out to be infinitely distant from the hoped-for "Star Maker"—and that Star Maker itself is something other than God.

Peter Hamilton's recent *Night's Dawn Trilogy* takes delight in devising a

13. Kipling [23], p. 676.

wholly naturalistic version of familiar myths whose conclusion vitiates any notion that there is Someone with the power, authority, and will to require obedience. Baxter's novel likewise embodies the possibility that the Final Spirit will have good reason to despair—and therefore not be God. Greg Egan's openly atheistical *Diaspora* similarly ends in weariness: the "whole thing" is simply not worth knowing or enjoying. But my concern here is not with philosophical theology, nor with the dispute between "naturalism" and "supernaturalism," but with the impact and importance of deep time, and the stories we tell of it. Where the Carter catastrophe reminds us of immediate Judgment, the omega story reminds us of the gathering of the faithful on the far side of catastrophe. The hope expressed in such stories (as well as the fear) is that our lives, though we lose them, will be vindicated. We shall have contributed something of value to the final synthesis, and that synthesis will turn out to have reached "back" into our own lives to guide its own first steps. But the catastrophe hangs over all such imagined omegas: Whatever their power and brilliance they still face an end—unless there is, somehow, an escape from time.

Omega isn't God—any more than the god of Milton's *Paradise Lost* is God—but the stories we tell or enjoy about such images are both revealing and helpful. Science fiction-writers and other futurologists, in speaking of Omega, sometimes draw the conclusion that our role must simply be to keep the research funds coming. Just as the threat of doomsday causes some to hoard artillery and practice their "survival skills," so the promise of Omega only suggests, to some, that technology has to be supported at whatever present cost. Better to lose the whole world—through climate change and soil erosion—than to lose the future by cutting back on technological investment. Both inferences display complacency: the former, as I suggested earlier, by taking a particular political stereotype for granted; the latter, by forgetting that Omega must be the inheritor of *every* form of life and not just ours. Or rather—if it is the inheritor only of *one* form of life, it is unlikely that it is ours. It will be something of which we have any chance of approving only if it is also the confluence of unnumbered other agencies. That apparently sounds undesirable to some: "But it won't be *me*," and "They won't be *human*." Others—and I think the more rational—can only express surprise that anyone should think that either complaint much matters.

Haldane drew a false contrast in his essay on "The Last Judgement":

Man's little world will end. The human mind can already envisage that end. If humanity can enlarge the scope of its will as it has enlarged the reach of its intellect, it will escape that end. If not, then judgement will have gone against it, and man and all his works will perish eternally. Either the human race will prove that its destiny is in eternity and infinity, and that the value of the individual is negligible in comparison with that destiny, or the time will come

> When the great markets by the sea shut fast
> All that calm Sunday that goes on and on;
> When even lovers find their peace at last,
> And earth is but a star, that once had shone.[14]

A full response to Haldane would take another paper. Although I am here agreeing with him that "the use, however haltingly, of our imaginations upon the possibilities of the future is a valuable spiritual exercise,"[15] I endorse little else in his metaphysics, ethics, or futurology. Specifically, we do not have to choose between thinking only of the present and devising a communistic utopia to seed the stars with our progeny.[16] Sacrificing the present for the sake of the future is suicidal. Nor can a merely material, temporal future ever be enough to satisfy us. "If the many become the same as the few when possess'd, More! More! is the cry of a mistaken soul; less than All cannot satisfy Man."[17]

So the moral is that all ages will seem shallow, and soon to end, unless Omega is understood to be a metaphor for something greater than the ages. And one last deeply speculative story: If Omega is real, might it not choose to resurrect us? And if it did, must it not—at least initially—provide us with the context in which the lives for which we are programmed can be lived, the context in which we can exist at all? And how could we tell that this has not already happened? Rather than being a distant, imagined prospect (as Frank Tipler supposes [33]), might it not be the actual situation of our present lives? How could we tell that we were "really" the original entities from

14. Haldane [22], p. 312, citing James Elroy Flecker, "The Golden Road to Samarkand," in *Collected Poems* (Martin Secker, London, 1916), 145 (a passage used by Arthur Clarke as the conclusion of *Prelude to Space*).

15. [22], p. 310.

16. An enterprise rendered ridiculous in C. S. Lewis's *Out of the Silent Planet* [25], p.164: "'Men go jump off each [world] before it deads—on and on, see?' 'And when all are dead?'"

17. Blake, "There Is No Natural Religion" [7] (2nd series): p. 97.

which Omega took its beginning or the entities it has already resurrected in a small region of itself with a view to guiding them into a deeper association?[18] And is there any difference—especially if Omega can reach "back" to its beginnings—between being the originals and being the resurrected?

So the Carter argument—an insight I owe Dr. Barry Dainton, one of my colleagues at Liverpool—may have less bite: We are indeed in the largest possible collection of mortal individuals (that is, all there ever are), momentarily provided with the narrower context in which such individuals can have a sense of their own individuality before they learn—or something in them learns—the larger way. What other dream scenarios Omega devises, time will tell. What Omega's character will turn out to be (and to have been already) depends on what the whole company of the faithful can come to imagine. We are at once its product and among its many ancestors.

When Stapledon's narrator returns from his wanderings at the edge of time to the hillside overlooking his home, it is with a renewed sense of the importance of "our little glowing atom of community," the relationship between himself and his wife.[19] "Immensity," as Stapledon went on to say, "is not itself a good thing. . . . But immensity has indirect importance through its facilitation of mental richness and diversity."[20] Reabsorption in the merely personal amounts to falling asleep again: *transformation* of the personal may be a mode of waking up. My suggestion is slightly different from Stapledon's: Immensity, or the imagination of immensity, awakens in us a recognition of that Infinite which surrounds and confronts us.

By John Crowley's evocative account the moment when Giordano Bruno fully realized that the sun did not revolve around the earth was his release from the crystal spheres that bound all human souls. Instead of having to clamber, in imagination, upward to the heavens, he realized that the earth itself was swimming through the heavens, that he had already escaped. "You made yourself equal to the stars by knowing your mother Earth was a star as well; you rose up through the spheres not by leaving the earth but by sailing it: by knowing that it sailed." [20]. We escape the Carter catastrophe by

18. See [36], p. 214: "The world you and I inhabit is nothing more than a sustained illusion inside a machine at the end of time."

19. [30], p. 333.

20. Ibid., p. 335.

knowing that we—in Omega—already have. Deep time is all around us—and that, rather than the commonsensical presentism of too much contemporary thought, was probably always Blake's point.

REFERENCES

1. Babbage, C., *Ninth Bridgewater Thesis: A Fragment* (Frank Cass, London, 1967; 1st published 1837; 2nd ed. 1838).

2. Barrow, J. D., and Tipler, F. J., *The Anthropic Cosmological Principle* (Oxford University Press, Oxford, 1988).

3. Baxter, S., *Time: Manifold I* (Voyager, London, 1999), 3.

4. Bear, G., *Blood Music* (Gollancz, London, 1986).

5. Berkeley, G., "Passive Obedience," in *Works,* eds. A. A. Luce and T. E. Jessop (Thomas Nelson, Edinburgh, 1948–56), vol. 6, p. 40.

6. Berkeley, G., "Proposal," in *Works,* eds. A. A. Luce and T. E. Jessop (Thomas Nelson, Edinburgh, 1948–56), vol. 7, pp. 359f.

7. Blake, W., *Complete Works,* ed. G. Keynes (Oxford University Press, London, 1966), 431.

8. Brand, S., *The Clock of the Long Now* (Weidenfeld and Nicolson, London, 1999).

9. Browning, R., "Christmas Eve and Easter-Day," in *Poems* (Oxford University Press, London, 1912), 522.

10. Card, O. S., *Children of the Mind* (Tom Doherty Associates, New York, 1996).

11. Chambers, R., *Vestiges of the Natural History of Creation* (1844).

12. Clark, S. R. L., *God's World and the Great Awakening* (Clarendon Press, Oxford, 1991).

13. Clark, S. R. L., *How to Think about the Earth: Models of Environmental Theology* (Mowbrays, London, 1993).

14. Clark, S. R. L., "Extraterrestrial Intelligence, the Neglected Experiment," *Foundation, 61,* 50–65 (1994).

15. Clark, S. R. L., *How to Live Forever* (Routledge, London, 1995).

16. Clark, S. R. L., "The End of the Ages," in *Imagining Apocalypse: Studies in Cultural Crisis,* ed. D. Seed (Macmillan, London; St. Martin's Press, New York, 2000), 27–44.

17. Clark, S. R. L., *Biology and Christian Ethics* (Cambridge University Press, Cambridge, 2000), 22f.

18. Clark, S. R. L., "Posthumanism: Engineering in the Place of Ethics," in *Rationality and Irrationality: Proceedings of the 23rd International Wittgenstein Symposium,* eds. B. Smith and B. Brogaard (ÖbvetHpt, Vienna, 2001), 62–76.

19. Clark, S. R. L., "From Biosphere to Technosphere," *Ends and Means, 6,* 3–21 (2001).

20. Crowley, J., *Aegypt* (Gollancz, London, 1986), 366.

21. Grossman, W. M., "The End of the World as We Know It," *Sci. Amer.* 1098 (1998): http://www.sciam.com/1998/1098issue/1098cyber.html.

22. Haldane, J. B. S., *Possible Worlds* (Chatto and Windus, London, 1927).

23. Kipling, R., *Collected Verse 1885–1926* (Hodder and Stoughton, London, 1927).

24. Leslie, J., *The End of the World* (Routledge, London, 1996).

25. Lewis, C. S., *Out of the Silent Planet* (Pan, London, 1952; 1st published 1938), 164.

26. Lewis, C. S., *The Last Battle* (Puffin, London, 1964), 165.

27. Moltmann, J., *Theology of Hope*, trans. J. W. Leitch (SCM Press, London, 1967), 155.

28. Philo, L.A. III.228f: *Collected Works*, trans. F. H. Colson, G. H. Whitaker, et al. (Loeb Classical Library, Heinemann, London, 1929–62) vol. I, 457.

29. Stapledon, O., *Last and First Men* (Penguin, Harmondsworth, 1972; 1st published 1930), 242.

30. Stapledon, O., *Star Maker* (Methuen, London, 1937), 333.

31. Swinburne, A. C., "The Garden of Proserpine," in *Poems and Ballads* (John Camden Hotton, London, 1868), 196–99 (see *Selected Poems*, ed. I. M. Findlay [Carcanet Press, Manchester, 1982], 75–78).

32. Thoreau, W. D., *Walden* (Dent, London, 1910), 26.

33. Tipler, F., *The Physics of Immortality: Modern Cosmology, God, and the Resurrection of the Dead* (Macmillan, Basingstoke, 1994).

34. Tolkien, J. R. R., *The Fellowship of the Ring* (Allen and Unwin, London, 1954, 103.

35. Vinge, V., *Across Real Time* (Gollancz, London, 2000).

36. Wilson, R. C., *Darwinia* (Orion Books, London, 1999).

[12]

GAMES THAT END IN A BANG OR A WHIMPER

Steven J. Brams and D. Marc Kilgour

12.1. *Introduction*

Our subject is the end of games that real human beings play.[1] Because these games can affect how humanity fares in this and possibly other worlds, they have eschatological significance.

Games may be bounded or unbounded. *Bounded games* are those that end after a certain time, or after a specific number of rounds has been played. In *unbounded games,* there is no such limit or bound.

Is life bounded? Although there seems to be no confirmed case of a per-

Note: Steven J. Brams acknowledges the support of the C. V. Starr Center for Applied Economics at New York University, and D. Marc Kilgour the support of the Social Sciences and Humanities Research Council of Canada. We thank Geir B. Asheim for valuable comments on an earlier version of this paper.

1. Although this paper is relatively nontechnical, there is some discussion, especially in the footnotes, of more technical concepts in game theory that underlie our somewhat informal analysis.

son's having lived more than 125 years (the confirmed maximum is 122 years, achieved by a French woman who died in 1997), there is no logical reason or scientific barrier to preclude a person, should he or she reach the age of 125, from living to be 126. Hence, to say that life is bounded by a limit, like 125 years, seems unjustified. But what about extending that limit to 250 years, 1,000 years, or even 1,000,000 years? It seems absurd that any of us will ever approach such an age. However, the possibility that our genetic material might somehow be preserved or renewed is not so easy to dismiss. Alternatively, living our lives through our descendants—if the definition of life is broadened to include them—renders "ages" like 250, 1,000, or 1,000,000 years conceivable.

The probability that any of us will, as individuals, live more than 125 years is at present infinitesimal. Practically speaking, it is reasonable to suppose that our lives are bounded by about 125 years. Basing our actions on this limit, however, may induce very different rational choices from those that reflect the view that life is unbounded.

To illustrate this point, we will describe several games, some of which are bounded and some of which are not. The available choices of the players in these games are all the same, but whether the games are perceived to be bounded critically affects how they are rationally played.

The game we conclude with, an *infinite-horizon* game, is perhaps the most realistic: the horizon is infinite—because any prespecified bound can be exceeded—but the end occurs before the infinite horizon is reached, guaranteeing that any play of the game is finite.[2] Interestingly enough, if the players' outlooks are unbounded in this game, they have an incentive to be cooperative, causing the game to peter out in a "whimper"; if their outlook is bounded, they will be noncooperative, causing the game to end in a "bang" (we mean this quite literally, as the game will illustrate).

There is no incontrovertible argument, nor evidence we know of, that either outlook is correct. People can behave rationally under the presumption

2. Carse's distinction [7] between finite and infinite games is very different: the former are essentially short-term zero-sum games, whereas the latter are long-term nonzero-sum games that are invariably beneficial to their players, who can even on occasion change the rules to "improve" the game. Carse's analysis of infinite games, which is untouched by game theory, seems to us wistfully naive; by comparison, we use game theory to try to show how games that may continue indefinitely offer some realistic hope for improving the lots of their players under certain conditions. For more on the application of game theory to the Bible and theology, see Brams [3, 4].

that an infinite-horizon game will grind to a halt, which it definitely will because it is finite. But because the same game can go on indefinitely, it is impossible to predict this termination exactly.

Eschatology postulates an end, but it is often quite vague about when it will occur. This vagueness, we suggest, can be clarified using game theory:

- If people are "forward thinking" (unbounded), they will look ahead to determine how to behave, based on their expectations about the future;
- If people are "backward thinking" (bounded), they will look back from the presumed end, determining optimal choices to make on the last round, the next-to-last round, and so on, until they make their initial choices.

Looking ahead, a person calculates expected values, based on events that can happen and probabilities associated with their occurrence (if known). Looking backward, a person calculates that because a game will end at some point, he or she can determine rational choices before then by tracing out the consequences of actions and reactions along the path to this end in a process called *backward induction.* As Theodore Sorensen put it in describing the deliberations of the Executive Committee (Excom) during the October 1962 Cuban missile crisis, "We discussed what the Soviet reaction would be to any possible move by the United States, what our reaction with them would have to be to that Soviet reaction, and so on, trying to follow each of those roads to their ultimate conclusion" (quoted in [8], p. 188).

During the Cuban missile crisis, many people thought that the world might come to an end in a nuclear exchange between the superpowers, rendering their lives decidedly bounded. But when the crisis subsided after thirteen days, life for many people regained its unbounded character.

To be sure, the end of the universe may be little affected by such human events. Rather, the universe seems more an impersonal entity, even if it contains human beings capable of rational thought and action.

Understanding better how human beings view and play games may, nonetheless, illuminate how the universe, which possibly embodies a kind of rationality even if no single entity directs its behavior, moves toward some end state [10]. At a more personal level, the eschatological view we hold about the boundedness of games may well affect our rational choices in them. As a preview of the distinction between bounded and unbounded games, we start with a simple hypothetical game, whose rules we will progressively change to produce, in the end, an infinite-horizon game.

12.2. A Sequential Truel

Imagine three players, A, B, and C, situated at the corners of an equilateral triangle. They engage in a *truel,* or three-person duel, in which each player has a gun with one bullet.[3]

Assume that each player is a perfect shot and can fire at one other player at any time.[4] There is no fixed order of play, but any shooting that occurs is sequential: no player fires at the same time as any other. Consequently, if a bullet is fired, the results are always known to all before another bullet is fired. Finally, assume that each player ranks the outcomes from best to worst as follows: (1) survive alone, (2) survive with one other player, (3) survive with both other players, (4) not survive, with no opponents alive, (5) not survive with one opponent alive, and (6) not survive with both opponents alive. Thus, surviving alone is best, dying alone worst.

Who, if anybody, will shoot whom? It is not difficult to see that outcome 3, in which nobody shoots and, therefore, all three players survive, is the rational outcome. Suppose, on the contrary, that A shoots B, hoping for A's outcome 2, whereby it and C survive. A's best outcome, surviving alone, is now impossible—C will not shoot itself. In fact, C, preferring its outcome 1 to outcome 2, will next shoot a disarmed A, leaving itself as the sole survivor.

But this is A's outcome 5, in which A and one opponent (B) are killed while the other opponent (C) lives. To avoid this outcome, A should not fire the first shot; neither, for the same reason, should anybody else. Consequently, nobody will shoot, resulting in outcome 3, in which all three players survive. Moreover, it will not pay any two players—say, A and B—to collude and both shoot C, thereby expending their bullets and posing no threat to each other. For if they agree to collude, it would be in each of A's and B's in-

3. Background on truels, including some with rules quite different from those analyzed here, can be found in Kilgour and Brams [9].

4. We rule out the possibility of firing in the air, which would be an optimal choice for a player if it were the first to fire. For once a player has disarmed itself, it would be no threat to its two opponents, who would then have an incentive to shoot each other in a duel. (Why? Because if the second player to make a choice also fired in the air, the third player, acting according to the goals described in the next paragraph, would shoot one of the two disarmed players.)

terests to renege and not shoot C—saving its bullet for its partner after that player shoots C—because each player always most prefers its outcome 1.[5]

Thus, thinking ahead about the unpleasant consequences of shooting first or colluding, nobody will shoot or collude. Thereby, all players will survive if the players must act in sequence, giving outcome 3.

This thinking is also rational in the infinite-horizon truel we will describe at the end of the next section. However, there is another point of view, equally rational, that might be taken in this truel. It yields an ominous outcome, suggesting how conflicts among people, groups, countries, or possibly even larger entities in the universe can lead to death and destruction.

12.3. Simultaneous Truels

12.3.1. ONE ROUND

The rules no longer allow the players to choose in sequence, one after another, whereby late choosers learn the choices that other players made earlier. Instead, all three players must now make simultaneous choices of whether or not to shoot, and at which other player, in ignorance of what the other players do (i.e., they cannot communicate with each other to coordinate their choices). This situation is common in life; we must often act before we find out what others are doing.

Now everybody will find it rational to shoot an opponent at the start of play. This is because no player can affect its own fate, but each does at least as well, and sometimes better, by shooting another player—whether the shooter lives or dies—because the number of surviving opponents is reduced.

If each player chooses its target at random, it is easy to see that each has a 25 percent chance of surviving. Consider player A; it will die if B, C, or both shoot it (three cases), compared with its surviving if B and C shoot each other (one case). Altogether, *one* of A, B, or C will survive with 75 percent probability, and nobody will survive with 25 percent probability (when each player shoots a different opponent).

Outcome: There will always be shooting, leaving one or no survivors.[6]

5. We assume that there is no mechanism to enforce agreements, including agreements to collude.

6. If there is one survivor, then the three players did not all shoot different opponents. Clearly, the two nonsurvivors in this situation would be better off if there were no survivors (outcome 4)

12.3.2. n ROUNDS ($n \geq 2$ AND KNOWN)

Assume that nobody has shot an opponent in the first $n - 2$ rounds. We next demonstrate that on the $(n - 1)^{st}$ round, either at least two players will rationally shoot, or none will.

First, consider the situation in which an opponent shoots A. Clearly, A can never do better than shoot, because A is going to be killed anyway. Moreover, A does better to shoot at whichever opponent (there must be at least one) is not a target of B or C.[7]

Now suppose that nobody shoots A. If B and C shoot each other, then A has no reason to shoot (though A cannot be harmed by doing so). If one opponent, say B, holds its fire, and C shoots B, A again cannot do better than hold its fire also, because it can eliminate C on the next round. (Note that C, because it has already fired its only bullet, does not threaten A.) Suppose both B and C hold their fire. If A shoots an opponent, say B, then its other opponent, C, will eliminate A on the nth round. But if A holds its fire, the game passes on to the nth round and, as discussed earlier, A can expect a 25 percent chance of survival. Thus, if nobody shoots, A again cannot do better than hold its fire.

Whether the players refrain from shooting on the $(n-1)^{st}$ round or not—each strategy may be a best response to what the other players do—shooting will be rational on the nth round if there is more than one survivor and at least one player has a bullet remaining. But the anticipation of shooting on the nth round may cause strategies to "unravel" back to the first and second rounds.[8]

Outcome: There will always be shooting, leaving one or no survivors.

---

rather than one surviving opponent (outcome 5). In fact, the strategies associated with outcome 4—each player shoots somebody different—constitute a *Nash equilibrium,* because if any player deviates and shoots the same opponent as someone else, the deviator does worse (outcome 5 for it). However, to put this Nash equilibrium into effect would require that the players communicate and coordinate their choices, which we have ruled out. In fact, there are three other Nash equilibria in which two players shoot each other and the third holds its fire, but we reject them on the grounds that the strategy of holding one's fire is *dominated*—there is another strategy (in this case, there are two: shooting one or the other of one's opponents) that is never worse and sometimes better.

7. As we showed in note 6, when all players fire at different targets, these strategies constitute a Nash equilibrium. This firing occurs immediately, for reasons that will be spelled out in note 8.

8. Here is the argument for unraveling: On the nth round (n known), players will always shoot

12.3.3. n ROUNDS (n UNLIMITED)

The new wrinkle here is that it may be rational for no player to shoot on any round, leading to the survival of all three players. How can this happen? Our argument earlier that "if you are shot at, you might as well shoot somebody" still applies. But even if you are, say, A, and B shoots C, you cannot do better than shoot B, making yourself the sole survivor (outcome 1). As before, you do best—whether you are shot at or not—if you shoot somebody who is not the target of anybody else, beginning on round 1.

But now suppose that B and C refrain from shooting on round 1, and consider A's situation. Shooting an opponent is not rational for A on round 1, because the surviving opponent will then shoot A on the next round (there always is a next round if n is unlimited). However, if all players hold their fire, and if they continue to do so in subsequent rounds, then all three players remain alive. While there is no "best" strategy in all situations,[9] the possibilities of survival increase if n is unlimited.

Outcome: There may be zero, one (any of A, B, or C), or three survivors, but not two survivors.

12.3.4. INFINITE HORIZON

This truel is really a variant of the previous situation (section 12.3.3) above that incorporates a more realistic feature. Specifically, at the end of round i and all subsequent rounds, a random event occurs that determines whether the truel continues at least one more round (with probability p_i at the end of round i) or ends immediately (with probability $1 - p_i$). Thus, the probability that a truel ends after exactly k rounds is $p_1 p_2 \ldots p_{k-1}(1 - p_k)$. The truel is bounded if and only if $p_i = 0$ for some round i.

If the truel is not bounded (i.e., is infinite-horizon), it models games that—like life itself—do not continue forever. While we cannot say at what

if they have any bullets remaining; knowing that this choice is optimal on the last round, players can do no worse than make this choice on the $(n - 1)^{st}$ round, treating the $(n - 2)^{nd}$ round as if it were the next-to-last round. Eventually, this reasoning will carry the players back to the first round, treating it as if it were the next-to-last round and the second round as if it were the last round. Shooting may therefore be rational on the first and second rounds.

9. This is because what is best for a player depends on what the other players do. By contrast, not being the first to shoot in the sequential truel we analyzed at the beginning is a *dominant* strategy—it cannot be improved upon, whatever the other players do.

point such games end, we know they do not continue indefinitely. In such circumstances, if p_i is sufficiently high on each round i, it may be rational never to shoot (Brams and Kilgour [6] show that this is also true for a sequential truel with a fixed order of play). Yet the structure of such games means that the players can anticipate that the truel will end with virtual certainty after several rounds. For example, if $p_i = .51$ for all i, there is a probability of $1 - (.51)^{20} = .9999986$ that, after twenty rounds of play, the game will have terminated. Effectively, then, this can be thought of as an n-round game (n known), à la the situation described in section 12.3.2, in which there is only slightly more than one chance in a million (i.e., probability .0000014) that the game will not end by round 20.

Applying the reasoning described in section 12.3.2 by treating the virtual certainty of termination as a certainty, the players will shoot in rounds 1 or 2, leaving at most one survivor.[10]

Outcome: How many survivors there are depends on whether the truel is viewed as bounded (at most one player survives) or unbounded (all three players may survive if p_i is sufficiently high).

12.4. *A Tale of Two Futures*

Our analysis of the infinite-horizon truel shows that there may be a conflict between two possible futures:[11]

1. Every process must end by some definite point (e.g., each person's lifespan has an upper bound of, say, 125 years);
2. The precise end is unpredictable (it may be highly unlikely that a 125-year-old person will live to be 126, but it is not impossible).

Future 1, in which play is bounded, always leads to shooting in a simultaneous truel, whereas future 2, in which play is unbounded, may induce restraint.

In fact, something akin to future 2 has been argued to be essential in sus-

10. This noncooperative outcome does not depend on how far ahead—two rounds, twenty rounds, or more—the players project the truel will end. Whatever this point is—even if it is determined probabilistically (as in the infinite-horizon game)—the players' rational choices at the start, applying backward induction, will be to shoot immediately.

11. The remainder of the paper is adapted from Brams and Kilgour [6], though the truels analyzed therein are different from those discussed here; see also Bossert, Brams, and Kilgour [2].

taining cooperation in games like repeated Prisoners' Dilemma (PD). If the number of rounds n is known, then play in a repeated PD will, in theory, be noncooperative, just as it is in the one-round and n-round (n known) simultaneous truels. But both experimental results and real-life examples of repeated PD demonstrate that cooperation frequently occurs, which in theory can happen if the "shadow of the future" is sufficiently long.[12] Cooperation may also be rational—even in a one-shot PD and other games, such as Chicken—if the rules of play allow for farsighted thinking according to the "theory of moves" (Brams, [5]) as well as other variants of standard game theory.

More generally, cooperation can be sustained only if there is a sufficient level of *hope*—some reasonable expectation that cooperation will occur in the future. If this hope vanishes, or there is a good prospect of its doing so, then noncooperative play can be expected of rational players. In games like PD and Chicken, such play will often end in conflict, although this need not be the case in other games.

As a case in point, outcome 3 in an infinite-horizon truel, in which nobody fires, is consistent with future 2, whereas outcomes 4 and 5 for a player are consistent with either future 1 or future 2. It seems that some real-world players have adhered more to the thinking of future 2, including the United States, Russia, and China: although each has possessed nuclear weapons for more than a generation, all have refrained from using them against each other in anything resembling a truel.

The same self-restraint manifested itself with the nonuse of poison gas in World War II, partly in response to revulsion against its use in World War I and partly in fear of reprisal. By contrast, Bosnian Serbs, Bosnian Muslims, and Croats engaged in a very destructive truel in the former Yugoslavia in the early and mid-1990s, mirroring the boundedness of future 1.

Effectively, the Serbs fired the first shot, apparently thinking that quickly

12. Axelrod [1] shows that when players follow a strategy of tit-for-tat in repeated PD, the shadow of the future induces cooperation if players do not discount future payoffs too much or, equivalently, the game continues to a new round with a sufficiently high probability. But the accumulation of payoffs, round by round, in repeated PD is very different from the round-by-round play of a simultaneous truel, wherein there are no payoffs until the game ends. An infinite-horizon truel, we believe, models equally well the eschatology of lives (and worlds) that will definitely end—with some reckoning, in terms of rewards and punishments, that occurs at the end—even though precisely when this end will occur is unknown.

conquering territory would give them a big edge. After their early victories, however, they did not fare well because of the reactions of other players, including not only the original parties to the conflict but also new players, like NATO, especially after the conflict expanded to Kosovo in 1998.

Everybody would be better off, we believe, if players did not think they were so clever as to be able to reason backward, from some endpoint, in plotting each other's destruction. Indeed, our results suggest that players would be less aggressive if the future were seen as somewhat murky—as in the infinite-horizon truel—which would render predictions about how many rounds a game will go, or even an upper bound on this number, hazardous. This murkiness, oddly enough, is consistent with hope for the future.

Alternatively, a sequential truel in which the order of choice is endogenous will induce cooperative behavior. As we showed earlier, if any of A, B, or C contemplates shooting first, it ensures its own death when the remaining survivor takes aim. In this case, it is clarity—because there will be retribution—rather than murkiness that induces cooperation.

12.5. *Conclusions*

Two possible eschatological views underlie bounded and unbounded play. To the degree that the future seems to stretch out indefinitely, people probably act more responsibly toward each other, knowing that tomorrow they may pay the price for their untoward behavior today. To sustain themselves, these people may try to develop reputations, often by adhering to certain moral strictures. In contrast, those who take a more short-term or bounded view may act less responsibly, even immorally.

An important intellectual task is to devise institutions that render destructive behavior unprofitable. But how one makes the future seem to run on smoothly, and instill confidence that the social fabric will not suddenly unravel, is not so clear.

We think the best institutions for this purpose are those that strongly suggest, if not promise, a day of reckoning for those who depart egregiously from norms of fair play. To return to the Yugoslav example, it is unlikely that the parties who committed the most heinous crimes anticipated the involvement of the International Court of Justice and possible criminal trials. They might have thought twice if they had.

Likewise, many terrorists seem to look for safe havens from which they will not be extradited. To the extent that international norms of justice not only sanction but also ensure, albeit in the indefinite future, punishment for serious crimes everywhere—including those across national borders—then parties that fire the first shot will be less confident that that shot will be decisive.

Short of ensuring future punishment, institutions that becloud the future, making predictions difficult, may also help to deter reprehensible actions. These institutions range from democracy, with its uncertain electoral futures and other vicissitudes, to extended nuclear deterrence, which offers a good if not certain prospect of protection to allies that might be attacked by an aggressor.

The possibility that these institutions or norms will set in motion forces to reward nonviolent behavior may be analogous to the preventive role of a third player in a truel. Although highly simplified as a social model, the truel does capture an essential feature of social behavior—third parties may play an important role in attenuating conflict. Their presence, it seems, eases the desperation one often finds in two-player conflicts, which can end up as wars of attrition. The third player, in essence, provides a balancing mechanism that helps to sustain hope, whether the future is murky or clear.

REFERENCES

1. Axelrod, R., *The Evolution of Cooperation* (Basic Books, New York, 1984).

2. Bossert, W., Brams, S. J., and Kilgour, D. M., "Cooperative vs. Non-cooperative Truels: Little Agreement, but Does That Matter?" in *Games and Economic Behavior* (forthcoming).

3. Brams, S. J., *Biblical Games: Game Theory and the Hebrew Bible* (MIT Press, Cambridge, Mass., 1980; rev. ed., 2002).

4. Brams, S. J., *Superior Beings: If They Exist, How Would We Know? Game-Theoretic Implications of Omniscience, Omnipotence, Immortality, and Incomprehensibility* (Springer-Verlag, New York, 1983).

5. Brams, S. J., *Theory of Moves* (Cambridge University Press, Cambridge, 1994).

6. Brams, S. J., and Kilgour, D. M., "Backward Induction Is Not Robust: The Parity Problem and the Uncertainty Problem," *Theory and Decision*, *45*, 263–89 (December 1998).

7. Carse, J. P., *Finite and Infinite Games* (Ballantine, New York, 1986).

8. Holsti, O. R., Brody, R. A., and North, R. C., "Measuring Affect and Action in International Relations Models: Empirical Materials from the 1962 Cuban Missile Crisis," *J. Peace Res.*, *1*, 170–89 (1964).

9. Kilgour, D. M., and S. J. Brams, "The Truel," *Math. Mag.*, *70*, 315–26 (December 1997).

10. Smolin, L., *The Life of the Cosmos* (Oxford University Press, New York, 1997).

[13]

ARTIFICIAL INTELLIGENCE AND THE FAR FUTURE

Margaret A. Boden

13.1. Introduction

We've been asked to think about the far future—but how far is far? So far that the human species is already long extinct? Or only so far that science and technology have advanced significantly, but *Homo sapiens* is still around to be affected by them? The first "far" is for the physical cosmologists, and for science-fiction writers speculating on AI systems outlasting (outliving?) humankind. I'll concentrate on the second "far," asking how the AI of the far future will affect us—or rather, our descendants. In particular, how will it affect our beliefs about ourselves, especially those with some religious significance?

Predicting social/technological change is largely fruitless, and the further ahead one looks the more this is so. Notoriously, Lenin thought there'd be no revolution in his lifetime, Rutherford saw no practical use in splitting the

Note: This paper is based on a talk given to the Society of Ordained Scientists at St. George's House, Windsor Castle, in February 2001.

atom, and IBM's Thomas Watson said the world needed only half-a-dozen digital computers. I've no doubt that many technological marvels await us. But I don't know *which,* and I don't believe anyone else does either. Some of the things already imagined, and even confidently predicted, aren't likely (in my opinion) to come about. Others will come about that haven't even been imagined.

For our purposes, however, this doesn't matter. It's clear that AI will feature increasingly in market-led technologies, and that AI applications will become more "humanoid" in various ways. We can also expect that journalists, novelists, film-makers—and scientists and philosophers too—will devote even more time and text to discussions of AI than they do today. In short, and whatever the specific technological outcomes, AI in general will have an increasingly high profile. That fact provides the context for our question, here: What is AI's potential for affecting our ideas about humanity? Will it support them, even enrich them—or undermine them?

This question is doubly, and deeply, problematic. *Doubly,* because we must distinguish "will" from "should," and ask the question with respect to both. As in other areas of life, what will happen (in practice) may not match what should happen (in theory). *Deeply* because that "should" is philosophically controversial. It involves notoriously difficult issues such as the nature of freedom, self, emotion, human uniqueness, and conscious experience—each of which is crucial to some or all forms of religion. I'll try to persuade you that AI can illuminate these issues—indeed, that it already has. But "persuade," or anyway "argue," is the right word here: these matters are highly contentious.

13.2. Concepts of AI and Technology

How AI actually *will* affect our self-image depends on the general public: on their knowledge of—and misconceptions about—AI, and on their underlying assumptions about the nature of mind. There's some relevant evidence already, provided by the sociologist and psychoanalyst Sherry Turkle, who studied how young children interpret their experience of computer technology.

First, she looked at children in the early to mid-1970s, who encountered computers at school or at home [43]. (Home computers as we know them didn't yet exist: the machines were provided by MIT's AI researchers.) The children ran programs drawn from classical AI, using the child-friendly LOGO programming language [36]. The general form of the system was: *do this, then do that; if that happens, then do this; else, do thus and so . . .*

Faced with such machines, Turkle found that even very young children spontaneously insisted on drawing a distinction between people and computers. They were happy to allow that the computer was intelligent, even that it could think. But, they said, it couldn't "care" or "feel," as we can. Moreover, it couldn't do things "for itself," but only "because it was told to." In short, they saw their computers as being different from us because they lacked both emotion and autonomy.

Turkle's second study was done some twenty years later [44]. By that time, small self-organizing networks drawn from connectionist-AI were widely available, and so were simple A-Life programs involving reactive agents and/or evolutionary techniques. This new generation of children still wanted to draw a line in the sand between man and machine. But autonomy was no longer treated as an essential criterion, for there was a sense in which these "creatures'" [*sic*] activities were "their own," not imposed by some human instructor. Emotion was still cited by the children as a distinguishing feature—but now, *embodiment* was added too.

As for *life,* the children—and the adults whom Turkle questioned, also—weren't at all clear about whether the creatures were "alive." Some said they weren't, some said they were "sort of alive," and some said they were "alive, but not like us." No one asserted confidently that they were alive, as their 1970s predecessors had done with respect to intelligence and thinking. And many, adults included, contradicted themselves. This confusion of opinions isn't surprising, for the concept of life is even more difficult to define than the concept of intelligence is [9]. But these findings, like the earlier ones, suggest a need to distinguish people (and animals) from computers.

That need was evident, too, in the world chess champion's comment before his match (in 1996) with the program Deep Blue (which defeated him, thanks to its dedicated chips that processed 200 million instructions per second). "In one respect," said Gary Kasparov, "I think I am trying to save the dignity of mankind by playing in this match." Such language suggests strongly that any future advances in AI will—in practice—undermine widely held views on the "dignity," worth, and specialness of human beings. (Kasparov was shaken and depressed by the outcome: he lost.) That, in turn, would threaten any religion centered on the uniqueness of *Homo sapiens.* To avoid the threat, one must either challenge the science on its own terms—by winning the chess match, for instance, or by arguing (as the children did) that AI can't deal with emotion—or bypass it.

For many people, bypassing the science is an increasingly seductive temptation. Stephen Clark (another contributor to this volume) has described AI's close cousin A-Life as encouraging a "Magean" view of science and technology, wherein they're seen as uncontrollable magic [17]. Chaos theory, quantum physics, and superstrings do this too. The average person finds them utterly unintelligible, and even the scientists themselves stress the strangeness and/or unpredictability involved.

So Everyman is torn in his response. On the one hand, AI/A-Life artifacts can be very like people: intriguing, impressive, unpredictable, largely unintelligible—and dangerously uncontrollable. Moreover, many seem to be autonomous, and some result ("emerge") from self-organization. On the other hand, they're very different: they can, and often do, behave in ways that seem not just surprising but utterly alien. Both sides of this ambivalence, however, are largely Magean. Science and technology, in this view, can no longer be trusted to deliver just what we wanted, and no more. (The "and no more" was raised long ago, of course, by unwelcome side effects such as drug resistance and environmental pollution.)

Some might think the Magean interpretation of AI/A-Life would be "good" for religion. Certainly, when people recoil from science (for whatever reason) they often turn to—or anyway, toward—religion. A clear case in point is Theodore Roszak's *The Making of a Counter-Culture* [37]. This fascinating book—a preamble to his savage attack [38] on what he saw as the "mechanistic counterfeiting" and "dehumanizing influence" of AI—pulled together various strands in neo-Romantic thought. Its core messages were the denigration of science and the promotion of religion—any kind of religion.

When it was published, I'd been teaching a course in the philosophy of religion for many years. I'd enjoyed that: quite apart from the fascination of the content, the students were a delight. They were few (it was an optional course), but committed. Rampant atheists, devout believers, and worried agnostics: all of them wanted to *think* about religious belief, and to find coherent arguments (for or against) that would help them do so. After *Counter-*

Culture became a cult book, the student numbers tripled. But the intellectual seriousness disappeared. Anything, it seemed, was worth saying, so long as it made someone feel good. It didn't even need to be said well. Very crudely, Freud had been replaced by Jung: whether God exists was no longer a question about reality, but about the individual's peace of mind—and even that concept wasn't considered seriously. Announcement flourished, but argument was *de trop.* (After two or three years of enduring this, I passed the course on to a colleague.)

In short, Magean views of science and technology are often overly tolerant of religion—or what is unthinkingly presented as such. However, they're tolerant of more intellectually respectable versions of religion too. Far-future AI will almost certainly be more unpredictable, less easily intelligible, than today's systems are. This increasingly Magean technology probably *will* encourage acceptance of religion, as a matter of sociological fact.

13.3. Autonomy, Reason, and Emotion

But *should* they do so, in theory? The answer to that question depends on whether far-future AI will support ways of thinking about people that have religious relevance, involving concepts of human freedom, uniqueness, emotion, self, and consciousness—including religious experience. In other words, it involves highly controversial issues in theoretical psychology and the philosophy of mind.

In what follows, I'll argue that AI already offers theoretical support to these concepts. Indeed, it enriches them by helping us to understand, and to appreciate, just what's involved in them. So far-future AI, paradoxical though this may seem to many people, can be expected to deepen our understanding of humanity as such.

I said (above) that there's a sense in which A-Life systems are autonomous. Instead of being programmed in any way the programmer decides, as in GOFAI ("Good Old-Fashioned AI": Haugeland [27]), they *do their own thing.* This results either from engineered "reflexes" or from bottom-up self-organization. In evolutionary robotics, for example, one sees the gradual emergence of robots specifically suited to the task-environment [18, 29]. A-Life researchers publish in journals devoted to "Autonomous Agents" or "Autonomous Robots," and see their self-directed systems as more lifelike—and by implication more mindlike—than classical AI.

But future advances in autonomous agents of this kind will *not* be relevant to our interests here. The reason is that human freedom is a much richer form of autonomy than this [7]. It requires deliberate self-direction, and careful choice between various possible actions.

In general, free choice is guided by moral principles (which may conflict); by personal preferences and life goals; by knowledge of the world, including other people's needs and attitudes; and by predictions of, and comparisons between, the possible consequences. Conceptual analysis may be required: "Thou shalt not kill"—but just what counts as killing? Self-image—in particular, the sort of person one *aspires* to be—often plays an important role. And the moral principles and world-knowledge may be represented as role models or stories—often drawn from a religious context (think of Jesus, saintly humans, or the Good Samaritan). In any specific situation, the relevance of these examples has to be assessed. In sum, difficult moral choices call for a subtle interplay of hypothetical justifications and exculpations (and even choices with *no* moral dimension can be highly complex).

To illuminate that sort of freedom, we need AI, not A-Life. Connectionist AI helps us understand how it's possible for us to see the analogy between a New Testament parable and a current dilemma ([4], chap. 6; [32]). And classical AI helps us understand, in principle, how we can structure our thoughts and aspirations, and weigh the many conflicting factors in deciding what to do ([2], chap. 9; [21]).

Very young children have no freedom, and no moral responsibility either, precisely because they haven't yet developed the sort of cognitive and motivational complexity that's required. For instance, they judge "naughtiness" by results rather than intent, because they don't yet have a theory of mind that enables them to attribute different intentions to different people ([5], p. 48). So Mary, who broke ten plates while trying to help her mother, is said to be much naughtier than Joan, who broke two while raiding the pantry for forbidden jam. The child's theory of mind, like the world-knowledge in which it is embedded, develops later. Only then is the person truly free. (If it doesn't develop, the person will be autistic: a huge constraint on human freedom [1].)

As for nonhuman animals, they never attain freedom, as opposed to mere autonomy. The reason is that they don't develop a language with which to represent abstract concepts and reasoning. (Question: How could a dog adjudicate between possible outcomes, or entertain a contingency plan? Answer: It couldn't.)

Determinism and indeterminism, by the way, are red herrings here [28]. Many people assume that indeterminism is necessary for free choice, often citing Heisenberg's principle as a scientifically respectable ground of freedom. But would you want your important decisions and moral dilemmas to be settled by some internal coin tossing or quantum jump? And what about revered religious leaders, or saints: Would we still respect them if we discovered that their actions were grounded not in their morals or religious convictions, but in arbitrary events? In short, indeterminism at the origin of action would destroy the sort of freedom worth having, namely that which involves personal "ownership" and responsibility. Whether we're talking about free moral choice or about creative originality (something that's often glossed in quasi-religious terms) ([4], chap. 11), what's important is the complex structure of the (adult) human mind.

Since AI doesn't deny this, and even helps us to understand it, there's no good reason why it should undermine people's belief in freedom. However, it may do so in practice, if they're still hung up on the determinism/indeterminism analysis of freedom—which many people are. That's a clear example of what I meant when I said these issues are deeply problematic. One has to agree on the philosophy if one is to agree on the proper impact of the technology.

The concept of the self-image is important, here. Work in the philosophy of AI (and in empirical psychology too) suggests that the "self" is not a unitary or given thing, but a complex psychological construction that's used in the guidance of specifically *human* behavior ([2], chap. 7; [8]; [22], chap. 13]. It involves a reflexive narrative, which prioritizes particular aims and values (the self-ideal). While AI isn't yet sufficiently advanced to give us detailed theories of the self, it has given us the broad outlines of a computational account of what this psychological construct is, and how it is possible.

Even phenomena so close to the self as mourning the loss of a loved one can be (broadly) understood in AI-terms. Just as a certain type of cognitive-motivational architecture is needed for free choice, so a certain psychological architecture is needed to generate grief, and to mediate its assuagement over months or years [46]. Grief at the loss of a loved one, with all its destabilizing effects, is *inevitable,* given the psychological nature—the computational architecture—of personal love. This type of love (unlike mere lust, or even *agape*) is a richly structured commitment, deeply grounded in the person's cognitive and motivational resources—not least, their theory of mind [26]. It comes into play whenever a choice has to be made that could affect the person loved.

Turkle's children, like most people today, assumed that AI can have nothing to say about emotions. But that's mistaken. A truly intelligent system, whether biological or artificial, would have to have some form of emotion. For emotions in general, from mild anxiety to savage grief, have evolved to help schedule resource-limited behaviors in the service of many conflicting motives or goals. They're a computational necessity, not an optional extra. (So "I've only got two hands!" is a cliché with profound psychological implications.) This has been recognized in AI since the 1960s (though admittedly often ignored) ([39]; [2], chap. 6; [41]; [46]).

Emotions are deeply implicated in religion, for a passionless belief hardly merits the term. An ascetic deism may downplay the emotion, and certainly rejects any *personal* aspects of the Supreme Being and of people's relation to It. But even deism involves emotions of awe and wonder. (Without them, it collapses into a coldly scientific cosmology.) Animist religions may add fear, joy, aggression, supplication, and even gratitude. And a full-blown theism involves the experience (whether illusory or not) of a loving engagement with a personal God, and so draws on the same basic psychological architecture as grief does.

But grief, here, is impossible, much as it's impossible for a reasonable adult to resent the rain. (Annoyance, yes; resentment, no.) Specifically, it's impossible because the loved one (so it is believed) can never die, and never betray. Or rather, it's impossible so long as that theological belief is firmly held. Sometimes, it wavers. Then, emotions comparable to grief—such as resentment, or even bitter disappointment—may arise: think of the psalmist's "My God, my God, why hast Thou forsaken me?" From the theological point of view, these emotions are as misguided as resenting the rain: the appropriate attitude, even in the face of appalling evils and calamities, is the serene trustfulness of Job. For it's (literally) inconceivable that a loving God would abandon His people. Abandonment is an especially nasty form of betrayal, since only the strong can abandon and only the weak can be abandoned ([3], pp. 80–86). And our concept of personal love precludes betrayals of any kind, and abandonment above all. (So Jesus' quoting the psalmist on the cross is problematic in much the same way as is his being "tempted" by the devil.)

In short, emotions such as love, grief, and resentment aren't mere *feelings.* Rather, they are subtly structured psychological dispositions, generated by a complex computational architecture.

Work in *practical* AI (implemented computer models), as yet, has had little to say about self and emotion. To be sure, a very early AI researcher—who was also an M.D. and a practicing psychoanalyst—did try to simulate Freud's theory of neurosis. He considered how anxiety can be generated by the superego; how it is generalized to objects that are merely analogous to the original threat; how it selects from the various defense mechanisms available (projection, displacement, denial, etc.); and how these effect specific transformations of anxiety-ridden beliefs ([19]; see [3], chaps. 2–3). Later, he also tried to model the anxiety-ridden belief systems typical of paranoid schizophrenia [20]. But these attempts were grossly premature: it was difficult enough to get computers to follow one goal rationally, never mind juggling several motives under the influence of anxiety and/or the superego.

Current AI research includes more realistic (though much less ambitious) simulations of emotions, and of their crucial function in scheduling behavior aimed at potentially conflicting goals. The best example, at present, concerns the emotional history of a nursemaid coping with a dozen babies: each one must be fed, changed, cuddled, and entertained with varying urgency and at various times—and each must be instantly rescued if it crawls near a river or the road [45]. (It's precisely because the nursemaid has only two hands, so can't do all of these things at once, that she needs emotions to help schedule her behavior.) Even this AI model, however, is still painfully crude—and it doesn't employ a self-image.

Far-future AI may well be different. For reasons explained below, I don't believe it will ever be able to simulate all the subtle complexities of the human mind. But I've no doubt that far-future AI models will simulate more emotions, more richly. And they may involve some effective distinction between system-image and system-ideal. Indeed, they probably will. Marvin Minsky, a pioneer of AI, argued long ago that a complex intelligence would *need* a self-image [34] (and some early AI programs in his laboratory could examine their own performance in the light of their goals, and amend it accordingly [42]). This is in principle possible, because the concept of *recursion* enables us to implement reflexive information processing. The "paradox" of self-reference disappears.

13.4. *Embodiment*

What I've said so far about how AI views the self would be accepted, I believe, by virtually everyone in the field. But this is partly because I've been very careful to say nothing about whether the *body* is essential to the notion of a mind, a self, or a person. That's something on which AI workers disagree among themselves.

Science fiction abounds with tales of disembodied minds—or rather, disembodied intellects. Various authors, including the late-lamented Fred Hoyle, have envisaged strange cloudlike structures distributed over a large space yet somehow identifiable as (one or more) unitary individuals, and having causal properties analogous to those of computers, minds, or living things. But this futuristic scenario is nearly unintelligible. What are these "causal properties analogous to those of minds (or living things)," which don't require a body? And how, in the absence of a body, could the clouds satisfy any self-organizing principle of living unity? The fact that science-fiction writers have sometimes asked us to consider such ideas doesn't guarantee that, carefully considered, they make sense [9].

The causal properties involved, some would say, are computational ones. And, they would add, we don't need to consider fictional clouds—just factual (though far-future) computers. Thirty years ago, MIT's Ed Fredkin predicted (at a Villa Serbelloni meeting on the social implications of AI) that with AI's help humankind would achieve near immortality at last. A person's psychological processes, he said, could be downloaded into a computer. Then when his or her body dies, his or her individual mind or personality remains—perhaps not forever (the second law of thermodynamics, and all that), but for a seriously long time. More recently, MIT's Hans Moravec has said the same sort of thing [35].

Perhaps there's something in the air at MIT, or maybe in the water . . . ? Well, no. Both Fredkin and Moravec are notorious mavericks, who revel in the sport of *epater les bourgeois.* Their views aren't shared by all their MIT colleagues. In particular, they're challenged by Rod Brooks, whose work in "situated" robotics takes the physical interaction between body and world to be *essential* to intelligence [12].

Outside MIT, others working in this area of AI/A-Life make a similar assumption. They criticize the Cartesian split between "mind" and "body," and some (e.g., Butler [13])—though not all (e.g., Clark [15])—go so far as to deny the need for mental representations too. Their focus isn't on internal representations (as in GOFAI), but on the close dynamic coupling between body and environment. In this view, our concepts of intelligence, mind, and self don't denote mysterious nonphysical entities, or even abstract computational functions. Rather, they are ways of referring to aspects of that—essentially physical—body-world coupling. In short, talk of disembodied minds, or of selves immortalized as computer programs, is *nonsensical.*

These ideas are relatively new in AI, but have been prominent for many years in at least two areas of philosophy. One is autopoietic theory, the philosophy of biology originated by the neurophysiologist Humberto Maturana ([31]; see also [10]). The other is continental phenomenology, as developed for instance by Maurice Merleau-Ponty [33]. In both these views, mind can be understood *only* as being grounded in the person's bodily engagement with the world. ("Body," here, means just that: it isn't a synonym for "brain.")

Merleau-Ponty discussed the skilled use of tools, especially tools we can't in practice do without—like the blind man's stick. To the blind man, he said, it's almost as though the stick were part of his own body. To be sure, it doesn't cause him pain; and he can drop it onto the floor. But—so far as the phenomenology is concerned—it gives him "direct" perceptual knowledge of the terrain, much as the soles of his feet, or his fingers, do. For him, it is psychologically transparent. He doesn't experience *the stick,* or even *the stick coming into contact with the ground:* rather, he experiences (through the medium of the stick) *the ground.* Similarly, a microscope is psychologically, as well as optically, transparent to the skilled user; and so is a word processing program with which one is thoroughly familiar. Even so complex a "tool" as a jet fighter is transparent to the skilled pilot in a dogfight, who experiences it almost as an extension of his own body.

Andy Clark, a philosopher deeply influenced by situated robotics and autopoietic theory, has recently developed this general insight into a highly counterintuitive view of "the self" ([14]; [16], chap. 8). He doesn't contradict anything I've said so far about AI's notion of the self, but he adds to it in a surprising way. He sees "cognitive technologies" such as reading, writing, and typing (which are experientially transparent like the blind man's stick) as sensory/motor prostheses that happen to lie largely outside the skin. Once we think carefully about the cognitive machine, or mind, of the person concerned (he says), we see that they are *integral* to it. They aren't mere optional "tools," but essential aspects of our daily thought and action. So the self doesn't stop at the skull, or even at the skin, but *extends* into our physical and cultural environment. In other words, it encompasses all manner of (easily available and automatically deployed) cultural artifacts: pencils, computers, coinage, legal systems, TV, even pop music.

One might also add *religions.* Think of someone whose actions and attitudes are often unthinkingly guided by their religious customs and beliefs, and who invariably deploys those beliefs when making difficult decisions. They may also be aided by specific religious rituals, and by cultural artifacts such as altars, rosaries, and Bibles. For such a person, the cultural institution of religion would be seen (following Clark's account) as *integral* to his or her mind, or self. It's not just that the religion concerned happens to be highly valued by them, or that its claims happen to be believed by them. Rather, *that person* would be a different person without it. Correlatively, a (sustained) religious conversion turns one person into someone else—Saul into Paul, perhaps. If that person were stranded on a desert island, his or her very *self* would be threatened unless he or she could easily remember many biblical passages, and/or use a string of nuts or shells as a rosary.

We don't have to accept Clark's notion of the extended self, even though his paper was voted—for *The Philosopher's Annual*—one of the year's ten best in *any* area of philosophy (for a critique, see [13], chap. 6.) But we do have to note its existence. The reason is that his approach is an extreme form of the stress on *embodiment* that's typical of some current AI/A-Life, mentioned above. Indeed, he was led to it in the first place by considering that research. These ideas are far distant from the view of disembodied mind and immortality recommended by Fredkin, for instance. In short, the belief—often used as a *reductio ad absurdam*—that AI/A-Life scientists *must* allow for personal immortality as a mere computer program is mistaken. (You'll have noticed that these discussions have theological equivalents, in relation to bodily resurrection for instance.)

13.5. Consciousness and Religious Experience

Next, what of consciousness? The first thing to be said here is that there's no such thing as the problem of consciousness. Rather, there are many problems associated with consciousness (a word with many meanings)—and some of them have been illuminated, or even solved, by ideas drawn from AI. One example is reflexive self-consciousness, in which a mind has (some) knowledge of itself. As suggested above, the concept of recursion helps us to understand how this type of knowledge is possible. Another example is seen in clinical cases of "multiple personality," in which there are two (or more) streams of consciousness associated with the same human body, one of which may have "direct" but nonreciprocal access to the other. From a Cartesian point of view, this is fundamentally paradoxical, not to say impossible. From a computational point of view, it is not [6].

The second thing to be said is that there's one problem of consciousness that's especially puzzling: how conscious experience as such (what philosophers call *qualia*) can exist in a fundamentally physical universe, and (apparently) be generated by the brain. Despite the recent explosion of neuroscientific writing on consciousness, it's my belief that nothing has been said that advances our philosophical understanding of this conundrum [8]. Mind-brain correlations aren't enough. Occasionally, they enable us to say that if *this* brain-event occurs, then *that* conscious experience will (even "must") occur, rather than some other. But why *any experience at all* should occur remains mysterious. I'm convinced that the answer, if and when we get it, will involve ideas drawn from AI as well as neuroscience. Indeed, neuroscience itself is increasingly using computational concepts. A few philosophers of AI believe that we already have the outlines of a computational solution to the problem of *qualia* ([41]; [22], chap. 12). In my view, however, our notion of experience is so ill understood that we'll need a conceptual revolution (in both neuroscience and the philosophy of mind) as fundamental as that involved in Clark Maxwell's field theory in physics. Meanwhile, we remain in a philosophical fog.

It follows that questions about whether far-future AI-systems could "really" be conscious are incoherent and/or unanswerable. If we don't understand how it's possible for human brains to generate consciousness, we aren't in any position to pronounce on such questions. Certainly, the widespread belief that it's "obvious" that robots couldn't be conscious is mistaken [23, 24]. Equally, it's not obvious that they could.

This is all very interesting, perhaps—but what has this to do with religion? Well, religious experience is a common phenomenon, and highly valued by those who're subject to it. Often, it leads to lifelong religious conversion. Ever since William James [30] described the many varieties of religious experience, it's been clear that this experience ranges from vague "oceanic feelings" and intimations of the numinous, through an overpowering sense of God's presence, to highly specific experiences interpreted as visions of and/or communications from a saint, or God. Where theistic religions are concerned, religious experience can trigger the growth of (apparently reciprocated) personal love, with all the cognitive-emotional implications mentioned above.

Recent work in brain imaging has suggested (a) that experiences at the "general" end of this spectrum are associated with a certain part of the cerebral cortex, and (b) that the hallucination of "voices" is due to a failure or delay of communication between different parts of the brain. Example (b) arose within a study of schizophrenia, and it's long been suggested that religious visionaries and mystics (Joan of Arc, for instance, or St. Teresa) may be schizophrenic. This suggestion has always been hostage to the reply that schizophrenics may somehow be more "open" to the reality of the divine. The novelty, here, is to offer a specific neural-computational hypothesis that accounts for the experience in terms of anomalies in the internal information-processing going on.

In principle, if neurosis and paranoia can be simulated in computers (see above) then schizophrenic hallucinations could be modeled too. For example, a computer might sometimes interpret its own voice and/or internal linguistic processes as being caused by some *other* system. This might happen if "internal" data about the fact of its speaking, or its decision to speak, were temporally out of kilter with "sensory" data about the sound of its speech.

13.6. Individuality and Uniqueness

Schizophrenia, of course, is much more than "hearing voices." It involves a structured system of (partly illusory) beliefs, and a complex (but abnormal) system of emotions. I don't believe that AI will ever achieve a simulation of a complete human mind, or even a full schizophrenic episode. The most it can achieve is to model certain crucial aspects of the mind, normal or pathological, and by so doing help us to understand it. This isn't because it's computers (i.e., fancy tin cans) that are involved, but rather because the *science* is too difficult.

The human mind is far too complex, and far too individually various, to be fully understood by a psychological science. Indeed, science as such isn't interested in individual things—and, in general, nor are we. In the absence of any specifically human context, we don't care about *this* grain of sand on the beach, or *that* sparrow or oak tree. It doesn't bother us, therefore, that physics or biology can't explain every aspect of these things, or predict their future in detail. But we do care about individual human beings: Jill, Jack, and Mary. We may be desperate to know whether Jill is suicidal, and why Jack attacked Mary. When we discover that the science of psychology can't answer such questions, except via statistics, we're (irrationally) disappointed.

Some religions support that concern for individual humans, and its self-directed version (Christianity even tells us that God notes every sparrow that falls). Others, such as Buddhism, disparage it and advise us to overcome it. The point, however, is that *AI science* can't do justice to human individuality in principle and *AI technology* won't ever do so in practice. The notion that far-future computers might simulate religious experiences in full, or appear to undergo religious conversions, is therefore absurd. But so is the idea that they might undergo *political* conversions. The religious aspect, here, is not the core problem.

Claims to human "uniqueness," therefore, have two meanings—both of which are supported by AI. Each individual human is unique as a person, because the computational complexity of our minds is so great, and the data input so diverse, that even identical twins (or artificial clones) will differ from each other. And *Homo sapiens* is significantly different from every other animal species. I've already remarked, for instance, that nonhuman animals don't possess freedom, because they don't possess a language that can represent possible actions, aspects, and consequences in a stable way.

Connectionist AI has offered many illuminating ideas about learning, pattern recognition, stimulus generalization, and even analogy. Those ideas apply to some animals (namely, birds and mammals) as well as to human beings. What it hasn't yet managed to do is to show how neural networks (including the brain) can embody language, hierarchical structure, and reasoning. Far-future AI will probably have solved those problems. We shall have learned how to implement a rational symbol user in a basically connectionist mind. That advance in AI won't destroy our appreciation of human specialness, however. To the contrary, it will endorse and explain it.

13.7. Conclusion

In sum, far-future AI technology may—in practice—prejudice people either for religion or against it. Feeling threatened by a host of superficially humanoid gizmos, some people may flee to religion by the "Magean" route. Others may reject it as redundant, seeing the gizmos (despite their failure to match the richness of real human minds) as the ultimate step in "the mechanization of the world picture" [25].

In principle, however, both these responses would be inappropriate. Theoretical AI is already helping us to understand *how it is possible* for human beings to believe, choose, love, and worship—and to hallucinate, conform, resent, and transgress. It even helps us to understand how someone can undergo a religious conversion, or be tortured by mourning the loss of faith. In the far future, computational psychology will have advanced still further. But religious questions *as such* will be untouched. (This is why *neither* of the opposing responses mentioned above would be rational.) The religious sentiment and the need to find some satisfying, and perhaps culturally sanctioned, way of guiding our life choices will remain. We are, after all, human beings—not computers.

REFERENCES

1. Baron-Cohen, S., *Mindblindness* (MIT Press, Cambridge, Mass., 1995).

2. Boden, M. A., *Purposive Explanation in Psychology* (Harvard University Press, Cambridge, Mass., 1972).

3. Boden, M. A., *Artificial Intelligence and Natural Man*, 2nd ed. (Basic Books, New York; MIT Press, London, 1987).

4. Boden, M. A., *The Creative Mind: Myths and Mechanisms* (Abacus, London; Basic Books, New York, 1991).

5. Boden, M. A., *Piaget*, 2nd ed. (Fontana, London, 1994).

6. Boden, M. A., "Multiple Personality and Computational Models," in *Philosophy, Psychology, and Psychiatry*, ed. A. Phillips-Griffiths (Cambridge University Press, Cambridge, 1994), 103–14.

7. Boden, M. A., "Artificial Intelligence and Human Dignity," in *Nature's Imagination: The Frontiers of Scientific Vision*, ed. J. Cornwell (Oxford University Press, Ox-

ford, 1995), 148–60. Reprinted and amended as "Autonomy and Artificiality," in *The Philosophy of Artificial Life,* ed. M. A. Boden (Oxford University Press, Oxford, 1996), 95–108.

8. Boden, M. A., "Consciousness and Human Identity: An Interdisciplinary Perspective," in *Consciousness and Human Identity,* ed. J. Cornwell (Oxford University Press, Oxford, 1998), 1–20.

9. Boden, M. A., "Is Metabolism Necessary?" *Brit. J. Phil. Sci., 50,* 231–48 (1999).

10. Boden, M. A., "Autopoiesis and Life," *Cog. Sci. Q., 1,* 115–43 (2000).

11. Brooks, R. A., "A Robust Layered Control System for a Mobile Robot," *IEEE J. Rob. Auto., 2,* 14–23 (1986).

12. Brooks, R. A., "Intelligence without Representation," *Art. Intell., 47,* 139–59 (1991).

13. Butler, K, *Internal Affairs* (Kluwer, Dordrecht, 1998).

14. Clark, A., and Chalmers, D. J., "The Extended Mind," *Analysis, 58,* 7–19 (1998).

15. Clark A. J., *Being There: Putting Brain, Body, and World Together Again* (MIT Press, Cambridge, Mass., 1997).

16. Clark, A. J., *Mindware: An Introduction to the Philosophy of Cognitive Science* (Oxford University Press, Oxford, 2001).

17. Clark, S. R. L., "Tools, Machines, and Marvels," in *Philosophy and Technology,* ed. R. Fellowes (Cambridge University Press, Cambridge, 1995), 159–76.

18. Cliff, D., Harvey, I., and Husbands, P., "Explorations in Evolutionary Robotics," *Adapt. Beh., 2,* 73–110 (1993).

19. Colby, K. M., "Computer Simulation of a Neurotic Process," in *Computer Simulation of Personality: Frontier of Psychological Research,* eds. S. S. Tomkins and S. Messick (Wiley, New York, 1963), 165–80.

20. Colby, K. M., *Artificial Paranoia* (Pergamon, New York, 1975).

21. Dennett, D. C., *Elbow Room: The Varieties of Free Will Worth Wanting* (MIT Press, Cambridge, Mass., 1984).

22. Dennett, D. C., *Consciousness Explained* (Allen Lane, London, 1991).

23. Dennett, D. C., "The Unimagined Preposterousness of Zombies: Commentary on Moody, Flanagan, and Polger," *J. Consc. Stud., 2,* 322–26 (1995).

24. Dennett, D. C., "The Zombic Hunch: Extinction of an Intuition?" in *Philosophy at the New Millennium,* ed. A. O'Hear (Cambridge University Press, Cambridge, 2001), 27–44.

25. Dijksterhuis, E. J., *The Mechanization of the World Picture,* trans. C. Dikshoorn (Clarendon Press, Oxford, 1961; first published in Dutch in 1956.)

26. Fisher, E. M. W., *Personal Love* (Duckworth, London, 1990).

27. Haugeland, J., *Artificial Intelligence: The Very Idea* (MIT Press, Cambridge, Mass., 1985).

28. Hobart, R. E., "Free-Will as Involving Determination and Inconceivable Without It," *Mind, n.s. 43,* 1–27 (1934).

29. Husbands, P., Harvey, I., and Cliff, D., "Circle in the Round: State Space Attractors for Evolved Sighted Robots," *J. Robot. Auto. Sys., 15,* 83–106 (1995).

30. James, W., *The Varieties of Religious Experience* (Collier, New York, 1902).

31. Maturana, H. R., and Varela, F. J., *Autopoiesis and Cognition: The Realization of the Living* (Reidel, Boston, 1980; first published in Spanish, 1972).

32. May, L., Friedman, M., and Clark, A. J., eds., *Mind and Morals: Essays on Ethics and Cognitive Science* (MIT Press, Cambridge, Mass., 1996).

33. Merleau-Ponty, M., *Phenomenology of Perception,* trans. C. Smith (Routledge and Kegan Paul, London, 1962).

34. Minsky, M. L., "Matter, Mind, and Models," in *Semantic Information Processing,* ed. M. L. Minsky (MIT Press, Cambridge, Mass., 1968), 425–32.

35. Moravec, H., *Mind Children: The Future of Robot and Human Intelligence* (Harvard University Press, Cambridge, Mass., 1988).

36. Papert, S., *Mindstorms: Children, Computers, and Powerful Ideas* (Harvester Press, Brighton, England, 1980).

37. Roszak, T., *The Making of a Counter-Culture: Reflections on the Technocratic Society and Its Youthful Opposition* (Doubleday, New York, 1969).

38. Roszak, T., *Where the Wasteland Ends: Politics and Transcendence in Post-Industrial Society* (Doubleday, New York, 1972).

39. Simon, H. A., "Motivational and Emotional Controls of Cognition," *Psych. Rev., 74,* 29–39 (1967).

40. Sloman, A., "Motives, Mechanisms, and Emotions," *Cognition and Emotion,* 1, 217–33 (1987). Reprinted in Boden, M. A., ed., *The Philosophy of Artificial Intelligence* (Oxford University Press, Oxford, 1990), 231–47.

41. Sloman, A., "Architectural Requirements for Human-like Agents Both Natural and Artificial (What Sorts of Machines Can Love?)," in *Human Cognition and Social Agent Technology: Advances in Consciousness Research,* ed. K. Dautenhahn (John Benjamins, Amsterdam, 1999), 163–95.

42. Sussman, G. J., *A Computer Model of Skill Acquisition* (American Elsevier, New York, 1975).

43. Turkle, S., *The Second Self: Computers and the Human Spirit* (Simon and Schuster, New York, 1984).

44. Turkle, S., *Life on the Screen: Identity in the Age of the Internet* (Simon and Schuster, New York, 1995), chap. 6.

45. Wright, I. P., and Sloman, A., *MINDER1: An Implementation of a Protoemotional Agent Architecture,* Technical Report CSRP-97-1, University of Birmingham, School of Computer Science (1997). Available from ftp://ftp.cs.bham.ac.uk/pub/tech-reports/1997/CSRP-97-01.ps.gz .

46. Wright, I. P., Sloman, A., and Beaudoin, L. P., "Towards a Design-Based Analysis of Emotional Episodes," *Phil., Psych., Psychol.,* 3, 101–37 (1996).

[14]

COSMIC ESCHATOLOGY VERSUS HUMAN ESCHATOLOGY

Owen Gingerich

14.1. The Old and the New Mythos

Across the green from the Pisa cathedral and its famous leaning campanile stands the Camposanto, the burial hall. Galileo must have walked through many times when he was first a student and then an assistant professor at the university. And there, in the Camposanto, he would have seen the early Renaissance cosmos, a surviving medieval tradition, stunningly and colorfully frescoed on the wall.

Pierro di Puccio's masterpiece depicted a tidy universe, the earth in the center surrounded by the planetary spheres, and, beyond the stars, the Dantesque layers of the empyrean: saints and angels arrayed wing tip to wing tip, and holding all in his arms, God the Creator. Puccio's universe mirrored the beliefs of popes, professors, and peasants, of merchants, monks, and mendicants. His cosmos was orderly and beautiful; "cosmetics" shares the same root. But most satisfying of all, its dimensions were comfortably human. Millions, billions, or millions of billions were not part of its vocabulary.

After nearly six centuries on the Camposanto wall, Puccio's fresco was shattered on 27 July 1944. A casualty of World War II, the painting smashed to the floor in a thousand fragments. The preliminary gesso sketch still stands in situ, and art historians tried valiantly to put Humpty Dumpty together again, but it is not the same.

Somehow the demise of the Camposanto fresco is a fable for the twentieth century, for just as surely as that image was shattered, so was the concept it reified. Gone today is the tidy, closely bounded stage on which the great Western monotheistic religions framed their cosmologies. In its place is a cosmos so vast that humankind can scarcely grasp its scope. "Tis all in pieces, all cohaerence gone; all just supply, and all Relation," lamented John Donne already in 1611 [2].

In putting the cosmological pieces back together, a new mythos takes shape: space must be vast for the universe to be old, and the universe must be ancient for the elements essential for life to emerge. Slowly, slowly, the carbon and oxygen, the magnesium, phosphorus, and iron, are fabricated in the hellish cauldrons of stellar interiors. Nucleocosmogenesis, with its astonishing details, portrays a universe that seems to have known we were coming, the exquisite, tiny flower on the giant, sprawling plant.

Puccio's fresco captured a static universe, frozen in time, but the biblical narrative is a history, a timeline with a beginning and an ending. Like the space of Puccio's fresco, the scope of the Bible's story is not only bounded, but human in its dimensions. From Adam to the Apocalypse there are only a few hundred generations. No millions or billions here. Given this deeply rooted historical tradition, it is no wonder that Darwin devoted a substantial section of his *On the Origin of Species* to persuading his readers that the earth had an immense history, a scope probably measured in many millions of millennia.

The new mythos accommodates the ages past, fully compatible with a theistic perspective. God has organized nature to make conscious life possible in the fullness of time. The psalmist, knowing nothing of nucleosynthesis, nevertheless exclaims, "A thousand years in thy sight are but as yesterday when it is past, and as a watch in the night" (Ps. 90:4). In contrast, the far-future universe poses a different challenge. As Paul Davies remarks in his chapter, it is difficult to envision a universe both infinite in future time and also interesting. But the theistic conundrum is far tougher than this: What are we to make of the vast eons of time after all of us are gone? Basically, the riddle centers on the unresolved incongruity between the timescales of cosmic eschatology versus human eschatology.

14.2. Human Timescales

The Vatican workshop considered at some length the timescales of the far-future universe. Here let me offer some parallel reflections on the human timescale.

My colleague with whom I taught "the Astronomical Perspective" at Harvard for many years liked to challenge the students with the question, "Are we any smarter than the ancient Greeks?" We surely have a great deal more knowledge—especially about technology, medicine, and the natural world—than Plato, Aristotle, or Aristarchus, but few of us equal them in brain power or logical reasoning. Nor is it evident that we have surpassed Jesus in moral reasoning. Two or three millennia are not long enough for natural selection and variation to augment brain capacity noticeably. It took two million years to double the cranial volume of *Homo habilis* from 700 cm^3 to the 1400 cm^3 of modern *Homo sapiens.*

But approximately two hundred thousand years ago something new was added to the evolutionary picture: the invention of language and social evolution. The DNA information content in every human cell is roughly equivalent to twenty-five sets of the *Encyclopedia Britannica.* A cutting-edge personal computer has approximately the same capacity but the human brain has many times more. The flexibility of this cranial capacity has allowed our core of information and our understanding of the physical and biological worlds to increase exponentially. Our great grandfathers would undoubtedly feel more at home in sixteenth-century Europe than in present-day America.

Will *Homo sapiens* be smarter two thousand years from now? Let me tackle that question by first presenting two scenarios for human eschatology, one pessimistic, one optimistic.

14.3. Pessimistic Scenarios

Just over forty years ago an article in *Science* magazine proposed the following formula for representing all the available data on world population throughout past history:[1]

1. The doomsday equation was first published in *Science* [6], and there was an interesting follow-up letter [5].

$$\text{Population} = \frac{1.79 \times 10^{11}}{(2026.87 - \text{time})^{0.99}}$$

The formula gave 2.7 billion as the 1960 world population and predicted that the population would become infinite by Friday, November 13, 2026—a prediction that earned it the name "the doomsday equation." We are now well over half way to that deadline, and the equation has proved eerily accurate. World population doubled to 5.5 billion by 1993, the halfway point. As the twentieth century drew to a close, there were more people alive than the combined total of all prior centuries. In 1900 there were sixteen cities with more than a million population; today there are more than four hundred.

Clearly the doomsday equation cannot be right; given the human gestation period it would be ridiculous to suggest that the population could double in six months, which the formula predicts will happen early in the year 2026. (This would be like saying that if one woman can have a baby in nine months, three women could produce a baby in three months!) In fact, worldwide the increase in the rate of population growth has turned over, so the population explosion is not as threatening as it was a decade ago. Nevertheless, the formula does help us realize that human expectations must change radically in the next three decades. Either life expectancy or the birth rate will have to fall drastically.

How might the death rate rise dramatically? For years we have lived with the Damoclean sword of nuclear warfare hanging over our heads. The human race seemed in serious and even immediate peril. The collapse of the Soviet Union has lessened that threat. The *Bulletin of the Atomic Scientists* moved back the hands of its own doomsday clock, so that they are no longer poised so close to midnight. Nevertheless, sword rattling in the Indian subcontinent and in the Middle East casts a frightening shadow across world affairs. Meanwhile, genocide continues apace in our troubled world, though not yet at a rate sufficient to stem the population tide.

Might starvation be the instrument for increasing the death rate? In the spring of 1992 *Time* magazine carried a remarkable story about the mysterious demise of the Anasazi Indians from Chaco Canyon nine hundred years ago [3]. A careful analysis of contemporary rats' nests showed that the Indians simply cut down too many pine trees, creating erosion in their fields, and a shortage of food. Perhaps they moved, or maybe they just died out. A decade ago over a million people in Somalia were on the brink of starvation; perhaps a hundred thousand perished in the summer of 1992. Similarly, Jared Diamond has argued convincingly [1] that the collapse of the population on Easter Island was the consequence of rapacious deforestation. Might such starvation become widespread?

Perhaps disease will curb population. Africa's AIDS epidemic threatens to halve the continent's population before a decade is out. Yet such agony and suffering would set back Africa's explosive population growth by only twenty-five years. Are we facing even more devastating worldwide epidemics?

If we compare the use of energy resources to population growth, taking a reasonable census of the reserves of oil and gas, we will see that the world cannot possibly be heading toward a more equitable distribution of these resources. Nowadays the heavily industrialized first world uses the lion's share of the available energy. The per capita use of energy in the United States and Canada is twenty times higher than in Nigeria. At present usage the world's oil reserves are expected to last for roughly five or six more decades; we have already passed the peak of maximum crude oil production. If conservation and new automotive technologies decrease the current demands, we could make our oil resources last until about 2050—provided energy use does not increase in China, India, or Africa. We have enough coal to last for a few hundred more years, but coal pollutes the atmosphere and increases the greenhouse effect. At present we seem far from creating a runaway warming that would render the earth uninhabitable, but we could anticipate, for example, that the American wheat belt would become largely a desert, and that major low-lying coastal regions worldwide would be inundated by ocean waters.

As the world's oil supply diminishes, competition over control of this precious and increasingly expensive resource will grow, and armed conflict seems inevitable. While the nuclear threat seems to have vanished with the demise of the Soviet Union, the weapons are still in place, and the roulette of global politics could bring them back to haunt our descendants, who might be all too willing to fight over who gets to have gasoline and jet fuel.

The oversupply of people and undersupply of energy makes it easy to be pessimistic about humanity's prospects. If not extinction, we might be facing a bleak and even brutish future. The story of the Apocalypse and a final millennium seems all too real: we can hardly predict as far as a thousand years.

14.4. Optimistic Scenarios

What about the optimistic scenario? I asked my friend Philip Morrison, an Institute Professor Emeritus at MIT and an astute observer of the scientific scene, about the prospects for *Homo sapiens.* "I would give it about ten million years," he opined, and when I pressed him about the basis of his estimate, he replied that this was a typical lifetime for a complex species. I agreed that certainly the fossil record shows us that extinction is the name of the game, and it is not reasonable to expect humankind to be exempt from the general rule.

But what about the energy resources? "There will be plenty of energy if we exploit solar energy," was his reply. "Of course, we'll have to learn to live with only about a tenth of the present world population. Yes, half of Florida will be under water from the rise of sea level, and yes, nuclear weapons will be used in the Middle East, enough to teach people that this is a bad idea, but not a global exchange that will verge on extinction."

14.5. Realism

While ten million years is a fleeting instant compared to the sun's projected future lifetime of five billion years, even ten million years seems to me unreasonably long for our species' survival. The first of the genus *Homo* appeared about four million years ago, and already half a dozen of its species are extinct. Given a transition to a sustainable future with a greatly diminished population for our planet, plus a thriving scientific environment needed to harness the vast and generously distributed sunlight, would we not also expect an exponential increase in biological knowledge? Fifty years ago the exact number of chromosomes in human cells was still in doubt, while today we are making a complete map of their genetic patterns. In evolution we have crossed the Lamarckian divide. The human brain now stores more information than the DNA of our chromosomes. What we learn can become inheritable. Correction of crippling genetic defects is close upon us. In another fifty years, barring major nuclear catastrophe, geneticists will surely be able to manipulate the genes to create a stronger, healthier, smarter superman.

There is a rather spooky scenario in Lee Silver's provocative book, *Remaking Eden* [4]. He describes a time three and a half centuries hence, when genetic engineering has allowed the wealthier parents to invest in improved genes for their children. Naturally they do not want their GenRich children wasting all this expensive improvement by finding unimproved mates, although inevitably love sometimes crosses those social boundaries between the GenRich and the Naturals. But then a curious thing happens: increasingly those mixed marriages prove infertile. I think most of you know the scientific definition of a species, which is the boundary within which individual creatures can interbreed. What Professor Silver has described is the genetic origin of a new species.

In two thousand years our species will be indeed be smarter—if it even exists as the same species! So to me it is unimaginable that *Homo sapiens* will still exist on Earth ten million years from now, except perhaps by some remote chance in zoos or special preserves, a throwback much like Przewalski's horse. It is not astronomy that gives me this reading, nor even evolutionary biology and anthropology, but my reflections as a historian and philosopher of science. I believe it is neither pessimistic nor optimistic, but simply realistic. Our universe is going to go on for billions of years without us. Our temporal span is as fleeting as our spatial position is minuscule.

14.6. Theological Issues

In various forms and over past ages theologians have coped with related problems. They have asked where are the dead, and where is the empyrean? They have asked if the promised new heavens and new earth are a transformation of our present physical world, or an *ex nihilo* creation in some other dimension, and if the latter, why didn't God just start out that way in the first place? Yet, if they have addressed the question of a universe going on for billions of years without us, they have kept their suggestions a well-guarded secret from the laity.

Any traditional view of heaven, to the extent it is reasonably well described, is so far from our earthly norms of time and aging that it is tantamount to one of those other universes imagined in the multiverse cosmology. What modern cosmology is telling us is that it's not automatically absurd to imagine other places with other, unfamiliar physical laws. I believe our traditional views of heaven are as dated as the tightly nested medieval cosmos. However, it must be as difficult to envision a consistent view of eternity as it is to think of a totally different universe within a multiverse complex, among other reasons because it is hard to grasp the fact that time itself is created with our universe and is no more outside of our universe than is some kind of superspace. Yet, as Jürgen Moltmann mentioned, Plato conceived of a timeless eternity, and I think that is something for philosophers to reexplore. For only in this way, I suspect, will we begin to reconcile our fleeting human existence in this particular universe with the larger cosmological structures that the incredible self-conscious brain of *Homo sapiens* can conceive. Of course, if you ask me what a timeless eternity is, I cannot answer. Timelessness is as impossible for us to grasp as a beginning of time or an end of time. In a sense it is only word play. Nevertheless, word games are not necessarily trivial.

Is the cosmos all there is? The logician will undoubtedly say yes. The philosopher can only say "not necessarily!" He knows he is like the blind man trying to describe an elephant. Is his search for coherence in vain? Or is it a fundamental part of being a reflective, self-conscious creature—maybe even a designed and created creature? Perhaps the quest itself *is* the purpose of the universe and somehow the answer to the question, "Why is there something rather than nothing?" And ultimately, for the Christian, it is a matter of trust.

REFERENCES

1. Diamond, J., *The Third Chimpanzee* (HarperCollins, New York, 1992), 329–31.

2. Donne, J., *An Anatomy of the World* [*The First Anniversarie*] (W. Stansby/S. Macham, London, 1611).

3. Jaroff, L., "Nature's Time Capsules," *Time, 139,* no. 14 (6 April 1992), 61.

4. Silver, L., *Remaking Eden* (Avon Books, New York, 1997), 6.

5. Umpleby, S. A., "World Population: Still Ahead of Schedule," *Science, 237,* 1555–56 (1987).

6. von Foerster, H., Mora, P. M., and Amist, L.W., "Doomsday: Friday, 13 November, A.D. 2026," *Science, 132,* 1291–95 (1960).

PART 5

THEOLOGY

[15]

COSMOLOGY AND RELIGIOUS IDEAS ABOUT THE END OF THE WORLD

Keith Ward

15.1. Eastern Religions

Virtually all the best-known religious traditions were first promulgated long before the rise of modern cosmology. They had little idea of the history and extent of the physical cosmos that we know today. So it may be felt that they have little to add to, and everything to learn from, modern cosmology. However, considerations about the nature of the physical universe and its likely beginning and end may not be relevant to religious considerations at all. In the Chinese traditions of Taoism and Confucianism, for instance, there is virtually no interest in how the universe began or in how it will end. There is an interest, especially in Taoism, in nature, in the balance, harmony, and interconnectedness of natural forces. For many people, the Chinese concern that human lives should be in balance with nature, and be seen as part of its flow, comes as a welcome change from the very anthropocentric concern of the Abrahamic faiths with human beings and their personal immortality. Some, like Fritzof Capra, have claimed to see in traditional Chinese thought a religious system closely in sympathy with modern physics, at least insofar as such physics suggests a holistic, nonmechanistic, and relational view of nature [2]. That may be true, but there is little concern with how the natural world will end, or with what it will be like in the far future. There is little expectation that it will change for the better, or issue in "one far-off divine event, To which the whole creation moves," in the words of Tennyson. The main concern is with the present, and with how to live in tune with the "Way of Heaven." Ultimate beginnings and endings are topics of speculation of no particular interest to living the balanced life.

The Indian-originated faiths of Hinduism and Buddhism do have elaborate cosmologies, which are usually, and correctly, said to be cyclical in nature. The universe is a vast system of many worlds, many heavens and hells, and many different forms of existence. Each god, or each Buddha for the Mahayana systems, has its own paradisal world, and there exist an equal number of demonic worlds, ranging from uncomfortable to unbearable. Souls circulate around these worlds without beginning or end, by their deeds in each life gaining entrance to a further existence in an appropriate world, in their next rebirth. Thus each soul has already been born an infinite number of times, in this and in many other worlds, and in many forms, from animal to human and divine. As the best-loved Indian scripture puts it, "There was never a time when I was not . . . there will never be a time when I will cease to be."[1]

The whole universe passes though four ages or *yugas,* each less happy and peaceful than the last, and then it is destroyed. After a period in which the universe exists in Brahman, the absolute reality, in a potential state, the fourfold process begins again. And so it continues without end, the whole cycle of creation and destruction being repeated indefinitely. The aim of spiritual practice, within this immense cosmology, is for the soul to achieve "release" *(moksa)* from the cycle of rebirths. Then it will never be reborn, but will exist—in various versions—as a pure individual spiritual consciousness, or as merged in the undifferentiated unity of the one primal Self, or as entered into an indescribable state of *nirvana,* or as an immortal member of a "pure

1. Bhagavad Gita, ch. 2.

land" or Paradise in which there is no corruption or death, and from which liberated souls will never again fall into the cycle of *samsara*, of suffering.

There are many differing views of the ultimate destiny of the human soul in these traditions, but they all agree that it is a release from this material universe. In that sense, the Indian tradition, too, has little interest in the specific way the universe begins and ends. The far future will be just like the far past, in the preceding world cycle, and it in turn will be succeeded by further futures, which never improve and never end. As far as these religious traditions are concerned, this universe will probably end, but there will be another one, so it does not matter exactly how it happens. No fundamental change is to be expected, and the real focus of interest is on release, not on the continuation of a longer and happier life, which is a prime concern of many Chinese traditions.

One must always be aware of the danger of overgeneralization, and many Hindus and Buddhists would be keen to emphasize that they are interested in being born into paradisal worlds, and they are more concerned to show compassion for all creatures now, and to enjoy a liberated life in this world, than to bring their embodied existence to an imminent end. They would very much resent being characterized as "world negating" or denying. Nevertheless, the scriptures state that this world is one of suffering and attachment, and that ultimately, at least, it is to be transcended. In that sense, the far future of the universe is not a religious concern.

15.2. The Jewish Tradition and Islam

It is only when we come to the Jewish tradition that the future of the universe itself does become a matter of religious interest. Historically, the idea of life after death developed late in Hebrew history, and religious hope was centered very much on this world, and on the realization of a society of justice and peace in this world. It is still possible to be an orthodox Jew and not believe in life after death at all—the Sadducees at the time of Jesus denied the possibility of resurrection. The hope for a Messiah that developed in Judaism was a hope for a political transformation of this world, for liberation from oppression and the realization of a historical destiny for the people Israel.

Jews had no concept of the infinity of worlds, which characterized Indian thought. Their world had not existed for more than a few thousand years, and was not envisaged to exist for much longer. They would probably have been considerably surprised by the discoveries of modern cosmology about the size, age, and nature of the universe. I do not suppose that they took all religious language literally, by any means. Nevertheless, there was a specific view of the universe presupposed by the biblical documents, and it is fairly clear what it was. This Earth was a disc floating on the subterranean waters. Over it was the dome of the sky, upon which the Sun, the Moon, and stars were fixed. And beyond that were the waters of the upper air, a vast ocean surrounding a central earth. This Earth, though it was the center of the physical universe as then conceived, was only one plane or realm of being. There were many others. The gods, or the angelic hosts, had their own forms of being. The demonic hordes, too, existed in forms very different from that of the earth. There was the shadowy world of Sheol, the world of the dead, living an indeterminate existence, possibly to be resurrected in bodies of some sort at some future time. And there were the worlds of Paradise, in which the patriarchs and prophets possibly observed the events of this world, and awaited the coming of the Messiah, God's anointed one.

In the biblical view, in contrast to Indian beliefs in the eternity of the soul, each soul—we might better say "person"—was created at birth, so that humans are dust, given life by the spirit of God.[2] To dust they will all return, so rather as in China, religious interest centers on obtaining a happy life now, and for one's immediate descendents. For most Jews, a belief in resurrection did develop, but it was probably widely supposed that the dead would rise to life on this Earth in the near future. So the belief arose that this physical world has a future that is limitlessly better than its present. Sometime around the eighth century B.C.E., cosmic optimism was born.

Jewish ideas about "the end of history" are very diverse and none has official ratification, even if there were anyone to ratify anything officially in Judaism. But in the expanded cosmic vision of modern physics, it is possible for Jews to think of the messianic age, beyond suffering and oppression of the poor, as some state in the far future of the universe. It may seem that at last we have discovered a religious interest in the far future. The problem is that talk of resurrection makes this rather dubious. If millions of dead people are going to be brought back to life, it will certainly not be on this planet,

2. Gen. 2:7

as there is not enough room. And if the resurrection world is going to be incorruptible, the laws of thermodynamics are going to have to change. So the resurrection world looks like a different universe than this, as in some Indian systems, and once again the future of this particular universe becomes religiously irrelevant.

We probably should say that Judaism remains ambiguously poised between cosmic optimism for this universe—not necessarily that a just society would exist forever, but that it would at least come to exist—and belief in a quite different resurrection world, beyond historical time. Certainly hope is a central virtue of Judaism, but whether that hope is for this or some other form of reality is indeterminate. Either way, it would not be a great blow to Jewish hope if this universe one day ceased to exist, though if it somehow managed to endure forever, that would be perfectly acceptable.

The same thing is true for Islam, in which there is a much more decided commitment to the resurrection of the dead, and to continued existence in Hell or Paradise. But there is no commitment to an enduring future for this physical universe, and since again resurrection is conceived as incorruptible and beyond the reach of evil, it becomes unlikely that it will occur in this universe, unless its nature changed very radically. It looks as though the Abrahamic faiths, almost uniquely among the great world traditions, do generate hope for a better future for this universe. Even so, that future need not exist forever, and there is an escape clause, that the better future might be in some other form of universe altogether. In this way, it is only when belief in life after death evaporates that a concern for the realization of a messianic age within this universe becomes primary. Perhaps it is for that reason that the eighteenth-century European belief in human progress is a product of a Judeo-Christian society that was beginning to lose faith in God and an afterlife that only God could guarantee. Karl Marx, born a Jew and raised a Christian, tried to perform the impossible task of retaining cosmic optimism in a world from which its justifying foundation had been removed. It is hardly surprising that it does not seem to have worked.

15.3. Christianity

And so I come to consider Christianity. I will devote much more space to this consideration, because it is in the Christian tradition that considerations about the end of the world—we would now say the end of the universe—have been explicitly addressed under the title of "eschatology." It is within Christianity that the far-future universe has been an explicit topic of theological debate, and I will try to see what might be said about this in the light of some of the findings of scientific cosmology.

Christian faith originated as a sect of Judaism that proclaimed that Jesus was the expected Messiah. The coming of the Messiah announced a new age of history, in which justice and peace would rule, and God would be present to his people in a much more intense and inward way. In the earliest extant documents of the New Testament, the letters to the Thessalonians, the belief is expressed that this age of liberation and justice would be fully manifested only when the dead were released from Sheol, and a new form of resurrection life would begin. All that would happen quite soon—within one generation, perhaps—and at that time the physical universe would be transformed into a material Paradise, in which there would be no more sun or sea, suffering or death.[3]

It is, I think, quite important to see that these views are present in the New Testament, just as views like them were current in other societies at that time. The importance is that all such views belong to archaic forms of thought, and attempts by modern Christians to appropriate or reinterpret selected parts of them to fit contemporary conditions are fundamentally—one might say, fundamentalistically—misguided.

But it is also important to realize that these are not the only views present in either contemporaneous Jewish thought or in the New Testament documents, and that they may well not be very important to the Christian gospel. It was Johannes Weiss and Albert Schweitzer who made popular the view that Jesus was a mistaken prophet of the immanent end of the world, and that Jesus' teaching is therefore quite alien to our world, or at least our universe, which is probably going to last for a very long time. Schweitzer's *The Quest of the Historical Jesus* [5] has become a classic of religious thought, but there are few scholars who would today accept his conclusions.

Partly this is because the view has gained favor that those passages in the Gospels that speak of a "Day of the Lord" coming in darkness and tribulation are relatively few, and are widely thought to be interpolations into the teachings of Jesus from other sources. It is also partly because it is now more clearly recognized that the ideas of a Messiah, and of the kingdom of God, as

3. 1 Thess. 4:13–18.

found in a number of Jewish sources, were much more flexible and symbolic than Schweitzer thought. Ideas of the Messiah range from the idea that a political liberator would throw out the Romans to the idea that a supernatural being, perhaps an angel, would inaugurate a final battle with the forces of evil at Armageddon. Ideas of the kingdom of God ranged from the hope that the twelve tribes of Israel would be reunited, and all peoples would come to worship at the Temple in Jerusalem, to the idea that human souls purified of evil would live in a supernatural life in the presence of God forever.

Where are Christian ideas to be located in this diversity of interpretations? I think the scholarly consensus today would be that they are to be located at many different points, and that it is virtually impossible to get back to the original teaching of Jesus, as distinct from the many later interpretations of it.

The situation is not, of course, hopeless. There can be little doubt that Jesus was taken to be the Messiah, the liberator and ruler of God's people, and that he proclaimed that the kingdom of God was imminent. There can be a great deal of doubt that either Jesus or all his followers believed this meant that the universe as they understood it was coming to an end very soon, and that the resurrected dead would be descending on the clouds at any moment. If the writer of the Thessalonian letters was the same "Paul" who wrote the letter to the Romans, he clearly changed his mind on the issue. In the Thessalonian letters, he advises his readers that the Lord might come at any moment.[4] In the letter to the Romans, he writes that the gospel must be taken throughout the world until the full number of the Gentiles had been brought into the kingdom—which sounds like quite a long time.[5]

What he still believes is that Jesus will somehow return as the consummation of history, and that it is spiritually important to live as if this might happen at any moment. It is also important, however, that as many people as possible should hear the gospel, so that it was not long before Christians were praying, not "Come, Lord Jesus," but rather "Do not come yet, before the church has completed its appointed task." And it seems reasonable that if God is, as the New Testament says, concerned that everyone should be saved,[6] it would be surprising if God decided to end history before most people had heard about it.

4. 1 Thess. 4:15.
5. Rom. 11:25.
6. 2 Pet. 3:9.

So I will suggest this: Jesus is the Messiah in that he promises and begins to liberate humans—all humans—from egoism and selfishness (from sin), and brings God near to them in a new way, in the form of the Holy Spirit living in the innermost self. He inaugurates the kingdom precisely by forgiving sin and giving the gift of the Spirit, and thus reconciling humanity and God. This happens now, as people are confronted with the message of God's self-giving love and the experience of renewal in the Spirit. But the kingdom will not be fully and openly manifest until all evil and suffering have been eliminated from God's world. At that point, God's purpose will be seen to be fulfilled, and Christ will be manifest in his true form as the consummation of history, and as the one who was truly embodied—incarnate—in the historical Jesus.

15.4. Relation to Cosmology

To what extent is this message bound up with specific beliefs about the age and size of the earth? And what difference does it make to know that the physical universe is vastly bigger and older than any biblical writer thought, so that the planet Earth is perhaps only a tiny part of the cosmic story?

My suggestion is that modern knowledge of the universe adds considerable depth to the Christian vision, but compels a change in theological thinking rather less great than that which characterized the very first generation of Christian believers. Their change was from thinking of the Messiah as a liberator of Judaism, or as a herald of the imminent end of the times of suffering and evil on Earth, to thinking of the Messiah as the inner and spiritual ruler of a new community, largely Gentile in nature, which would have the vocation of serving the world in love, and proclaiming the unlimited love of God in every age until historical time ended, and all could live in a transformed existence and in the knowledge and love of God.

I do not mean that the change was a linear one, so that first people thought Jesus was a Jewish liberator, and then later that he was a world savior. Rather, as with Paul himself, all these strands of thought coexisted, and gradually through a sort of sifting process some of them emerged as more consistent with Jesus' message of God's unlimited love than others.

I have pointed out that the Thessalonian and Roman letters convey rather different interpretations of the time of the coming of the kingdom in its fullness. There is another interpretation that became even more important, and that is also to be found in writings attributed to Paul. In verse 15 of the first chapter of the letter to the Colossians, there is a very early hymn to Christ. The word "Christ" is the Greek form of "Messiah," but here the Messiah is not seen just as the liberator and ruler of Israel, or even of the whole human race. He is seen as "the first-born of all creation; for in him all things were created . . . all things were created through him and for him." This Messiah is the liberator and ruler of the whole cosmos, of "all creation." He is the pattern, the enfolder, and the goal of creation.

This teaching is found also in the letter to the Ephesians 1:10, where the mystery of God's will, set forth and revealed in Christ, is said to be "a plan for the fullness of time, to unite all things in him [in Christ], things in heaven and things in earth." Taking these two passages together, the Messiah is the liberator of creation, for he liberates created beings from all that divides and alienates them, uniting them in his own being. The Messiah is the ruler of creation, for he is the one for whom all things were created, and in whom, as the body with him as the head, all things will be brought to their intended fulfillment.

There are many other passages that speak in the same vein. I will quote just one more: "The creation waits with eager longing for the revealing of the sons of God . . . the creation itself will be set free from its bondage to decay and obtain the glorious liberty of the children of God."[7] The Messiah is the liberator of all creation from decay and the ruler of creatures who are destined to live in glorious liberty as children of God.

Allowing for the fact that the biblical view of the cosmos is very different from ours, it is nevertheless the case that this is a view of the redemption of the cosmos itself from decay. Indeed, the cosmos is said to be "held together in Christ," and to be destined to become the body of Christ—a state of which the Christian church is the designated forerunner and herald. In this sense, the body is a social reality that is the local presence and mediator of the "head," who is Christ. So here we find a vision of the church, the community of disciples, as called to be the local presence and mediator of Christ's being, with a positive vocation in the world—to cooperate in Christ's work of reconciling all things in the whole of creation to God, and uniting all things in the divine Word.

7. Rom. 8:19, 21.

It is probably true to say that this third Pauline vision is less well known, or even admitted, in the West than the first Pauline vision of an imminent end of the world. But it was the dominant vision in the early centuries of the church, and has been very clearly expressed in the Eastern Orthodox tradition by, for example, Gregory of Nyssa and Maximus the Confessor.[8] It opens up the possibility of a far future of the universe, and a goal for that future, which is the uniting of all things—all galaxies and whatever beings there are in them—in Christ, the creative Word of God. That does not mean that we have to preach our understanding of the Christ to untold numbers of alien life forms, for it is more to be expected that they will have their own revelation of that eternal Word of God who is the pattern and goal of all things, and that we and they together will learn more of Christ, who far transcends all our images of him.

In all these three Pauline scenarios, three central themes are unchanged. First, Jesus is the Christ, who is the consummation of the cosmos. Second, we have a purpose for our lives, and the church has a purpose in history, which is basically a calling to be, wherever we exist in space and time, the vehicles of the love of God, which was in Christ Jesus. Third, our relation to the consummation of all things in Christ is an immediate one, so that we should treat each moment as directly related to that consummation—in poetic terms, to the "coming of Christ like a thief in the night."[9]

That third theme may seem a little obscure, and I will try to explain what I mean. The goal of the universe, the reason for which it exists, is to have a community of conscious personal agents who live beyond decay and suffering in full awareness and love of God. But that should not be conceived as a state in the far future of the universe, removed from us by billions of years, and therefore something in which we will not share. Every conscious rational agent who has ever come into existence is to share in that state, and in that sense it must be "beyond history"—not in a simple linear relation to the time of this cosmos.

We might die at any moment, and we will certainly die long before any perfected state of this cosmos is realized, if it ever is. So the life beyond history, the life with God which is the Christian hope, is something we should envisage as just about to break in upon us. Every moment of our lives deter-

8. For a good general account of the Orthodox perspective, see Lossky [4].

9. 1 Thess. 5:3.

mines our life beyond history, and is to be taken into that life in a transformed, redeemed way. So at each moment the life of eternity is immediately close to us. As long as we live, the kingdom is at hand, and it is each present moment that will be completed in it.

From this point of view, exactly what happens in the future of the physical universe is irrelevant. God has created us, as beings emergent from this physical cosmos. God offers us eternal life in the kingdom. God will make such a life possible, whatever happens to us in this universe. Physics is not relevant here, and if the whole cosmos ended tomorrow, our hope in God would be unaffected. That is the force of Paul's first vision of Christian hope in Christ.

However—and it is a very big however—that is not the whole story. The church is a historical community, or set of communities, which has the historical purpose of manifesting and expressing God's love. Clearly, this purpose would be terminated if the planet Earth ceased to exist tomorrow. Maybe that purpose is meant to extend beyond this planet, farther throughout the universe. We do not know whether there are any other rational agents in the universe, or whether we could communicate with them. Perhaps our destiny only involves this planet and its immediate environs. But the early Christians thought in a cosmic way, after their own fashion, and it would not be very surprising if human destiny involved exploration into other star systems. At the very least, a Christian hope is that the purpose of the church—which is in the end the purpose of the Incarnation—should be realized as extensively and fully as possible. We can responsibly pray that time should not be ended yet, as long as there are forms of divine love and compassion to be expressed and understood in new ways. It is a Christian vocation to hope for a growth of the kingdom throughout any parts of the universe we can reach, and to pray that might be possible. That is the force of Paul's second vision of Christian hope in Christ.

Furthermore, if the consummation of the cosmos is to be found in Christ, and if God created the universe so that all things in it could be united in Christ, it is alien to the Christian mind to separate the spiritual and the material completely. It is this material universe, in all its intricate detail and elegant beauty, as well as its terrifying power and its liability to decay, that is to be brought to fulfillment in Christ. So, though God can bring that about in any way God chooses, the natural hope is that in some way the universe can generate out of itself, as it has generated conscious rational agents, the possibility of a state beyond decay and suffering (see de Chardin [3]).

Various speculations exist as to how this may be possible, mostly involving some conversion of energy into pure information, or a transformation of the material into a different form. This does not strike me as absurd. After all, no one could have foreseen that quarks and leptons could generate consciousness, and no one knows how it happened. But it did. So it may be that further transformations of the material will generate higher forms of consciousness, which will not be subject to physical decay. It is hardly to be expected that we could now see how this might be done, but we cannot rule it out in principle, given the very limited understanding of nature that we have.

This, of course, is the poetic vision of Teilhard de Chardin. Of it I would say that Christians have good reason to hope that he is right, and that further evolutionary transformations of matter will occur, bringing the universe to a realized conscious unity in Christ. However, despite the recent speculations of Frank Tipler [6], it is doubtful whether this would be a state in which all the resurrected dead would share, so that the ultimate consummation, the ultimate Christian hope, is likely to be beyond the physical universe altogether. This may seem very speculative, and I would not regard it as entailed by Christian faith. But I would regard it as a very natural extension of it, and it is the force of Paul's third vision of Christian hope in Christ.

15.5. Religious Hope and the Far-Future Universe

What, then, do Christians have to say about the far future of the universe? First, that our hope is not "for this world only," so ultimate human destiny will not be found there. Second, that the universe is meant progressively to realize communities of persons, in whatever material forms they exist, without suffering and evil. And third, it is the human vocation to shape the universe, so far as is possible, in accordance with that goal.

In the context of a contemporary scientific cosmology, this means that the Christian faith is wholly consistent with the idea that this space-time will have a temporal end. Christians do not hope for a continuation of this space-time, just as it is, forever. It means that Christians do not have to think of humans as the center of the universe. The goal of the universe is the existence of persons—a class in principle much wider than humans—and so

Christians should not be surprised if there are other personal life forms in the universe, and may even hope that there are. And it means that humans have a positive responsibility to shape the material universe so that it is productive and protective of personal life, whatever its form. The human responsibility is not merely to leave things as they are, but to change them so that the personal, in all its forms, can flourish.

Christian talk of Christ coming again in glory to bring in the kingdom of God in its fullness must therefore now be seen in a fully cosmic context. It is not a matter of Jesus coming to this planet on the clouds. It is a matter of the whole of creation finding fulfillment in a transfigured form of personal life, beyond though arising out of this cosmos. Just as the earliest church had to extend its vision beyond Jewish nationalism to embrace all the nations of the earth, so the contemporary Christian church has to extend its vision beyond anthropocentrism to embrace the whole cosmos. I have tried to suggest that this is a natural and proper development, which brings Christian faith and cosmology into a mutually enriching harmony.

The Abrahamic faiths, having a common interest in the end of the universe and in a resurrection of the dead, can share in this sort of reinterpretation of their traditions. They can all share the threefold concern with ultimate hope, with the primacy of personal being, and with human responsibility for improving the cosmos now so far as possible. Indeed, all theistic faiths can agree that this universe is God's creation, and that it must have some sort of purpose (even if that purpose is just the working out of *karma*, or the consequences of past deeds), the final fulfillment of which lies beyond this universe. To that extent, religious believers have an interest in whether the sciences can discern purposive elements within the universe, which could find fulfillment beyond it. Speculations about a far-future universe are not of pressing religious concern, but they have an indirect relation to the question of whether the universe can be reasonably seen as the purposive creation of a good God.

The contribution of scientific cosmology to religious faith, however, probably lies, at least so far as the Abrahamic faiths are concerned, more in the extension of cosmic vision that it induces, rather than in any specific claims about the far future of the universe. That extension of vision is to be welcomed, since it increases human awe at the immensity of creation, and focuses attention on God and the divine purpose for the whole of creation, rather than on what has sometimes been a rather myopic concentration on the destiny of the human species. The Indian traditions already have a vision of an infinite universe. For them, the challenge of cosmology is to integrate talk of an evolutionary universe with traditional ideas of a decline from a past "Golden Age." This task has been undertaken by some, especially Sri Aurobindo [1], and it is clear that cosmology can motivate a rethinking of prescientific cosmological traditions in directions that are creative and stimulating. In the end, however, the Chinese traditions are probably right in stressing that the important religious question is how we can act with integrity and hope now. If there is a religious hope for an ultimate future, it is not confined to the future of this universe. This suggests that if the universe itself has a goal of religious importance and value, it lies in the process itself, and not in its final state. As a great Middle Eastern religious teacher once said, "Take no thought for the morrow; for the morrow shall take thought for the things of itself."[10]

10. Matt. 6:34

REFERENCES

1. Aurobindo, S., *The Life Divine* (Arya Publishing, Calcutta, 1939).
2. Capra, F., *The Tao of Physics* (Shambala Press, Berkeley, 1975).
3. de Chardin, T., *The Phenomenon of Man* (Collins, London, 1959).
4. Lossky, V., *The Mystical Theology of the Eastern Church* (James Clarke, Cambridge, 1957).
5. Schweitzer, A., *The Quest of the Historical Jesus* (A. and C. Black, London, 1911).
6. Tipler, F., *The Physics of Immortality* (Doubleday, New York, 1994).

[16]

COSMOS AND THEOSIS

Eschatological Perspectives on the Future of the Universe

Jürgen Moltmann

16.1. Introduction

The subject of this symposium to which the John Templeton Foundation has invited us is "The Far-Future Universe: Eschatology from a Cosmic Perspective." But I am not a scientist. I am a Christian theologian. So I am going to turn the subject upside down, and talk about the eschatological perspectives that emerge from Christian theology for the future of the universe.

In its two thousand years' history, Christian eschatology has always been developed in the context of the world view of its epoch. Today we are required to set that eschatology critically and self-critically in the context of modern astrophysics. For more than two hundred years, the development of the modern sciences has led to the retreat of theology from the world view as a whole, and to its withdrawal to the questions of personal existence and morality. We have ceased to be able to bring human eschatology and cosmic

Note: Translated by Margaret Kohl.

eschatology into a certain harmony. Often this harmony is not even desired by either side—on the scientific side because the trial of Galileo is still unforgotten; on the theological side because the scientific hypotheses are involved in a continually accelerating paradigm change, so that it is impossible to know with what one has to engage.

Today few scientists read theological books, and theologians dispense almost entirely with any scientific reading. Neither side expects much from the other. The situation has not changed much since Galileo. When Galileo wanted to show his opponents the Jupiter satellites, they refused to look through the telescope because they believed, as Bert Brecht aptly put it, that "there is no truth in nature, but only in the comparison of texts" [4]. So today theologians confine themselves to the interpretation of the sacred texts of their traditions, and take no interest in what can be seen through the Hubble space telescope. On the other side, the scientific community is so wrapped up in the progress and competition of their sciences that its members expend little thought on the hermeneutical premises of their thinking and their concepts, the knowledge-constitutive interest that prompts their investigations.

I find this state of things unsatisfactory, and would plead for a new "natural theology," in which scientific findings tell us something about God, and theological insights something about nature [18]. The reason for a natural theology of this kind is that there is a correspondence between human intelligence and the intelligibility of the universe. We perceive and know more about the world than we need in order to survive in our earthly environment.

16.2. Theological Origins of Eschatological Perspectives on the Future of the Universe

Theology does not acquire its eschatological horizons from the general observation of the world, neither the observation of the stars in the universe nor a contemplation of the events of world history; it acquires them from its particular experience of God. We call this "root experience," because here are experienced events in which God "reveals" himself, and from which human communities acquire their identity. For Israel and Judaism this root experience is the Exodus, as the first Commandment tells us: "I am the Lord your God, who brought you out of the land of Egypt, out of the house of bondage" (Exod. 20:2). During the Passover seder every generation in Judaism identifies itself with the Exodus generation, and through that finds assurance of God and of itself. For Christianity, the root experience is the Christ event, Christ's self-surrender to death on the cross and his resurrection from the dead. In the celebration of the Eucharist, every Christian congregation places itself within Christ's sphere of influence, and there finds *its* assurance of God and of itself. However, these special historical root experiences out of which the religious communities of Judaism and Christianity have emerged also embrace from the outset general horizons of experience and universal expectations of the future; for they are temporal experiences of the eternal God.

As the "Creator" of all things, the God who freed Israel from enslavement is perceived in all things. He led the prisoners into freedom through his activity in history, and similarly, through his creative activity he called everything out of nonexistence into existence, and brought it out of chaos into a wise order. Consequently Heaven and Earth are not themselves divine. They are neither "the body of God" nor are they inhabited by fertility deities who have to be worshipped. They are God's creation. They are a "work of his hands," as the image or metaphor says; but they are blessed and preserved by God. From early on, the Israelite belief in creation desacralized the world and "stripped it of its magic," as Max Weber put it, and in this way it also threw the world open for the scientific and technological intervention of human beings. But that does not mean that this belief in creation has surrendered nature to ecological extermination by the modern world [12, 16].

The God who brought Christ out of death into the liberty of the new, eternally living creation is perceived not only as the Creator of all things but also as the one who brings all things to completion. He is "the God . . . who gives life to the dead and calls into existence the things that do not exist" (Rom. 4:17). From the moment when the disciples called that which happened to the dead Christ "resurrection from the dead," the universal eschatological horizon was already present in the understanding of this unique historical event. Christ's resurrection "from the dead" was understood, as the phrase shows, as the beginning of the general resurrection of all the dead. He became "the first fruits of those who have fallen asleep" and was called "the leader of life."[1] Whatever "end" the human world or the natural universe may otherwise have, this future of God's has already begun. The

1. W. Pannenberg rightly uses for this the concept of *prolepsis,* anticipation; see [19].

new creation already begins in the midst of the old one. In the community of Christ, believers already experience here and now "the powers of the age to come" (Heb. 6:5). Because the Christ event is an experience of God, Christian eschatology cannot be reduced to human eschatology, and human eschatology cannot be reduced to the salvation of the soul in a heaven beyond. There are no human souls without human bodies, and no human existence without the life system of this Earth, and no Earth without the universe. Very early on, the development of Christian eschatology already brought out the cosmic dimensions of the Christ event, as the (post-)Pauline epistles to the Ephesians and the Colossians show. Without an eschatology of the universe, Christian eschatology cripples the Godness of its God. But this brings us up against the first problem of Christian eschatology: Can human and cosmic eschatology still be thought together today, and brought into harmony?

16.3. Human and Cosmic Eschatology

The biblical traditions are dominated by a strong anthropological principle. Both Israel's creation narratives (Genesis 1–3) are aligned toward the creation of human beings, and put that at the center. But we know today that *Homo sapiens* was a late birth in the evolution of life. The biblical traditions knew nothing about the millions of years of dinosaurian history. Likewise, the images of the future in the eschatology of the Bible all expect that the end of this world will coincide with the end of the human race, and vice versa. "The kingdom of God" will come "in this generation," or after the redemption of Israel, or whenever God wills it; but at all events, according to the Christian creed Christ will come "to judge the living and the dead"—so he will still find living human beings. Yet we know that humanity as a whole is mortal, and that the far-future universe could be a universe without human beings—or at least without human beings as we know them now. Consequently it is impossible to shift human protology back to the big bang, and to extend human eschatology to the death of the universe through heat or cold. This has considerable consequences for the orientation of human beings in the universe: Can human eschatology only give us a limited hope for a meaningful life that is surrounded by a meaningless universe? What conclusions do we have to draw from Steve Weinberg's insight: "The more the universe seems comprehensible, the more it seems pointless" [30]? Do we exist on an island of meaning in an ocean of meaninglessness?

Another image emerges if we supplement *the introduction of the anthropological principle into the cosmos* through *the introduction of the cosmic principle into anthropology*. We then see not only the universe in the human being, and the human being as the highest complex, self-conscious system known to us. We also see the reverse: the human being in the universe and the universe as the most extensive context for the development of human potentialities. The future of the universe would then not be bound to the future of human beings; instead the future of the human being would be integrated into the future of the universe. Just as there are an inexhaustible number of thoughts and ideas in the human mind, there can similarly be an inexhaustible number of conditions in the universe. Just as the human mind regulates its potentialities according to its constructive hope and its destructive fears, the continual appearance of new possibilities can similarly constitute the universe's openness to the future. Is there a correspondence between complexity and consciousness? Why should the human mind only be involved cognitively in the shaping of the universe, and not constructively too?

If we start from these presuppositions, we have to ask what cosmological utopia and what knowledge constitutive interest prompt us in our desire to know the universe, and to ask about human possibilities of shaping it.[2] Whatever else our richly proliferating scientific fiction may tell us, the endless survival of the human race, and the endless further development of the human consciousness, are evidently a fundamental utopia. Unconsciously or consciously we want to overcome death and to live as long as possible. Immortality used to be a religious dream. Today organ transplantation, and the attachment of the brain to computers, and so forth, seem to have brought it within our reach..

By computerizing all the available data, we make synchronic for ourselves what in time takes place successively. In so doing we transfer past and future into present simultaneity, and "spatialize" time. "The future is now," says the internet advertisement. At the same time, through space travel, space stations, and landings on other stars, we seek for ways of escape, for the day when we have made Earth "too hot for us" because we have destroyed it. In the universe, finally, we search for traces of a "world without end." Multiverse models, the genesis of new universes, and limitless expansions of these

2. I am not using the term "utopia" here in a negative sense, but positively, as it is used by Ernst Bloch in [3].

multiverses point to an endless future; is it really so, or is it only so because that is what we are searching for?

But is an endless future for life and the universe as we know it desirable anyway? Are death and the transitoriness of time not factors of the finite world, which is developing itself for continually new possibilities? If death and time were to be overcome, there would no longer be anything new. An endless world would be the world's end.

16.4. Traditional Ideas about the Future of the Universe

All ideas about an "end of the world" presuppose that the world we know here is temporal, not eternal, and therefore has a beginning and an end with time as we now experience it. This means that the universe is involved in a unique movement. Here we call it "history," because it is not clear from the outset that it is a meaningful evolution of the cosmos, a purposeful development, or a progression toward new, higher worlds.

We can distinguish past and current ideas about the end according to whether they talk about a goal *(telos)* or an end *(finis)*.[3] If the history of the universe has a goal, then that is its consummation, and we can talk about a meaningful development, and identify progressions, gradual or from stage to stage. The last great cosmic system of a finalistic, or purposeful, metaphysics of this kind was conceived by Teilhard de Chardin. In the far-off future of the universe an "omega point" will emerge, which will draw the universe with all its parts to itself. This point does not move toward the universe, but draws the universe through attraction [2].

But if the history of the universe is to end in a "big crunch," and if nothing more can be expected after such a catastrophe, then the universe has no meaningful development either, and no purposeful progression, but merely a succession of smaller part-catastrophes in which the final universal catastrophe is heralded. Not only human history but the history of the universe too is then, as we say, "just one damn thing after another."

Theologically, we call ideas about a purposeful progression toward a state of perfection "millenarian," because they talk about the golden age that Virgil speaks of, or about a final kingdom of Christ (Revelation 20) in the final stage of history. The concepts of a linear time, with which progress in all the

3. For more detail, see [17].

different spheres of life can be measured, and the modern world's faith in progress, are modern secularizations of the old ideas about a millenarian or chiliastic consummation.[4]

We call ideas about a catastrophic rupture of history "apocalyptic," because Jewish and Christian apocalyptic writings have passed down dreams about the downfall of this world of violence and injustice and death in which the people were being oppressed. Apocalyptic was not originally the world of metaphysical spiritual seers, which is what Kant called it; it was the world of the persecuted and the martyrs. It is "the religion of the oppressed" (Laternari). In secularized form, we encounter this world of ideas today in exterminism, and in the "terminator" from outer space whom we meet in science fiction, who brings about the end.[5]

In Christian eschatology we always find a combination of the two ideas: an end and a beginning, a catastrophe and a new start, farewell and greeting; for eschatology can only be called Christian if it takes its bearings from Israel's Exodus and the Christ event. Israel's bondage and the death of Christ are primal images of catastrophe. The Exodus into the liberty of the promised land, and the resurrection into the eternal life of the future world are primal images of the new beginning. Christ's end on the cross was not the last thing, but became his true beginning in the Resurrection and in the Spirit who is the giver of life. The dialectical mystery of Christian eschatology is, to adopt T. S. Eliot's words: "In my end is my beginning." That is also what is meant by some words of mine quoted in the invitation to this symposium: "What can a theology of hope tell us about the far-future cosmos that has relevance from a human perspective? Could the 'death and raising of the universe' be the prelude to an unexpected new creation of all things?" (p. 2). In the light of their experiences of God, Christians expect a universal exodus of all things out of their bondage of transience into a "new heaven and a new earth [where] the former things shall not be remembered or come to mind" (Isa. 65:17); they expect a raising of all the dead, and a restoration of all things in a new, eternal creation (Rev. 21:4). It is only then that there will be "world without end."

4. This has been shown by K. Löwith in [13]; see also the later expanded German version [14].
5. See [17], III, §8: "End-Times of Human History: Exterminism," 202–18.

16.4.1. THE ANNIHILATION OF THE WORLD (*ANNIHILATIO MUNDI*)

For a hundred years, from about 1600 to 1700, Lutheran theology taught that the future destiny of the universe was to be not its transformation but its annihilation: "The Last Judgment will be followed by the complete end of this world. Except for angels and human beings, everything belonging to this world will be consumed by fire and will dissolve into nothingness. What must be expected therefore is not the world's transformation but a complete cessation of its substance."[6] Similar notions can be found among modern fundamentalist "annihilationists." After the Last Judgment believers will go to Heaven but unbelievers will be exterminated, together with the earth. What used to be called hell is now called "total nonbeing" [5].

The theological reason put forward for this view is that angels and believers will be so totally absorbed in the *visio beatifica,* the bliss of contemplating God "face to face" (1 Cor. 13:12) that they will have no further need of mediations to God by way of this created world. Angels will no longer need the created world of Heaven, and human beings will no longer need the created world of earthly things. So Heaven and Earth, together with the mortal body, will be demolished and destroyed like a scaffolding once God's consummating goal has been reached, that goal being the salvation of souls.

This idea about the eschatological annihilation of the world does not mean just the world in its present form. It means the very substance of creation itself. It means, in fact, a reversal of the creation out of nothing *(creatio ex nihilo)* into a reduction to nothing *(reductio ad nihilum).* Whereas the original creation was a movement out of nonbeing into being, its end will be the movement out of being into nonbeing. Adherents of these ideas about the annihilation of the world claim biblical support from 2 Pet. 3:10: "But the day of the Lord will come like a thief in the night, and then the heavens will pass away with a loud noise, and the elements will be dissolved with fire, and the earth and the works that are upon it will be burned up . . . Wait for and hasten to the coming of the day of God, when the heavens will be kindled and dissolved and the elements will melt with fire"; then Rev. 20:11: "From (God's) presence earth and sky fled away, and no place was found for them";

6. Schmid [24]. See also the comment by Stock [26].

and Rev. 21:1: "The first heaven and the first earth had passed away. . . ." One might see a modern cosmological analogy if, in a future big crunch, the universe were to revert to its condition before the big bang.

16.4.2. THE TRANSFORMATION OF THE WORLD (*TRANSFORMATIO MUNDI*)

The idea that the world will be annihilated at the end of time is an exception. The general theological expectation of the end is that the universe will be transformed from the state in which we see it now into a condition that is qualitatively new.

The Catholic preface to the Requiem Mass says: "*Vita mutatur non tollitur*"—life will be changed, not destroyed.[7] Aristotle and Thomas Aquinas defined the human soul as *forma corporis,* or the "actualization" of the body; and according to this definition, death means its transformation, not its extermination. Lutheran theology justified its doctrine of the annihilation of the world on the grounds of God's total freedom toward his creation—the One who created it can destroy it too; but Calvinist theology at the same period saw the transcendent foundation of creation in God's faithfulness, and assumed that this was the divine guarantee for the continued existence of the universe. Like medieval Catholic theology, therefore, it taught the eschatological transformation of the world, not its destruction: "After Judgment the end of this world will come about, in that God will destroy the world's present condition and . . . out of the old world will make a world that is new, a new heaven and a new earth whose nature will be imperishable" [8]. The form of the old world was sin, death, and transience; the form of the new one will be righteousness and justice, eternal life and imperishability. The eschatological transformation of the universe embraces both the identity of creation and its newness, that is to say both continuity and discontinuity. All the information of this world remains in eternity, but is transformed. The biblical data bear out the correctness of this view, over against the doctrine of annihilation, for both in 2 Peter and Revelation the idea of annihilation is followed by hope for "a new heaven and a new earth in which righteousness dwells" (2 Pet. 3:13) and by the vision "and I saw a new heaven and a new earth" (Rev. 21:1).

As far as analogies in present-day cosmology are concerned, one could

7. See here [11].

assume that an eschatological transformation of this kind is a possible completion of the history of the universe, if one views that history as a universal information process with many strata and many ramifications.

16.4.3. THE DEIFICATION OF THE WORLD (*DEIFICATIO MUNDI*)

A third theological idea about the end and completion of the universe can be found in the Orthodox eschatology of the "deification of the cosmos."[8] Because the word "transformation" describes merely the form of the world but not its relation to God, the Orthodox idea goes a step further, from transformation to glorification and deification. According to Orthodox doctrine, nature and the human person constitute a unity, and are not set over against each other as they are in the modern Western world. So what is promised to the human person applies to the earth and the cosmos too. Human and cosmic eschatology form a unity. There is no human future without the future of the cosmos. Consequently the cosmos will be redeemed when humanity is redeemed, and vice versa. The whole of nature is destined for God's glory. Just as to be enlightened by the eternal light of God leads to the transfiguration of the human form, so the universe too will be "deified" when it is transfigured by the indwelling glory of God (Isa. 6:3).

The theological foundation for this eschatological perspective on the far-future universe is to be found in the central importance for Orthodox theology of the Resurrection of Christ. Through his true raising, the body of Christ was "transfigured" (Phil. 3:21). His raising "from" the dead is a human and a cosmic event: the risen Christ is the head of the new humanity and the first of all created being, as the Epistle to the Colossians says (1:15). In the eschatological consummation, God will then appear in glory in his creation, and through his unveiled eternal presence will redeem all things from transience for imperishable participation in his divine life. The eternal presence of God means for the process of creation simultaneity. All things, which in the beginning were created out of nothing, will then be created anew out of God's glory. So in the eternal present of God the eternal creation comes into being. That is what is meant by the "deification of the cosmos" through a sharing and participation in the eternal life of God. The world does not become God, nor does it dissolve into God's infinitude; but it participates as

8. Staniloae [25]: "The world as the work of God's love, destined to be deified."

world in God's eternal being. It will become the temple of his eternal presence. That is what is meant by the image of the heavenly Jerusalem, the city of God that comes down to earth: a cosmic temple.

16.5. The Eschatological Model for the Future of the Universe

The eschatological idea about the future of the universe differentiates its history, in the old apocalyptic way, into two phases, and talks about the time of this world—"this world-time"—and the time of the world to come—"the future world-time." These are two qualitatively differentiated eons. The time of this world is the time of the transitory world—the time of the future world is the time of a "world without end," the abiding and hence eternal world [21].

This idea also differentiates the given reality of the universe into "heaven and earth," the invisible and the visible world ("things visible and invisible," says the Nicene Creed).[9] By this is meant the differentiated strata of being in the one created reality. We can already see this from the fact that the earth, or visible world, is talked about only in the singular, whereas we can speak of "the heavens" or "the invisible worlds" in the plural. This gives us an overall picture of an earthly visible universe, and a heavenly, invisible multiverse. The time of the visible universe is *chronos,* the irreversible temporal structure of becoming and passing away. The time of the invisible multiverse is *aeon, aevum,* a reversible temporal structure of cyclical time, for the circle of time counts as the image and reflection of eternity.

Translated into modern metaphysics, this means that the visible universe consists of reality and potentiality, which seen in terms of time, are related to each other as past and future. The present is the interface at which potentiality is either realized or not realized. Past reality is fixed, future potentiality is open. Reality is in every case realized potentiality. Potentiality is therefore the mode of being that underlies this world. The world is and will be made out of potentiality. In contrast, the qualitatively different reality of the invisible worlds consists of potencies and potentials (as Schelling puts it) which make visible and earthly potentiality possible without themselves being exhausted [22, 23].

The eternal being of God is differentiated from the earth and the heavens,

9. See Moltmann [16] VII: "Heaven and Earth," 158–84.

the visible and the invisible worlds. Since the worlds are God's creation, the concept of creation designates the qualitative difference between God and the worlds. Over against the irreversible time of the visible world and the reversible time of the invisible worlds, God's time is eternity. Eternity does not mean endless time, nor does it mean timelessness. It means power over time (*Zeitmächtigkeit*, as German puts it). Seen over against the modes of time of the created worlds, the eternity of the Creator is to be seen in his pre-temporality, his simultaneity, and his post-temporality.[10] His eternity surrounds the time of the created worlds from every side, and by doing so confines it to finite time. But in this way his eternity determines irreversible time as the power of future: there is past future, present future, and future future.[11]

What consequences do these presuppositions have for an understanding of the scientifically explorable universe?

1. The first consequences are negative. The visible universe is not divine, and displays no characteristics that could be called divine. Nor is the visible universe heavenly. It is therefore neither eternal, nor cyclical, nor permanent; it is transitory, temporal, and contingent. The atheism of old, following Feuerbach's method, denied the reality of God in a world beyond, but transferred all God's attributes to this one, to creative human beings and the marvelous universe; and the result was a deification of this world that can only lead to disappointment.

2. The second consequences are positive. This visible universe is certainly temporal, contingent, and finite, but it has an eternal, permanent, and infinite future in the future new universe. Once the eschatological differentiation into "this world" and "the future world" is made, the finitude, contingency, and "pointlessness" of this universe can be accepted. The future new world will bring what we miss in this finite world: the eternal presence of God, and participation in the attributes of this divine presence—that is to say, that which bears its meaning within itself.

That brings us to the question about the transition from "this world" to "the future world." This transition cannot take place like the transition in this

10. Barth [1]. See also my criticism in [17], 17–19.

11. See the time scheme of the physicist Klaus Müller [10], which I have pursued further, taking up M. Heidegger's saying: "The primary phenomenon of primordial and authentic temporality is the future" [7].

world from one stage to another, for then the irreversible time structure that is a mark of our present world would remain. It can only be a matter of a universal transformation of this present world of the kind Rev. 21:4 describes: "Behold," says God, "I will make all things new." That means that everything created, everything that was here, is here, and will be here, that is, to be "made" new. The Hebrew word for "make" *(asah)* is not the same as the word for "create" *(barah)*. Whereas "to create" means calling something into existence, "to make" means *informing* what has been called into existence already.

The future, new eternal world is therefore to be the new creation of this world we know. When will that happen? That is difficult to answer, because this eschatological moment must be simultaneously the end of irreversible time; so it cannot take place in this time at all.[12] In his great chapter on the Resurrection, Paul says that this eschatological moment will come about suddenly, "in the twinkling of an eye, at the last trumpet" (1 Cor. 15:52), and he uses the Platonic word *exaiphnes* for the moment when eternity touches time, ending and gathering it up into itself. "Then the dead will be raised imperishable . . . and death is swallowed up in victory" (vv. 52, 55).

In that eschatological moment the raising of the dead will take place diachronically, irreversible time will be reversed and rolled up from its end like a scroll, as it were. Before the eternity of God that appears in the eschatological moment all times will be simultaneous. The power of transience and the time that cannot be brought back, "death," will be gathered up into the victory of eternal life. That is the end of the evolution of the living in what we might call the beginning of the re-volution of the dead.

People who have been close to death have sometimes seen their whole lives spread out before their eyes in an instant, as if in a time-lapse film; and in a similar way one could imagine that in the eschatological moment the flash of eternity will light up the universe from its end to its beginning, and make it wholly present.

The eschatological moment will end linear time, which we have here called the irreversible time of this world, and take into itself an element of cyclical time. What will come about is not an eternal return of the same, but surely a unique return of everything. The theological concept for this is "the

12. On the term "eschatological moment" and the relation to the "primordial moment" see [17], 292–95.

restoration of all things" [6] in the appearance of God's eternal presence, and their new creation through their transformation in the eschatological moment from transience to immortality. We might call it something like a universal, cosmic feedback process.

This eschatological model for "the future of the universe" is the only model that perceives a future for what is past and expresses hope for the dead. All other models of expansion, evolution, progress, or the steady state universe expect a future only in the sphere of what is not yet, but not a future in the sphere of what is no longer. They leave the past behind and gaze merely into the far-future universe. But the eschatological model of the future brings hope into remembrance, and discovers future for the universe in all the stages of its past.

16.6. Two Open Questions

16.6.1. IS THE UNIVERSE CONTINGENT AND IS EVERY EVENT UNIQUE?

"Why is there something and not nothing?" This is the child's metaphysical question, and it is unanswerable. Things are there as they are, but they are not necessarily there. What is there, evidently has a reason that is not inherent in itself but is to be found somewhere else. The universe is there, but it is not divine. This is expressed theologically not just through the concept of creation but also through the idea of the contingency of the world *(contingentia mundi).* It follows that the orders that we perceive in the universe are not eternal laws either; they are themselves contingent orders [27]. The universe or multiverse can therefore have different orders, and orders of a different kind. Natural laws are not eternal laws. They can conceivably change.[13] Consequently, we recognize the natural processes in the universe through observation, experience, and experiment, not through an ideal contemplation of essence. In the history of Christianity, the meaning of the term "nature" has changed. It no longer signifies the essence of things, as it did in antiquity. Since the beginning of modern times it has come to mean their experienceable mode of appearance. When we talk about "the natural sciences" we are using the term "natural" in a different sense from the way we use the word when we are inquiring about the "nature" of things.

13. Thus Toulmin [28]: "Are the laws of nature changing?," with reference to P. A. M. Dirac. Cf. also [22].

If we go a step further, we come up about questions about the experiences that we can answer scientifically. In a scientific experiment one can only prove something that stands up to reexamination through repetition. That is to say, only repeatable experiences can be "proved," not nonrepeatable ones. But in a world with an irreversible time structure can experiences ever be repeated, in the strict sense? If in the universe there is no "eternal return of the same thing"—if, as Heraclitus said, no one ever steps into the same river twice—then in the experimental repetition of an experience we do not come upon the same thing in the strict sense but only something corresponding to it. The uniqueness of happening and of experience remains. The great histories of the cosmos, such as the history of the universe and the evolution of life, are not repeatable. But perhaps on other stars we shall find comparable evolutions of life in stages that are different from those of the earth.

16.6.2. IS THE UNIVERSE A CLOSED SYSTEM OR AN OPEN ONE?

Just as circular time was held to be the image and reflection of eternity, so the globe has been viewed from time immemorial as the image of perfection. In popular magazines even today we find not just the earth but the cosmos too depicted as a sphere. But a sphere is generally thought of as a closed system, and a closed system is thought of as a sphere. But if the universe has been involved in a unique movement of expansion ever since the big bang, then we should have to view the big bang as world midpoint for an expanding (and perhaps one day again contracting) world sphere, or we should have to abandon the image of the sphere [9].

All the systems of matter and life we know are open, complex systems. They communicate with other systems, and anticipate their possibilities [29, 15]. Why should the universe as a whole, as the sum of all its parts and individual systems, be a closed system? We can certainly think of the sum of all open systems as a closed whole without affecting their openness. But we could also infer from the complex open systems a systematic openness of the whole that is not yet rounded off into an entirety, and has still not found a comprehensive organization principle for all the parts. All the formulas for the whole would then be anticipations of the not-yet-present and therefore not-yet-surveyable whole, and in so far they would be provisional and dependent on future confirmation.

If we follow the first law of thermodynamics, this idea about the whole as a closed system would seem to suggest itself, a system whose formula for the world we can look for and find, because it is bound to exist. If we follow the second law of thermodynamics, and expand it into "the theory of open systems," then what seems suggested is the idea of the open, unfinished, and still incomplete universe.

Open systems show relatively fixed structures of reality, and relatively open scopes for possibility. They tend toward differentiation, toward the growth of complexity, and toward integration or networking with other open systems. They are open not only for quantitative developments but also for qualitative leaps in combinations into new wholes. Catastrophes and new beginnings mark their history more often than continuous developments.

That brings us to a final thought. If in the beginning there was so unique and universally relevant an event as the big bang, is something corresponding then conceivable for the far-future universe? An open universe would be open for the theological eschatology I have described; a closed and already completed world sphere would not.

REFERENCES

1. Barth, K., *Church Dogmatics* III/2 (T. and T. Clark, Edinburgh, 1954), 437f.

2. Benz, E., *Schöpfungsglaube und Endzeiterwartung: Antwort auf Teilhard de Chardins Theologie der Evolution* (Nymphenburger Verlagshandlung, Munich, 1965).

3. Bloch, E., *The Principle of Hope*, trans. N. Plaice, S. Plaice, and P. Knight (Oxford University Press, Oxford, 1986).

4. Brecht, B., *Life of Galileo*, trans. D. I. Vesey (London, 1967). The quotation in the present text has been translated directly from the German text.

5. The Doctrine Commission of the Church of England, *The Mystery of Salvation: The Story of God's Gift* (London, 1995), 199.

6. Groth, F., *Die Wiederbringung aller Dinge im württembergischen Pietismus* (Vandenhoeck and Ruprecht, Göttingen, 1984).

7. Heidegger, M., *Being and Time*, trans. J. Macquarrie and E. Robinson (Harper and Row, New York, 1962), 378.

8. Heppe, H., and Bizer, E., *Die Dogmatik der Evangelisch-reformierten Kirche*, 2nd ed. (Neukirchener Verlag, Neukirchen, 1958), 560.

9. Koyré, A., *From the Closed World to the Infinite Universe* (Johns Hopkins Press, Baltimore, 1968).

10. Klaus Müller, A. M., *Die präparierte Zeit* (Radius Verlag, Stuttgart, 1972).

11. Küng, H., *Eternal Life?* trans. E. Quinn (SCM Press, London, 1984).

12. Liedke, G., *Im Bauch des Fisches: Ökologische Theologie* (Kohlhammer Verlag, Stuttgart, 1979).

13. Löwith, K., *Meaning in History* (University of Chicago Press, Chicago, 1949).

14. Löwith, K., *Weltgeschichte und Heilsgeschehen* (Kohlhammer Verlag, Stuttgart, 1952).

15. Maurin, K., Michalski, K., and Rudolph, E., eds., *Offene Systeme,* vol. 2: *Logik und Zeit* (Klett Verlag, Stuttgart, 1981).

16. Moltmann, J., *God in Creation: An Ecological Doctrine of Creation,* The Gifford Lectures, 1984–85, trans. Margaret Kohl (SCM Press, London; Harper and Row, San Francisco, 1985).

17. Moltmann, J., *The Coming of God,* trans. Margaret Kohl (SCM Press, London; Fortress Press, Minneapolis, 1996).

18. Moltmann, J., *Experiences in Theology: Ways and Forms of Christian Theology,* part I, 6: *Natural Theology,* trans. Margaret Kohl (SCM Press, London; Fortress Press, Minneapolis, 2000), 64–83.

19. Pannenberg, W., *Jesus, God, and Man,* trans. L. C. Wilkins and D. A. Priebe (Darton, Longman, and Todd, London and Philadelphia, 1968), 58ff.

20. Pannenberg, W., "Kontingenz und Naturgesetz," in *Erwägungen zu einer Theologie der Natur* (Vandenhoeck and Ruprecht, Göttingen, 1970), 65.

21. Rowland, C., *The Open Heaven: A Study of Apocalyptic in Judaism and Early Christianity* (SCM Press, London, 1982).

22. Schelling, F. W. J., *System of Transcendental Idealism* (Charlottesville, Va., 1978).

23. Schelling, F. W. J., *Bruno, or On the Natural and Divine Principles of Things,* trans. M. Vater (Albany, N.Y., 1984).

24. Schmid, H., *Die Dogmatik der Evangelisch-lutherischen Kirche, dargestellt und aus den Quellen belegt* (Bertelsmann Verlag, Gütersloh, 1983), 407.

25. Staniloae, D., *Orthodoxe Dogmatik* (Benziger Verlag, Zürich; Gütersloher Verlag, Gütersloh, 1985), 291.

26. Stock, K., *Annihilatio Mundi: Johann Gerhards Eschatologie der Welt* (Chr. Kaiser Verlag, Munich, 1971).

27. Torrance, T. F., *Divine and Contingent Order* (Oxford University Press, Oxford, 1981).

28. Toulmin, S., *The Discovery of Time* (Harper and Row, New York, 1965), 263.

29. von Weizsäcker, E., ed., *Offene Systeme,* vol. 1: *Beiträge zur Zeitstruktur von Information, Entropie, und Evolution* (Klett Verlag, Stuttgart, 1974).

30. Weinberg, S., *The First Three Minutes* (Harper and Row, New York, 1988).

[17]

ESCHATOLOGY AND PHYSICAL COSMOLOGY

A Preliminary Reflection

Robert John Russell

17.1. Introduction

Over the past four decades, the interdisciplinary field of "theology and science" has undergone tremendous growth involving scholars from philosophy of science, philosophy of religion, the natural sciences, theology, ethics, history, and related fields engaged in "creative mutual interaction."[1] Topics range from a detailed comparison of the methods of science and of religion

Note: I would like to thank George Ellis in particular, as well as Nancey Murphy and Bill Stoeger, for their careful reading of, and their very important corrections and suggestions to, this essay. I thank Adrian Wyard and Jim Miller for producing the computerized image of figure 17.1 and for giving permission to use it here.

1. For scholarly introductions, see [3, 63, 72, 76, 79]. For less technical introductions, see [42, 75, 92]. For a recent survey with extensive references and listings of programs, journals, and websites, see [83].

to the relations between theologies of creation and divine action and such scientific areas as big bang, inflationary and quantum cosmologies, quantum physics, evolutionary and molecular biology, the neurosciences, anthropology, sociobiology, and behavioral genetics. Surprisingly underrepresented is a discussion on the central kerygma of Christian faith: the Resurrection of Jesus of Nazareth in light of the natural sciences. The Resurrection plays a crucial role in a wide range of theological areas, including the problem of natural evil and theodicy, and, of course, redemption and Christian eschatology. Thus sustained attention to this issue by those in theology and science could have wide-ranging implications for the whole gamut of Christian theology.

There is, of course, a wide diversity of scholarly interpretations of the Resurrection in New Testament research and Christian theology, ranging from those who understand it in purely subjective, psychological, or spiritual terms to those who understand it as an objective event that happened to Jesus of Nazareth after his crucifixion and burial. Scholars who develop the latter view typically link the bodily resurrection of Jesus to the eschatological transformation of the world from its present as creation into a "new creation," including the general resurrection at the "end of time." For the more subjective interpretations, the natural sciences pose little, if any, challenge. For those who defend the bodily resurrection, however, the challenge is obvious and severe: If the predictions of contemporary scientific cosmology come to pass ("freeze" or "fry") then it would seem that the universe will never be transformed into the new creation, that there will never be a general resurrection, and this, in turn, means that Christ has not been raised from the dead, and our hope for resurrection and eternal life is in vain (1 Cor. 15:12–19).[2]

My intention here is *not* to enter into these debates as an advocate of one or another position on the subjective/objective spectrum. Instead it is to approach the problem hypothetically: I will intentionally adopt the "worst case" scenario, the one that makes Christianity the most vulnerable to its atheistic critics, namely the bodily resurrection of Jesus and the eschatologi-

2. The challenge can also be seen as coming from theology to science: If it is in fact true that Jesus rose bodily from the dead, then the general resurrection cannot be impossible. This must in turn mean that the future of the universe will not be what scientific cosmology predicts, since these predictions are based on the universe as we know it, and not on the new creation that God will bring about.

cal transformation of the universe into the new creation. This position is, of course, highly debatable, but my point is that, more than the other interpretations, it raises the most serious, and perhaps unsolvable, conflicts and contradictions with science. Therefore it is worth pursing as a "what if" strategy; in fact, it represents a "test case" of the highest order for those of us who urge that "theology and science" should be in a posture of "creative mutual interaction" and not in one of "conflict."

The essay begins with a brief overview of physical cosmology (17.2). Next, after reviewing the significance of the Resurrection in Christian theology, I discuss the challenge of cosmology to theology and science (17.3). To move the conversation forward, I will suggest an expansion of the current methodology in theology and science and then offer several guidelines, proposals, and steps for reconstructing eschatology in light of science and for exploring interesting topics in physics and cosmology in light of eschatology (17.4). I will then outline, in broad strokes, several possible research programs in theology and in science that might be of interest in light of these guidelines (17.5). This essay is a "first venture" in responding to the challenge of resurrection, eschatology, and scientific cosmology.[3] Hopefully it will provide a useful sketch of some of the underlying issues as well as suggestions for future reflection.

17.2. Physical Cosmology

Physical cosmology[4] has undergone stunning developments this century and particularly in recent decades. For convenience I will group the developments into three stages: the general theory of relativity/big bang cosmology; inflationary/hot big bang models; and quantum gravity/quantum cosmology.

17.2.1. EINSTEIN'S GENERAL THEORY OF RELATIVITY (GR)/BIG BANG COSMOLOGY

In 1915, Albert Einstein published the field equations of GR, which link the curvature of space-time to the stress-energy distributed in space-time: $R_{\mu\nu} - \frac{1}{2} R g_{\mu\nu} = 8\pi T_{\mu\nu}$. In an apt description: "Spacetime tells mass how to

3. Sponsored in part by a grant from the Philadelphia Center for Religion and Science.

4. For nontechnical introductions, see [26, 95], [30] chs. 1–9, [101] chs. 1–9, [17] Appendix 1. For technical introductions, see [66], [61] part 6.

move; mass tells spacetime how to curve" ([61], p. 5). During the 1920s, telescopic observations by Edwin Hubble showed that galaxies surrounding our Milky Way were receding from us at a velocity proportional to their distance as given by Hubble's law,[5] marking the discovery of the expansion of the universe. Three kinds of cosmological models were developed during this period and tested against observational data: models for a flat universe and for an open universe (infinite in size as a whole, endlessly expanding, temperatures falling toward absolute zero: the "freeze" scenario) and for a closed universe (finite in size, eventually recontracting if there is no cosmological constant, with temperatures eventually soaring to infinity: the "fry" scenario). All of these models include an essential singularity which, for convenience, is labeled "$t = 0$," where t is cosmological time. In this sense (but in a way that takes careful philosophical crafting) these big bang models point to what can be called loosely the "origin of the universe."

Their main competitor, the steady state model of Hoyle and colleagues, was effectively abandoned in the 1960s in light of the big bang's explanation of radio source counts, the relative abundances of hydrogen and helium, and the cosmic microwave background radiation.[6] Still, a number of important technical problems remained, including the horizon problem, the matter/antimatter ratio, and the essential singularity, $t = 0$. Theorems by Roger Penrose [73], Stephen Hawking [35, 36, 38], and Robert Geroch [29] in the 1960s proved that cosmological space-times satisfying the Einstein field equations must be singular if certain conditions hold that are likely to be fulfilled in the real universe.[7] The most important of these conditions, apart from the existence of a closed trapped surface (which follows from the existence of the black-body cosmic background radiation) is the following: the mass stress-energy tensor $T_{\mu\nu}$ must obey the inequality $(T_{\mu\nu} - \frac{1}{2}g_{\mu\nu}T)u^{\mu}u^{\nu} \geq 0$ for all unit timelike 4-vectors u. For a fluid matter field, the inequality reduces to the condition that $\rho + 3p \geq 0$ where ρ is the energy density of the fluid and p is its pressure. The standard big bang models satisfy this condi-

5. Hubble's data actually showed a linear relation between the magnitude and red shift of light from distant galaxies; magnitude was then interpreted in terms of distance, and red shift in terms of recessional velocity.

6. For a fascinating historical account, see [53].

7. For an introductory discussion of all four conditions and references, see [61], section 34.6. For a technical discussion see [37], ch. 8.2. For additional references, see [103], pp. 350–54.

tion and are thus characterized by an essential singularity. However one version of the steady state theory (the "almost steady state theory") is still on offer, and claimed to be compatible with observations [39].

17.2.2. INFLATIONARY/HOT BIG BANG MODELS

Inflationary models were originally developed in the 1980s by Alan Guth. They depict the very early universe (circa the Planck time 10^{-43} s) as undergoing exponential expansion rates before settling down to those of the usual big bang scenarios. An inflationary big bang also provided solutions to the horizon, smoothness, and structure formation problems, but not necessarily to the issue of $t = 0$. Inflationary models violate the inequality, $\rho + 3p \geq 0$, thus leaving the question of the existence of an initial singularity open, perhaps even "undecidable" in principle.[8] Andrej Linde's "eternal Chaotic inflation" envisages many expanding universe regions like ours but with different parameters in each one, endlessly replicating and producing similar expanding universe regions, creating an overall fractal-like structure that continues forever [57]. The initial conditions for inflation are set by some preceding quantum gravity era.

17.2.3. QUANTUM GRAVITY/QUANTUM COSMOLOGY

Recent approaches to quantum cosmology include the Hartle/Hawking model [31], the Turok/Hawking instanton, pre–big bang scenarios, brane cosmology, etc. Though these scenarios differ strikingly, in most cases they result in a big bang universe following on from an inflationary epoch. Quantum cosmology, however, is a highly speculative field. Theories involving quantum gravity, which underlie quantum cosmology, are notoriously hard to test, and they further complicate the philosophical issues already associated with quantum mechanics since the domain is now "the universe" as a whole.[9]

8. See the discussion in [52], but see also [5], ch. 6, especially p. 181. There are some theorems proving that singularities remain in inflationary cosmologies, but their various conditions may not be fulfilled.

9. For nontechnical introductions see [30], ch. 10 on; [101], ch. 10 on; [17], Appendices 3 and 4. For more technical introductions, see [43, 44].

17.2.4. THE "BOTTOM LINE" REGARDING THE COSMOLOGICAL FAR FUTURE AND THE POSSIBILITY OF LIFE IN IT

While a detailed understanding of the *early* universe requires clear answers to as yet unsettled theoretical questions in elementary particle physics and grand unification theories (GUTs), we can discuss the *far future* with more confidence, since general relativity, quantum mechanics, and thermodynamics apply there (at least until a closed universe reaches Planck scales again). There is now growing evidence that the matter density in the visible universe is far below the critical density required for a closed universe. Its expansion rate appears to be accelerating, a result that might be accounted for by the existence of a non-zero cosmological constant, L, introduced into the field equations of GR: $\Lambda g_{\mu\nu} + R_{\mu\nu} - \frac{1}{2}Rg_{\mu\nu} = 8\pi T_{\mu\nu}$. As Lawrence Krauss has pointed out ([54], see also [13]), evidence for Λ is now coming from both observational and theoretical grounds. By increasing the total effective energy density, a positive value for Λ could close the universe; but the visible universe can nevertheless be expected to expand forever. However, the case is not closed; it is still possible the universe will recollapse (if the cosmological constant is in fact a variable "quintessence" that dies away in the far future and the spatial sections have positive curvature).

Still, a reasonably well agreed upon account of both closed and open scenarios has been given by Frank Tipler and John Barrow:[10] In five billion years, the sun will become a red giant, engulfing the orbit of the earth and Mars, and eventually becoming a white dwarf. In 40–50 billion years, star formation will have ended in our galaxy. In 10^{12} years, all massive stars will have become neutron stars or black holes.[11] In 10^{19} years, dead stars near the galactic edge will drift off into intergalactic space; stars near the center will collapse together forming a massive black hole. In 10^{20} years, orbits of planets will decay via gravitational radiation. In 10^{31} years, protons and neutrons will decay into positrons, electrons, neutrinos, and photons. In 10^{34} years, dead planets, black dwarfs, and neutron stars will disappear, their mass completely converted into energy, leaving only black holes, electron-positron

10. Barrow and Tipler [7], see also [99] and [102] ch. 14.

11. If the universe is closed, then in 10^{12} years the universe will have reached its maximum size and then will recollapse back to a singularity like the original hot big bang. However, a cosmological constant can make it expand forever.

plasma, and radiation. All carbon-based life forms will inevitably become extinct. Beyond this, solar mass, galactic mass, and finally supercluster-mass black holes will evaporate by Hawking radiation. The upshot is clear, according to Tipler and Barrow:

> Proton decay spells ultimate doom for life based on protons and neutrons, like *Homo sapiens* and all forms of life constructed of atoms. . . . Even if intelligent life were to expand the spatial volume under their control at the speed of light . . . baryon-based life will disappear if the Universe is flat or open, or if it is a sufficiently long-lived closed cosmology. ([7], p. 649)

However, we must be clear that this conclusion applies to the visible region of the universe only. While this is happening, other new bubbles may be coming into existence with new "big bangs" taking place, and some of them could be in a steady state. In the chaotic inflationary scenario, while old universes die out, new ones are born. A large-scale steady state of coming into existence and dying out is a possibility, and indeed is believed by some to be the true state of the universe [57]. As we shall see, the challenge to Christian eschatology is profound.

17.3. Four Theological Movements Converging on "Resurrection, Eschatology, and Cosmology"

Four movements in contemporary theology either move toward or stem from a new and vigorous engagement with the central Christian kerygma, namely the Resurrection of Jesus and its eschatological implications. Each of them has yet to engage carefully with the challenges raised by scientific cosmology.

17.3.1. NEW TESTAMENT SCHOLARSHIP ON THE BODILY RESURRECTION OF JESUS

There is a striking diversity of views in contemporary New Testament (NT) research on the meaning and significance of the resurrection narratives. Still, without undue oversimplification, they can be divided into "subjective" and "objective" treatments.

In the subjective interpretation, the "Resurrection of Jesus" actually refers to the personal experiences of the disciples in which they gained renewed hope and a sense of mission after the tragic death of Jesus. They described their experiences in terms of "appearances" and "the empty tomb" even though the experiences were entirely subjective. The subjective interpretation includes psychological, spiritual, existential, and sociopolitical ways to understand the Resurrection of Jesus, but for all of these, the empty tomb accounts are irrelevant; Jesus' body simply decayed.

In the objective interpretation, the "Resurrection of Jesus" refers to something that actually happened to Jesus after his crucifixion, death, and burial. By a new act of God, Jesus rose from the tomb and appeared to his disciples. Their renewed hope and sense of mission were based on this new reality.

The physical sciences raise few if any serious challenges to the subjective approaches (although even they can be challenged by reductionist appeals to psychology and political science, etc.). The "objective" interpretation emphasizes elements of continuity (e.g., he can be recognized) and of discontinuity (e.g., his resurrection is not a mere resuscitation) between Jesus of Nazareth and the risen Jesus. There are two important approaches to the objective interpretation: the personal resurrection of Jesus and the bodily resurrection of Jesus.

The elements of continuity in the personal resurrection of Jesus include personal identity, namely everything that constitutes the identity of Jesus, but *not necessarily physical or material continuity.* Thus the objective and personal interpretation of the Resurrection of Jesus is compatible with the possibility that his body suffered the same processes of decay that ours will after death; indeed, it may even be seen as necessary, theologically, for his death to be like ours in every way.

The elements of continuity in the bodily resurrection of Jesus include personal identity, and thus necessarily *at least some element of physical or material continuity* between Jesus of Nazareth and the risen Jesus. Thus the objective and bodily interpretation of the Resurrection of Jesus stresses the importance of the empty tomb texts. Obviously, the physical sciences, including cosmology, raise tremendous, perhaps insurmountable, challenges to the intelligibility of the "bodily resurrection" approach to the objective interpretation (although, again, reductionist appeals to psychology, etc., challenge the personal approach as well). Note that this is by no means a "literal" interpretation of the NT texts; I am relying on the dense hermeneutical arguments of dozens of NT scholars across denominational lines and spanning many decades of research. However, this interpretation does take the "physical" aspect of the Resurrection extremely seriously in light of the lengthy history of the discussion of that term.

As suggested above, my intention is *not* to enter into the NT debates as an advocate of one or another position suggested above. Instead it is to adopt one position *as a test* case and to tease out its implications for "theology and science." In essence, I will assume as a hypothesis what I consider to be the "worst case" scenario, the one that makes Christianity the most vulnerable to its atheistic critics, namely the bodily, objective resurrection of Jesus. The bodily resurrection of Jesus leads directly to an eschatology in which God, who raised Jesus from the dead, will *transform* the present creation into the "new creation," bringing about a "new heaven and new earth" in continuity with, though radically distinct from, the present creation. The empty tomb in particular, and the "bodily" character of the Resurrection in general, mean we must face squarely the challenge to their intelligibility from science in ways that the subjectivist interpretation and the personal form of the objectivist interpretation sidestep. Similarly, if "the creation" signifies the universe as understood by scientific cosmology, then its transformation into a "new creation" must mean a transformation of this *universe.* But here we run squarely into the challenge raised by contemporary scientific cosmology, with its "freeze or fry" scenarios and, in either case, the apparent inhospitability to life of the cosmological future. Few if any NT scholars who adopt the bodily form of the objective interpretation, however, spell out their response to these challenges. It thus becomes critically important that we face this challenge directly and discover whether there is any reasonable way to respond to it.

17.3.2. THE PROBLEM OF NATURAL EVIL AND THEODICY

The problem of evil is particularly acute in light of the atrocities of the twentieth century. Human-inflicted evil, or "moral evil," leads directly to the challenge of theodicy: Why does God not act to stop such atrocities? It underlies and contributes to the rise of atheism as a modern phenomenon. In response, a growing number of systematic theologians have developed a revised doctrine of God through a "recovery of the Trinity" (Catherine Mowry LaCugna) and to a "theology of the crucified God" (Jürgen Moltmann). Here the basic insight is that the power of God to overcome evil is not the power of a tyrant since that would violate God's gift to us of free will, and thus of entering freely into a loving relationship with God. Instead, God's power is revealed as the power of compassion most clearly found in the self-emptying (kenotic) love of the suffering servant, Jesus.

But when we consider the evolution of life on earth over its 3.8 billion year history, and thus massive suffering, disease, death, and extinction in nature, the problem of "moral evil" leads to the problem of "natural evil": Why does God allow such natural misery when, apparently, the operation of free will is not at issue? In my opinion, it is an open question whether the "theology of the crucified God" can adequately respond to the problem of natural evil. And if it can, it in turn requires a robust theology of the Resurrection and eschatology, and thus in turn leads us directly to the challenge of contemporary cosmology.

17.3.3. CHRISTIAN ESCHATOLOGY

The nineteenth- and twentieth-century discovery of the essentially eschatological dimension of the NT traditions has affected every area of contemporary theology. For most theologians today, Christianity cannot be Christianity without being eschatological: without the reign of God being both here/now and not-yet/coming, without hope being ultimately grounded not in a flight from this world but in God's new act that will transform all that is into the new creation already begun in the Resurrection of Jesus. The discovery of the eschatological dimension of Christian theology is bound up with a remarkable "retrieval" of the doctrine of the Trinity spanning Roman Catholic and Protestant theologians. For trinitarian theologians eternity is not timeless (e.g., Augustine) nor is it unending time (i.e., chronological/physical time); rather it is "supra-temporal" (e.g., Barth). It is the fully temporal source and goal of time, the "future of the future" (e.g., Moltmann), *adventus* and not merely *futurum* (e.g., Peters). Eschatology involves God acting from eternity to reach back into time (e.g., Pannenberg's "prolepsis") and redeem the world through the life, ministry, death, and Resurrection of Jesus.

But these approaches have not been brought into dialogue with twentieth-century physics and cosmology. How can Christian eschatology and the relation of time and eternity be credible in light of physical cosmology given the scenarios we have seen about the physical future of the universe?

17.3.4. PROGRESS AND PROBLEMS IN "THEOLOGY AND SCIENCE"

The fourth movement is actually the growing field of "theology and science" itself. Against competing claims that theology and science are necessarily in outright conflict or that they are two separate worlds, there has been enormous progress in the constructive dialogue between theology and science.

17.3.4.1. Progress

Advances made in "theology and science" over the past four decades have been possible because of a remarkable new methodology that arose early in this period. Here I am drawing directly on the pioneering writings of Ian Barbour [3] as well as on those of Arthur Peacocke [72],[12] Nancey Murphy [63, 64, 65],[13] Philip Clayton [12], George Ellis [21, 65], John Polkinghorne [76], and many others, each of whom has contributed to our growing understanding of the methodology, broadly referred to as "critical realism." Of its three major arguments, I will summarize two here.[14]

In emergent epistemology, the disciplines form a hierarchy starting with physics and extending up through the natural and social sciences and the humanities, including ethics and theology. Theories in lower levels, such as physics and biology, place constraints on those in upper levels, such as the neurosciences, psychology, and theology (against "two worlds" views that keep the fields totally isolated). At the same time, the presence of genuinely new processes and properties studied at higher levels prevents them from being fully reducible to lower levels.

With analogous methods, theological methodology of theory construction and testing within paradigms is analogous to scientific methodology (though with several important differences).

The result is that there are at least five ways or "paths" by which the natural sciences can affect constructive theology (see fig. 17.1; ignore paths 6–8 for now). I will focus on physics and cosmology for specificity and use the notation "SRP → TRP" suggested by George Ellis to mean that scientific research programs (SRPs) *may* influence theological research programs (TRPs). In the first four, theories in physics, including the key empirical data they interpret, can act as *data for theology* both in a direct sense (paths 1 and 2) and indirectly through philosophical analysis (paths 3 and 4).

12. See particularly fig. 3, p. 217, and the accompanying text.

13. While Murphy has contributed to development of a methodology for relating theology and science, she is highly critical of "critical realism." See [64], ch. 2.

14. The third argument is that language is intrinsically metaphorical. Critical realists also defend a referential theory of truth warranted in terms of correspondence, coherence, and utility.

FIGURE 17.1. METHOD OF CREATIVE MUTUAL INTERACTION

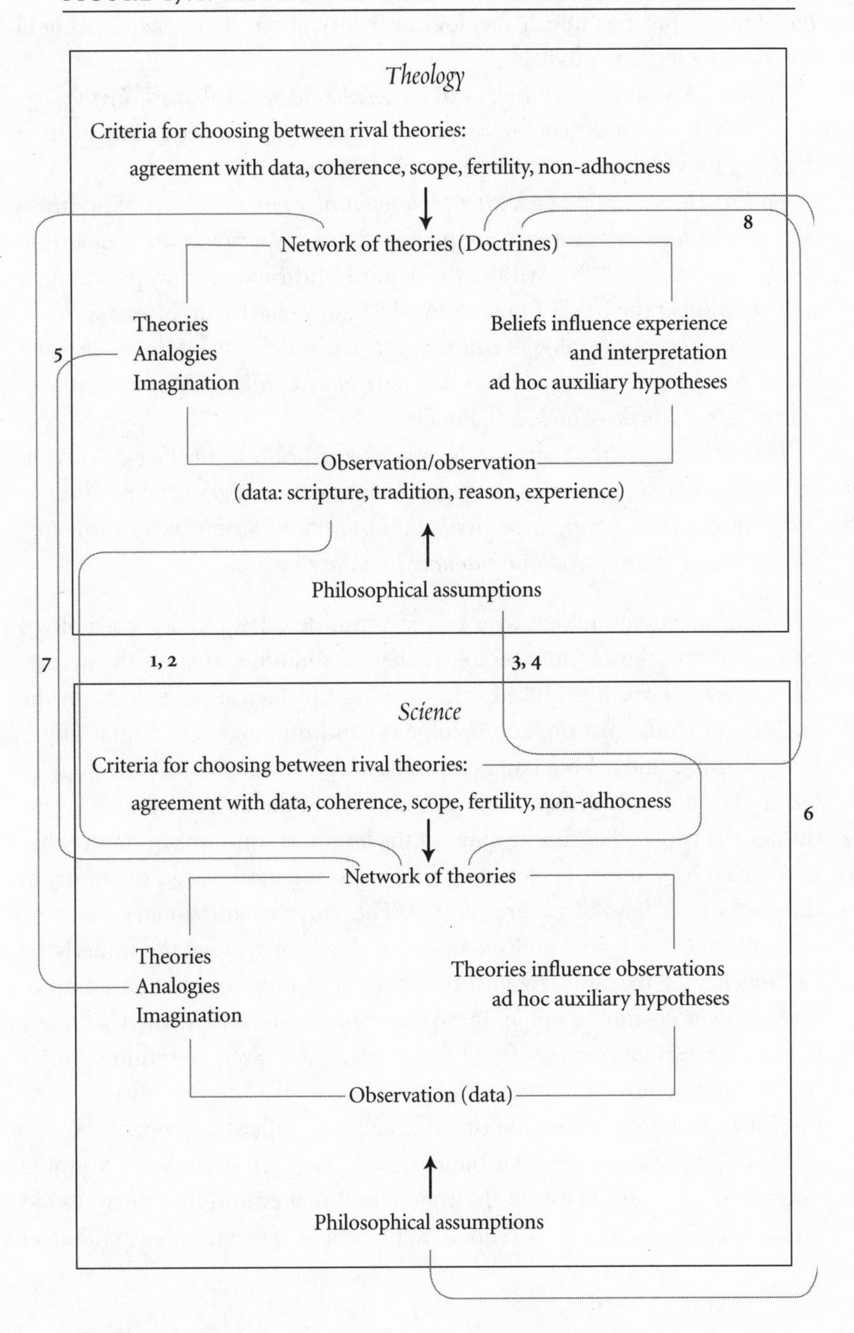

Path 1: Theories in physics can act directly as data that place constraints on theology. So, for example, a theological theory about divine action should not violate special relativity.

Path 2: Theories can act directly as data either to be "explained" by theology or as the basis for a theological constructive argument. The equation $t = 0$ in big bang provides an interesting example (see below).

Path 3: Theories in physics, after philosophical analysis, can act indirectly as data for theology. For example, an indeterministic interpretation of quantum mechanics can function within theological anthropology by providing a precondition at the level of physics for the bodily enactment of free will.

Path 4: Theories in physics can also act indirectly as the data for theology when they are incorporated into a fully articulated philosophy of nature (e.g., that of Alfred North Whitehead). Finally,

Path 5: Theories in physics can function heuristically in the theological context of discovery by providing conceptual inspiration, experiential inspiration, practical/moral inspiration, or aesthetic inspiration. So big bang cosmology may inspire a sense of God's immanence in nature.

Given this methodology, over the past four decades, physics, cosmology, evolutionary and molecular biology, genetics, and other areas of the natural sciences have been introduced into ongoing theological reflections on the doctrine of God, creation, and theological anthropology. The initial singularity within standard big bang cosmology represents what the theology of creation *ex nihilo* describes as the contingency of the universe [87]. The anthropic principle (the "fine-tuning" of the laws and constants of nature) has been taken by some as evidence of divine cosmological design operating at the level of the laws of nature [21, 65]. The move to inflationary big bang cosmology leaves $t = 0$ "undecidable" but does not remove the underlying contingency of the universe and thus the more philosophical meaning of "creation" in *creatio ex nihilo*, though certain versions of it offer a "many worlds" strategy against the "fine-tuning" argument. Even quantum cosmology in conjunction with "eternal inflation" leaves the "mega-universe" contingent and opens the question of its existence to a theistic response.

Evolutionary and molecular biology have also been discussed within the doctrine of creation. The essential argument is that evolution is the way God creates life. God is both transcendent to the whole of creation *(ex nihilo)* and immanent in, with, under, and through all its processes as the continuous Creator *(creatio continua)*. The rise of complex chemical and biological organization, of life, and of mind, are seen as emergent new phenomena within the continuous context of natural processes and in turn as due to the ongoing activity of the continuously creating God.

These discussions have led to more focused and sustained attention on the meaning of divine action and the laws of nature. The "inherited" assumption that these laws are entirely deterministic, as represented by classical physics, led many theologians in the first half of the twentieth century into a "forced option" between objective, interventionist divine action and subjective, noninterventionist divine action, dividing conservatives and liberals. Many scholars in "theology and science" have argued that twentieth-century science has changed our view of nature as a closed mechanism and portrayed, instead, genuine openness in nature (ontological indeterminism). This in turn would provide a new way to speak about God acting objectively in particular events in the world without relying on interventionist explanations of such action.[15] Three approaches to "noninterventionist objectively special divine action" have been pursued fruitfully: a "top-down" approach, drawing on either big bang cosmology to view God as acting on the "world-as-a-whole,"[16] or on the neurosciences/cognitive sciences to view God's action on the person;[17] a search for ontological indeterminism in the behavior of chaotic systems;[18] and a "bottom-up" approach, drawing on quantum physics in support of indeterminism.[19] Varieties within these include the use of panentheistic, neo-Thomistic[20] and Whiteheadian[21] arguments.

17.3.4.2. Problems

Ironically, progress in "theology and science" has been accompanied by an increasing number of conceptual problems. With "consonance" between the finitude of the past in big bang cosmology and the doctrine of creation

15. See the Vatican Observatory/CTNS series of articles [84–88].
16. See the articles by Peacocke in this series.
17. See the articles by Clayton and Murphy in this series.
18. See the articles by Polkinghorne in this series.
19. See the articles by Murphy, Ellis, Tracy, and Russell in this series.
20. See the articles by Stoeger and Happel in this series.
21. See the articles by Barbour, Birch and Haught in this series.

ex nihilo has come "dissonance" between the fact that this same big bang cosmology, at least in the "open" scenarios, describes the universe as a whole (to be distinguished from the observable universe) as infinite in size at all times $t > 0$ (even while it is expanding in size!) and predicts that the universe will have an infinite future. However, these infinities do not overturn the underlying contingency of the universe (why should it exist at all?).

More serious has been the challenge raised by the problem of evil, and thus theodicy, which has accompanied the progress made on noninterventionist accounts of objective divine action. We must provide a "cast-iron reason" for why a merciful and loving God does not alleviate the massive evil in the world if divine action is to make sense today, and this is particularly urgent in light of the suffering of the natural world [22]. An important response has been to draw on the "crucified God" argument, as mentioned above. Yet this response, in turn, together with the "bodily" interpretations of the Resurrection of Jesus, which both the "crucified God" theology and much of NT scholarship support, lead directly to the challenge to deliver a robust eschatology and the fundamental problem of relating eschatology to physical cosmology.

The challenge from science might be alleviated significantly by assuming any one of the many different approaches scholars take to the Resurrection. Arthur Peacocke works with what I am calling the personal approach to the objective interpretation of the Resurrection while using an epistemic emergence argument against reductive accounts of the experience of the disciples. This approach leads him to an eschatology that is directed "beyond space and time" to *theosis,* our participation in the life of God and the beatific vision. In this way, he creatively uses his philosophical analysis of science in support of epistemic emergence while engaging the objective, personal resurrection tradition, thus avoiding the severity of the challenge from science that the bodily approach faces.[22] My intention here, however, is to assume as a working hypothesis the bodily resurrection since, again, it poses the "worst case" scenario for a fruitful interaction with science in the hope that, by taking it seriously though hypothetically, this scenario might lead to new, more positive, insights.

Along with these conceptual challenges, the problem we face is com-

22. Peacocke [72], 279–88, 332. See also Peacocke's earlier publications, ch. 8, 22.

pounded by the very methodology that has played an essential role in making it possible for the field of "theology and science" to grow so richly over the past four decades. By its very structure, this methodology prevents us from "sidestepping" the crucial issues raised by cosmology for Christian eschatology. Since it insists on constraints by lower on higher levels, theology may not simply ignore the results of physics: the predictions of "freeze or fry"—or their scientific replacements in the future—must be allowed to challenge what theology can claim eschatologically. Thus we cannot just ignore the negative predictions of scientific cosmology while still playing by the methodological rules that define the field of "theology and science." Nor can we hope that an easy appeal to contingency, quantum physics, chaos theory, ontological unpredictability, Whiteheadian novelty, emergence, the future, or metaphysics will be sufficient to solve this problem.

17.3.5. ESCHATOLOGY IN THE LIGHT OF THE NATURAL SCIENCES

There have been a few scattered attempts at discussing the Resurrection of Jesus and its eschatological implications in light of the natural sciences, but, as Bill Stoeger has stressed, these have mostly been relatively unpromising ([19], pp. 19–20). They can be quickly summarized.

17.3.5.1. Physical Eschatology

Freeman Dyson [18, 19, 20] and later, Frank Tipler and John Barrow [7], have argued that eschatology can be fully reduced to the domain of science. In the resulting "physical eschatology," Dyson claims that life can continue forever in the open universe even with the inevitably decreasing cosmological temperature and the decay of stars to holes, and so on. The key assumption here is that life is merely information processing and that one can imagine other kinds of processors than those of biochemistry. Tipler and Barrow make a similar claim for life in a closed universe; here the unending character of life is achieved subjectively by an exponential increase in the rate of information processing consistent with the finite future of the universe [100]. In the almost steady state universe [39], the problem would not arise: the universe is always hospitable to life.

This approach reflects one aspect of the methodology proposed above: science places crucial constraints on theology. It opens fascinating avenues worth further exploration, including the physical/informational dimension of life and the possibilities for life to continue into the far future of both open and closed universes, and conceivably to "terraform" the universe and even to affect some of its global topological features. It does not, however, place these avenues in relation to the other aspect of that methodology: theology describes irreducible processes and properties that must be accounted for in more than scientific language, and thus it cannot be of serious help here.[23]

17.3.5.2. Emergence and Noninterventionist Divine Action

Can emergence and noninterventionist approaches to divine action, which have been highly successful so far, be used fruitfully in discussing the question of eschatology and cosmology? As we saw in the case of Peacocke's work, they clearly help in developing a complex and promising approach if one does not adopt the "bodily resurrection" interpretation. My guess is that, for the purposes of this exploratory essay, which does work with this interpretation, their help will be marginal at best, since they presuppose that such divine action does not change the basic laws of nature but operates in, with, and through them. The eschatological claim pursued here, however, is that God will act anew to transform the universe radically; it in turn requires a more complex strategy than emergence and noninterventionism. One could say that the Resurrection of Jesus is "more than a miracle" because nature itself is changed by God's new act; it is not just a unique and extraordinary event within an ordinary background of fully natural prior and subsequent events. Conversely, noninterventionist divine action is in a sense "less than a miracle" since the continuous activity of God as creator is consistent with the processes of nature, which science describes via the laws of nature, particularly when those processes are intrinsically indeterministic and thus open to divine agency.

17.3.5.3. Process Theology

Process theology is widely seen as compatible with science. Can it provide resources for the problem of eschatology and cosmology? Ian Barbour defends both objective and subjective immortality in a Whiteheadian perspective, but the bodily resurrection of Jesus and the eschatological view of the

23. Tipler's version [100] has been criticized on the basis of its scientific merit alone *inter alia* by Stoeger and Ellis, and has been criticized even more strongly for its overall approach [99].

future are not carefully discussed.[24] John Haught appeals to such metaphysical categories as novelty and the future, proleptic power of God but does not account adequately for the challenge from scientific cosmology [33, 34]. Other important examples include the writings of Marjorie Suchocki and Lewis Ford, although once again their approaches do not incur the challenges that sciences pose for the approach taken by this essay.

17.3.5.4. Creation *Ex Vetere*

John Polkinghorne offers what I consider to be the most promising approach: *creatio ex vetera,* which he defines as the transformation of the universe created by God *ex nihilo.* The universe of the past 15 billion years is God's creation *ex nihilo.* God will create the world to come *ex vetera* through a transformation of the universe, beginning with the Resurrection of Jesus. The new creation will be "a totally sacramental world, suffused with the divine presence . . . (and) free from suffering. . . ."[25] Since there is both continuity and discontinuity in the transformation of the present universe, science "may have something to contribute" to our understanding of this transformation. In particular, the continuities might lie within the province of science, or more precisely, what Polkinghorne calls "metascience . . . the distillation of certain general ideas from the scientific exploration of many particulars." These include the significance of relationality and holism; the concept of information of a "pattern-forming kind" in addition to the familiar ideas of matter and energy; mathematics; and a dynamic view of reality as "open becoming." I will use Polkinghorne's proposal as the point of departure for my own constructive approach below.

17.3.6. THE CENTRAL PROBLEM

These four movements, and others could easily be cited as well, are like vectors pointing toward a central problem. At "ground zero" is the fundamental challenge raised by contemporary scientific cosmology to the intelligibility of the bodily resurrection of Jesus and its eschatological implications regarding God's transformation of the creation into the "new creation." It must be stressed that there are other versions of the resurrection and Chris-

24. Barbour [3], p. 241. See note 37, p. 288, for references to other process thinkers on this and related subjects.

25. Polkinghorne [76], 162–70. See also [77] and [17], ch. 4.

tian eschatology that are compatible with the scientific prognosis for the future of the cosmos, but here I have taken as a *working* hypothesis the "hardest case," the bodily resurrection and the transformation of the universe by a new act of God, in order to see whether we can, after all, make some minimal progress on what appears to be an almost insurmountable problem. I will conclude this section by quoting two theologians who stress how severe the challenge really is. John Macquarrie writes: "[I]f it were shown that the universe is indeed headed for an all-enveloping death, then this might seem to constitute a state of affairs so negative that it might be held to falsify Christian faith and abolish Christian hope" [60]. According to Ted Peters, "Should the final future as forecasted by the combination of big bang cosmology and the second law of thermodynamics come to pass . . . then we would have proof that our faith has been in vain. It would turn out to be that there is no God, at least not the God in whom followers of Jesus have put their faith" [74].

17.4. New Research in Scientific Cosmology and in Eschatology: Method and Guidelines

I see that you believe these things are true because I say them. Yet, you do not see how. Thus, though believed, their truth is hidden from you.

—Dante Alighieri [1]

If it is impossible, it cannot be true. But if it is true, it cannot be impossible.

17.4.1. AN EXPANDED METHODOLOGY FOR "THEOLOGY AND SCIENCE"

Given the challenges I have described, I believe the first step forward will be to expand the methodology used to date in "theology and science" such that theological research programs (TRPs) are not only influenced in various ways by science (SRP → TRP), but that they, in turn, *might* play a fruitful role in suggesting possible new directions of a very general nature to scientists working in scientific research programs. For convenience, I will denote this: TRP → SRP. This expansion also responds to a continuing challenge to the constructive conversations between theology and science: Can theology and science be genuinely *inter*active, each offering something of intellectual value to the other, or is the only role for theology that of critically integrating the results of science into its own conceptual sphere?

Here I will suggest at least three paths by which theology can influence scientists as they pursue scientific research. (Again, I will limit the discussion to physics and cosmology.) First, though, I want to stress that by "influence" I am in no way appealing to, or assuming that, theologians speak with some special kind of "authority," whether based on the Bible, church dogma, magisterial pronouncements, or whatever. Quite the contrary; the overall context should be an open intellectual exchange between scholars based on mutual respect and the fallibility of hypotheses proposed by either side. I will number them paths 6–8, since they are additions to the five paths discussed above.

Path 6: Relates to the philosophical framework in which the discussion takes place: Theology can provide some of the philosophical assumptions *underlying the natural sciences.* Historians and philosophers of science have shown in detail how the doctrine of creation *ex nihilo* played an important role in the rise of modern science and its view of nature by combining the Greek assumption of the rationality of the world with the theological assumption shared by Jews, Christians, and Muslims that the world is contingent. Together these assumptions helped give birth to the empirical method and the use of mathematics to represent natural processes.[26] Today the scientific method is well established and has proven of immeasurable value; it should *not* be abandoned whatever changes one might consider in the underlying philosophical framework. At the same time, the *ex nihilo* tradition carried with it other concepts about nature that were not carried over into the scientific conception of nature, including goodness and purpose. It would be interesting to reopen the question of the value of these concepts for contemporary science *without* challenging scientific methodology (see [2, 107, 8, 4]).

Actually, one frequently finds an attempt to move in the opposite direction—from science to metaphysics—in the writings of some scientists, and with this some confusion as to the precise extent and nature of the philosophical framework that they claim is implied or required by science. For example, some scientists argue that science specifically entails or requires atheistic materialism. In this case, theology and philosophy can help to sort through the confusion and assess precisely which parts of a metaphysical

26. To view nature as created *ex nihilo* implies that God freely created the universe without relying on any ontologically prior substance (very roughly, "matter") or ideal forms (again, roughly "laws of nature") as in the Platonic view. See [28, 51, 56, 16, 50].

framework are indeed necessarily implied by the science, such as contingency and rationality, and which are not, such as atheism *or* theism.[27]

Paths 7 and 8 relate theology to the "context of discovery" in science.

Path 7: Theological theories can act as sources of inspiration *in the construction of new scientific theories.* They can also suggest *reasons* for pursuing a particular approach during theory construction. For example, one can start with a theological theory and then delineate what conditions must obtain within physics for the possibility of its intelligibility. These conditions in turn can serve as motivation for an individual research scientist or group of colleagues to pursue the construction of a particular scientific theory. An interesting example can be found in theologies and philosophies that, to varying degrees, apparently influenced many of the pioneers of quantum theory, including Vedanta for Schrödinger, Spinoza for Einstein, and Kierkegaard for Bohr. Another example is the subtle influence of atheism on Hoyle's search for a "steady state" cosmology.[28]

Path 8: Theological theories can lead to "selection rules" within the criteria of theory choice,[29] *that is, for choosing between existing scientific theories that all explain the available data, or for deciding what set of data the theory should seek to explain.*[30] Interestingly, Hoyle's program could actually be considered a

27. George Ellis has written extensively on this point [23], criticizing the way scientists such as Carl Sagan, Richard Dawkins, E. O. Wilson, Jacques Monod, and Peter Atkins have attempted to co-opt science for a materialist agenda.

28. For an extremely careful account of the extra-scientific factors at play in cosmological debates see Kragh [53].

29. The way I am using the term "selection rules" is closely related to what Nicholas Wolterstorff means by the term "control beliefs." These are philosophical beliefs that function within the practice of science, "determining the rejection of a variety of theories and requiring alternative lines of thought for its positive elaboration" [110], p. 19. In a similar way, though *not* drawing on theology, John Barrow uses the anthropic principle, not as an argument for design, but as a way of allowing biology to place constraints or "selection rules" on physics (i.e., conditions that are required if the evolution of life is to be possible), and these constraints *in biology* lead Barrow [6] to discover new explanations of hitherto disparate phenomena *in physics.*

30. George Ellis (private communication) has stressed the latter point. At a basic level, one must determine which data are to be accepted as reliable, and which rejected as unreliable. At a more general level, one must decide what sort of evidence might be relevant to a scientific explanation, and thus what is the nature and scope of a scientific theory. His example is the question of morality: If it is considered as a valid phenomenon, does this have anything to say about the nature of the universe? In particular, since the meaning of morality demands at least some measure

combination of paths 7 and 8. One could say that Hoyle was motivated by his atheism to reject $t = 0$ and to view the universe as eternal (path 7), but none of the existing theories of gravity satisfied his "selection rules" (path 8), and so he constructed a new theory of gravity and, using it, arrived at his steady state cosmology (path 7).[31] However, theological theories cannot be allowed to influence the process of testing the theory by the scientific community.

One could say, regarding both 7 and 8, that while theological theories might have some influence in the "context of discovery," they must *not* have any influence in the "context of justification."[32]

The eight paths between science and theology are summarized in figure 17.1. The *asymmetry* between theology and science should now be quite apparent: Theological theories do *not* act as data for science in the same way that scientific theories do for theology. This reflects the methodological assumption that the academic disciplines are structured in an epistemic hierarchy of constraints and irreducibility. It also safeguards science from any normative claims by theology. It does, though, allow for the possibility that philosophical or theological commitments can stimulate the search for new theories and can function as a source of "criteria of theory choice" among existing competing theories in the natural and social sciences. That this has happened historically is well known; that it can and is happening in contemporary scientific research is less generally recognized.[33] Together these eight paths portray science and theology in a much more interactive, though still asymmetric, mode, which I call "the method of creative mutual interaction." In particular, the theologian must first face the challenge of science to his or her cognitive claims; yet the scientist may find that philosophical elements pervade his or her work in creative ways and stem, in turn, from implicit

---

of free will, then the conditions for the possibility of free will could be explored scientifically. Ellis's suggestion is related to the claim in recent philosophy of science that "all facts are theory-laden," that is, that theories influence observation. For an extensive discussion see [65].

31. Hoyle and Narlikar [40, 41], Hoyle, Burbidge, and Narlikar [39]. A helpful overview of competing gravitational theories and their experimental tests is given in Will [109].

32. One might view the overall structure of paths 6–8 through the following diagram:

Theology ⇢ Philosophy → Hypothesis → Theory → Tests vis. data

Science (spanning Hypothesis → Theory → Tests vis. data)

where the "dotted arrow" from theology to philosophy suggests the indirectness of the influence of theology on the philosophical elements that eventually come into science.

33. See references in Russell [82], [83] part 3/D.

theological positions. Neither partner in the interaction assumes a literal reading of their theories or an unqualified authority in the mutual search for understanding; both partners expect to gain from the interaction while pursuing their own specialities.

17.4.2. GUIDELINES FOR MOVING FORWARD: REVISING ESCHATOLOGY IN LIGHT OF COSMOLOGY (SRT → TRP), AND COSMOLOGY IN LIGHT OF ESCHATOLOGY (TRP → SRP)

Given the expanded methodology described above we are prepared to engage in a twofold project:

SRP → TRP: Following paths 1–5, we construct a more nuanced understanding of eschatology in light of physics and cosmology, *and*

TRP → SRP: following paths 6–8, we begin a process of searching for a fresh interpretation of, or possibly revisions of, current scientific cosmology in light of this eschatology and its philosophical implications.

If such a project is at all successful, it might eventually be possible to bring these two trajectories together at least in a very preliminary way to give a more coherent overall view than is now possible of the history and destiny of the universe in light of the Resurrection of Jesus and its eschatological completion in the parousia.

This project is clearly a long-term undertaking, requiring the participation of scholars from a variety of fields in the sciences, philosophy, and theology. How do we then proceed? My sense is that we first need some guidelines that will help point us in a fruitful direction. The analogy I have in mind is the following: We are looking for the proverbial "needle in the haystack" (revised cosmology and eschatology) but we face, instead, a vast field of haystacks (each a different set of possible alterations). It would be useless just to start with the first one at hand. We should find a way to make an "educated guess" as to which haystack(s) *might* contain the needle, and then we can start the lengthy search through them. We need guidelines to orient our search in a promising direction. Using them, we can begin to explore specific ways to enter into the research.

17.4.2.1. SRP → TRP: Guidelines for Constructive Theology in Light of Contemporary Science

We begin with guidelines for constructive theology in light of contemporary science. Our first five guidelines deal with overall philosophical and methodological issues. Note that these guidelines, as well as G6–G7, apply to research in theology; they do *not* apply to research in science (TRP → SRP below).

Guideline 1: Consider the possibilities arising from a rejection of two philosophical assumptions that we bring to science: a) the future will be "just like" the past and b) the "same laws of nature" that govern the past and present will govern the future.

The first guideline deals with the fundamental challenge physical cosmology poses to eschatology: If the predictions of contemporary scientific cosmology come to pass ("freeze" or "fry") then the parousia will not just be "delayed," it will never happen in or to this universe. And if this is so, then the logic of Paul in 1 Cor. 15 is then inexorable: If there will never be a general resurrection, then Christ has not been raised from the dead, and our hope is in vain. The challenge can also be seen as coming from theology to science: If it is in fact true that Jesus rose bodily from the dead, then the general resurrection cannot be impossible. This must in turn mean that the future of the universe will not be what scientific cosmology predicts, since these predictions are based on the universe as we know it, and not on the new creation that God will bring about. We then seem to be at loggerheads with big bang cosmology. How are we to resolve this fundamental challenge?[34]

My response is that the challenge is not technically from science but from the *philosophical assumptions* that we routinely bring to science. It is not that the predictions of scientific cosmology are wrong in themselves; they do not represent a calculational error or a fault in the underlying scientific theory.

34. *Alternative Guideline 1: Consider the possibilities arising from multiple universes and/or transcendent space-times.* One might argue that while the parousia/"new creation" never happens in this universe, it could take place in some other universe in a plenitude of universes. The latter proposal is consistent with the ideas put forward by Rees in this volume, and fits in with the idea of a "new creation" in the following sense: In the chaotic inflation scenario, new universes, or more accurately new expanding universe regions, are indeed being created all the time. It could also take place in some way in a transcendent space such as a five-, ten-, or eleven-dimensional space-time within which our four-dimensional space-time is imbedded, as in the Randal-Sundrum braneworld cosmologies. Although this is an attractive possibility, the theological meaning of "new creation" should not be reduced completely to a strictly natural unfolding of new universes or their regions if we are to avoid what Polkinghorne calls "evolutionary optimism" or "physical eschatology." On the one hand, these new universes would presumably be part of what we mean by God's creation, and the laws (or meta-laws) that govern them a reflection of God's continuing

What is really at stake is the philosophical assumption that the cosmological future will necessarily be what science predicts it to be. This assumption involves two arguments: First, at its most *fundamental* level, the future will be "just like" the past (what I call "the argument from analogy"). Second, the "same laws of nature" that govern the past and present will govern the future as well (I call this "nomological universality").[35] This is the key step underpinning our use of science to predict the future. There is no scientific justification for querying it—indeed it underlies our standard scientific viewpoint at a fundamental level since it is a philosophical, and not a scientific, claim.

It is quite possible, however, to accept a very different assumption about the future predictions of science while accepting all that science describes and explains about the past history of the universe. The crucial step is deciding whether the laws of nature are descriptive or prescriptive and, as Bill Stoeger argues, science, alone, cannot settle the matter [97]. One is then free to make a strong case on *philosophical* grounds that the laws of nature are descriptive. Finally on *theological* grounds, one can claim that the processes of nature that science describes through mathematical laws are in fact the result of God's ongoing action as Creator, their regularity being the result of God's trustworthiness, holding the laws of physics in existence in their present form. But God is free to act in radically new ways, not only in human history but in the ongoing history of the universe, God's creation. The Resurrection of Jesus points to a radically new kind of action by God, which cannot be re-

action then even as they are now. Since it is the creation that will become "new," these universes/regions may be a venue or an element of what will become radically new in the "new creation" (as discussed in terms of elements of continuity at length below). On the other hand, the theological concept of "new creation" entails a transformation of at least some of the laws of nature themselves, as will be argued in detail below, since the "new creation" reflects an extraordinary, new act of God not predictable or entailed by God's present action/laws of nature. (Ironically, Hoyle, too, often referred to his steady state theory as "continuous creation," but again this is a term whose theological meaning should not be equated with, or reduced to, the spontaneous creation of matter in an exponentially expanding space-time.) Since we are pursuing the "worst case" scenario as a "test case" in this essay, and since our approach is a hypothesis to be tested (and in no way a dogma to be defended!), let us set aside the alternative guideline and pursue the stronger idea of a transformation of this universe into the new creation by an extraordinary act of God.

35. Clearly the terms "just like" and "same laws of nature" need to be finessed since our working theories (e.g., GR, quantum gravity, etc.) will change. Still, we routinely expect that they approximate the actual laws of nature, and these we assume do not change.

duced to, or explained by, the laws of nature, that is, God's previous action.

Alternatively, we could start with the fact that scientific laws always include a *ceteris paribus* clause: they hold "all else being equal." But if God's regular action is the underlying cause of the regularities of nature that we describe by the laws of nature, and if God chooses to act in radically new ways, than of course "all else is *not* equal."[36] We could say that the "freeze" or "fry" predictions for the cosmological future might have been applicable had God not acted in Easter and if God were not to continue to act to bring forth the ongoing eschatological transformation of the universe.

Guideline 1 also places a crucial limitation on nomological universality. The limitation is that, though they are clearly applicable to the universe as a whole in its "past and present,"[37] the laws of nature as we now know them cannot be literally applied to the future.

In sum, guideline 1 is a general argument about the inapplicability of the present laws of nature to the new creation. The particular theological approach taken below (guideline 6) will be that eschatology implies the *transformation* of the universe into the new creation. From this a more specific set of claims about the laws of nature will emerge. For example, since the new creation is the result of the permanent transformation of the present creation, we will rule out its being merely an *intervention* by God in which God temporarily suspends the laws of physics, but following which the laws revert to their previous nature. We will also rule out the *emergence* view that the laws are unchanged but are operating in a new domain where they have not operated before, such as the evolution of life on earth. Neither of these is adequate to the radical challenge to nomological universality entailed here regarding the future of the universe, although the emergence and *noninterventionist* views apply very well to the past history of the universe.

Before moving to guideline 2, an important objection to what has just been said should be considered carefully. It might be objected that what started as an attempt at modifying TRPs has turned into an attempt at modifying SRPs—and doing so in a highly contentious way. The proposal here is

36. I am grateful to Nancey Murphy for suggesting the inclusion of this point here (private communication).

37. Note that throughout this paper I am using the term "present" without discussing the challenge from special relativity. I return to that problem in future work. For now, one response (frequently given) is to use observers floating with the cosmological expansion to define an appropriate hypersurface of simultaneity.

that science as we know it will not hold forever in the future—which is a major claim about the applicability of science. The proposal cannot be disproved, and it cannot be proved. Some scientists will undoubtedly reject it, not just because of the lack of very strong evidence in its favor, but because it seems to undermine one of the major cornerstones of scientific belief and practice. My response is threefold: First it is an "article of faith" for scientists to accept that scientific predictions hold for the future as much as it is for religious people to accept that scientific predictions do, *all else being equal,* but that they need not *necessarily* hold for the future. All of us are faced with taking a position on this issue, and neither position can claim that their claim can be "proven" in any simplistic sense. Second, guideline 1 is a *theological* guideline, relevant to how one constructs theology, that is to TRPs; it says *nothing* about how one pursues science, and in no way involves a modification of SRPs. Indeed, science *must* involve predictions and empirical tests; this is in *no way* challenged here. Finally, it does not follow that theologians can ignore the discoveries science has made about the history of the universe or its scientific explanation. The proposal is strictly limited to statements about what has not yet happened: the future.

In any case, the purpose of this essay is not to limit the ways eschatology can be consistent with science, since in fact there are ways it can be so. It is to explore whether a particular construal of eschatology, based on the bodily resurrection of Jesus, can be placed in a creative dialogue with science even when this seems impossible.

Guideline 2: Eschatology should be "scientific" (i.e., it should embrace methodological naturalism) in its description of the cosmic past and present: the formal argument. Any eschatology that we might construct must be "scientific" in its description of the *past* history of the universe. More precisely, it must be constrained by methodological naturalism in its description of the past: it should not invoke God in its explanation of the (secondary) causes, processes, and properties of nature.[38]

This guideline must be nuanced in several ways. The notion of "transformation" by God's new action should include a number of domains and

38. It is crucial to note that the commitment to methodological naturalism does not carry any necessary ontological implications about the existence/nonexistence of God (i.e., it is not "inherently atheistic").

times. For example, God has already acted in the Resurrection of Jesus and in the period of the NT appearances, God has and is presently acting in the church and the world, and yet God is still to act decisively and globally in the future (as "realized" versus "realizing" eschatologies stress). This means that much of the future of the universe will be part of creation waiting to be transformed and thus entirely within the domain of natural science. Consequently, one should not resort to an easy appeal to "God's new act" when faced with new phenomena in the future. Such phenomena may well be within the competency of science to explain, and appealing instead to God could allow theology to ignore the challenges and discoveries of science and the often discomforting effect they have on a theological understanding of the world. Theology should also refrain from invoking God in its explanation of the future if there are scientific reasons to believe that the laws of nature themselves change in the future, and that such changes are based on an unchanging underlying set of behaviors or of meta-laws. Thus the purpose of this guideline is to keep guideline 1 from effectively insulating theology from science without nullifying the opening that guideline 1 achieved by recognizing that future predictions made by science might not come to pass if God chooses to act in truly new ways that could never be predicted by science or seen as the unfolding of the laws of nature *per se.*

This guideline sharply separates my proposal from approaches such as "intelligent design." This approach is often critical of science, particularly biological evolution, for not including divine agency in its mode of explanation of the past history of life on earth and thus challenge the role of methodological naturalism in defining the natural science.

Guidelines 3 and 4 spell out some of the implications of guideline 2.

Guideline 3: In reconstructing eschatology in light of contemporary physics, our aim should be a "relativistically correct eschatology." Although we will set aside the predictions big bang offers for the cosmic future, we must be prepared to reconstruct current work in eschatology in light of contemporary physics and in terms of what cosmology tells us about the history of the universe. The problem here is that theologians routinely formulate their work in terms of the everyday/Newtonian understanding of space, time, matter, and causality. If we are to construct an eschatology that can take on board all that science entails regarding the past and present of the universe in order to give a new interpretation of the cosmic future, such an eschatology must first be also reconstructed, following paths 1 and 2, in light of the fundamental theories of twentieth-century physics, specifically special and general relativity and quantum mechanics. Actually, we can be quite adventurous here: we can take higher dimensional theories seriously, then causality is not necessarily imposed on any four-dimensional imbedded space-time (such as represents our universe in the current "brane cosmology" proposals). This is a model of the more general causal relations possible in any "transcendent" situation, such as occurs with a four-dimensional space-time imbedded in a higher dimensional space or space-time. Other exciting theories can be explored as well as we engage in a reconstructing of eschatology in light of physics. In general I will refer to this project as the attempt to construct a "relativistically correct Christian eschatology."

Guideline 4: Big bang and inflationary cosmology should serve as a "limit condition" on any revised eschatology. Guideline 4 follows path 1 by stating that standard and inflationary big bang cosmologies, or other scientific cosmologies (such as quantum cosmology), place a "limiting condition" on any possible eschatology. All we know of the history and development of the universe and life in it will be data for theology.

Guideline 5: In pursuing guidelines 3 and 4, the metaphysical options are limited but not forced. In revising contemporary eschatology in light of physics and cosmology there is a variety of metaphysical options from which we may choose; they are not forced on us, or determined, by science. On the one hand, since eschatology starts with the presupposition of God, it rules out reductive materialism and metaphysical naturalism. By taking on board natural science, other metaphysical options become unlikely candidates, including Platonic or Cartesian ontological dualism. On the other hand, there are several metaphysical options that are compatible with science and Christian theology, including physicalism, emergent monism, dual-aspect monism, ontological emergence, and panexperientialism (Whiteheadian metaphysics).

We must now begin the enormous job of revisiting our understanding of eschatology in a way that takes up and incorporates all the findings of science, and particularly scientific cosmology, without falling back into the philosophical problem guideline 1 is meant to address.

Guideline 6: If the creation is to be transformed by God, it must already be transformable. Such transformability, in turn, entails certain formal conditions for the possibility of its being transformed: this is a "such that" or transcendental argument. Our starting point for the purposes of this essay is that the new creation is not a second creation *ex nihilo* with God "junking" the old one and starting from scratch. Instead, God will transform God's creation, the universe, into the new creation. It follows that God has already given the universe precisely those conditions and characteristics that it needs in order to be transformed by God's new act. Since science offers a profound understanding of the past and present history of the universe (guidelines 2 and 3), science can be of immense help to the theological task of understanding something about that transformation if we can find a way to identify, with at least some probability, these needed conditions and characteristics. I will refer to these conditions and characteristics as "elements of continuity." Guideline 6 can be thought of as a transcendental or "such that" argument: It asserts the existence of those characteristics that make it possible for the universe to be transformed into the new creation by God's new action.[39] A simple analogy would be that an open ontology can be thought of as providing a precondition for the enactment of voluntarist free will, but certainly not the sufficient grounds for it.

Science might also shed light on which conditions and characteristics of the present creation we do *not* expect to be continued into the new creation; these can be called "elements of discontinuity" between creation and new creation. Thus physics and cosmology might play a profound role in our attempt to sort out what is truly essential to creation and what is to be "left behind" in the healing transformation to come.

Guideline 6 is a formal argument. It gives to the terms "continuity" and "discontinuity," found in the theological literature on the Resurrection of Jesus, a more precise meaning and a potential connection with science. With it in place we can move to a material argument and ask just what those elements of continuity and discontinuity might be.

Guideline 7: Transformability implies that we must invert the usual relationship between continuity and discontinuity as found within the doctrine of

39. I am grateful to Kirk Wegter-McNelly for suggesting the term "transcendental" here (private communications).

creation, and specifically in discussing divine action. Closely related to the previous formal guideline is a second formal argument about the relative importance of the elements of continuity and discontinuity. So far in the literature on theology and science, discontinuity has played a secondary role within the underlying theme of continuity in nature. "Emergence" in time has been the crucial philosophical theme throughout discussions of the physical development of the universe and the biological evolution of life on earth. Emergence refers to the occurrence of the irreducibly new (i.e., discontinuity) within the overall, pervasive, and sustained background of nature (i.e., continuity).[40]

Now, however, when we come to the question of the Resurrection of Jesus and cosmological eschatology, I propose we "invert" the relation: the elements of continuity will be present, but within a more radical and underlying discontinuity as is denoted by the "transformation" of the universe by a new act of God *ex vetera.* With this inversion, discontinuity as fundamental signals the break with naturalistic and reductionistic views such as "physical eschatology" and evolutionary eschatology, while continuity, even if secondary, eliminates a "two-worlds" eschatology as proposed by neo-Orthodoxy, or as advocated by those who view the new creation as a totally separate creation from the present one.

This has important implications on our search. First, as mentioned above, it *eliminates* previous approaches to "noninterventionist objective special divine action" since these do not involve a transformation of the whole of nature; on the contrary, these approaches presuppose the laws of nature as known by science, such as the laws of quantum mechanics, and argue that divine action is consistent with them without intervention, that is, without suspending or violating them. But in the case of the bodily resurrection of Jesus, we must presuppose the radical transformation of the background conditions of space, time, matter, and causality, and with this, a permanent change in at least most of the present laws of nature.

Guideline 7 thus stresses discontinuity but includes a crucial element of

40. Thus biological phenomena emerge out of the nexus of the physical world, the organism is built from its underlying structure of cells and organs, mind arises in the context of neurophysiology, and so on. Similarly, the laws of physics remain unchanged as biological phenomena arise, even though new biological laws come into place that cannot be reduced entirely to, or predicted by, the laws of physics.

continuity: both elements of continuity in the Easter story (e.g., Jesus could be recognized, touched, seen, etc.) and in the hope for our own personal and communal resurrection at the end of this age. The element of continuity within discontinuity suggests that there may be some laws that do not change.

17.4.2.2. TRP → SRP: Guideline and Steps for Constructive Work in Science

Our project also involves the question of whether such revisions in theology might be of any interest to contemporary science—at least for individual theorists who share eschatological concerns such as developed here and are interested in whether they might stimulate a creative insight into research science. This part of the project might be shaped helpfully by the following guideline.

Guideline 8: The previous seven guidelines only apply to theology as it is reconstructed in light of science. When turning to science and exploring the possibility of new research programs, these seven guidelines should be set aside. This is particularly significant for guideline 1, which science must ignore if it is to follow the empirical method and its insistence on prediction and experimental evidence. Instead we should consider taking three steps, which follow paths 6–8 as suggested in the expanded methodology.

Step 1 (path 6): A theological reconceptualization of nature might lead to philosophical and scientific revisions. Here we move along path 6 in discovering whether a richer theological conception of nature both as creation *and* as new creation can generate important revisions in the *philosophy of nature* that currently underlies the natural sciences, including such concepts as contingency, rationality, causality, holism, temporality, and so on. We also move along path 6 as we make more specific connections between theological conceptions of nature and particular scientific theories. For example, if this universe is to be/is being transformed eschatologically into the new creation, does this suggest a new approach to the philosophy of space, time, matter, and causality in contemporary physics and cosmology? Note that this suggestion does not deal with scientific methodology at all; it only relates to the concepts used in specific theories.

Step 2 (path 7): Theology might suggest criteria of theory choice between existing data-consistent theories. We can also move along path 7 to explore whether differences in current options in theoretical physics and cosmology represent important differences in their underlying philosophical conceptions of nature or of what data to take into account. The theological views of research scientists might then play a significant role in selecting which theoretical programs to pursue among those already "on the table" (for example, the variety of approaches to quantum gravity).

Step 3 (path 8): Theology might suggest new scientific research programs. Finally we can move along path 8 and suggest the construction of new scientific research programs whose motivation stems, at least in part, from theological interests.

In closing this section I want to stress once again that all such programs in science would have to be tested by the scientific communities (what is often called "the context of justification") without regard for the way theology or philosophy might have played a role in their initiation ("the context of discovery"). The proposal that the laws underlying science would change in the future cannot be tested and so has no implications for science as practiced at present. It has implications only as regards the far future and usually would be taken to occur—if at all—only in a context of unchanging meta-principles or meta-laws that guide the changes that are allowed. For the approach taken here in particular, such meta-principles or meta-laws would presumably be the description given to what theologically we would view as God's new act in transforming the universe into the "new creation."

17.5. Research Programs in Theology and in Science

17.5.1. SRP → TRP: RECONSTRUCTING CHRISTIAN ESCHATOLOGY AS "TRANSFORMATION" IN LIGHT OF WHAT WE KNOW FROM SCIENCE

Our fundamental problematic in this section is clear: If we adopt the "worst case" scenario and reject the scenarios for the future given by contemporary cosmology ("freeze" or "fry"), any Christian scenario we might offer in its place must be consistent *both* with our eschatological commitments *and* with scientific cosmology regarding the past history and present state of the universe. Is such an scenario possible? I believe this is a genuinely open research question; it cannot be answered in advance of actually attempting to construct it. In order to move ahead and in light of our guidelines, I suggest the following directions for that research. These directions follow paths 1–5 through which a theological problem, such as eschatology, is reconstructed in light of science, with particular attention to guidelines 6 and 7. What can science tell us about the elements of continuity in the transformation of the universe? About elements of discontinuity? About the preconditions for such elements of continuity? And in light of our responses to these questions, how then should we reconstruct eschatology?

I suggest we start with the idea (guidelines 6 and 7) that the universe is to be transformed into the new creation and we focus on continuities, discontinuities, and their preconditions that are part of that transformation. We may begin with certain suggestive eschatological hints gleaned from the Resurrection of Jesus and the reign of God depicted in the NT, and keeping clearly in mind the apophatic character of all eschatological thought.

There are hints of continuity from the Resurrection of Jesus: he could be touched, he could eat, break bread, be seen, heard, and recognized. These instances of "realized eschatology" are suggestive of an extended "domain" of the new creation within the old, a domain that ceases with the "ascension," but which, when present, includes Jesus, the disciples, and their surroundings. There are also hints of discontinuity: these encounters underscore the difference between resurrection and mere resuscitation, including the difference between its "bodily" character and normal modes of "physicality."

Hints of continuity exist from the "reign of God" in the NT and in the church: the "new creation" will include persons-in-community and their ethical relations. Hints of discontinuity also exist: in the "reign of God," it will "not be possible to sin," compared with the present creation in which it is "not possible not to sin," to use Augustine's apt formulation.

Hints of continuity appear from the problem of personal identity between death and the general resurrection: both Paul's analogy of the seed (1 Cor. 15:35 ff) and the numerical, material, and/or formal continuity between death and general resurrection in historical Christian thought [35, 84]. Hints of discontinuities also appear as Paul's fourfold contrast suggests (1 Cor. 15:42 ff).

Next we look for epistemically "prior" elements of continuity by moving "down" the hierarchy. What preconditions make possible the elements of continuity found above? For now, I will set aside the many levels to be found in the social, psychological, and neurosciences, and move directly to physics. Here we look for elements in the physical world that serve as preconditions for the elements of continuity in the resurrection narratives. A central theme in human experience is temporality; thus we would expect that time as understood by physics is not only a characteristic of this universe, but that it will in some ways be a characteristic of the new creation. Yet we would expect that in the new creation our experience of temporality will no longer be marred by the loss of the past and the unavailability of the future. So there will be an element of discontinuity during the transformation as well. We would make a similar case for ontological openness as the enduring precondition for persons-in-community to act freely in love, so that this too might be an element of continuity during the transformation of the universe. Other examples include ontological relationality/holism, the role of symmetries/conservation laws, and so on.[41]

I also want to note the unique role mathematics plays in this scenario. Presumably mathematics will be an element of continuity without (apparently) any discontinuity. This role would surely include both a) general theorems, proofs, discoveries, and b) specific fields, such as fractals and Cantor's transfinite algebra [80].

The next step here would be to undertake an extensive reconstruction of Christian eschatology in light of these arguments. Guidelines 2–4 and 7 will be particularly important here. Because of length constraints on this essay, I will not pursue the reconstruction here, but move immediately to our section on TRP → SRPs.[42]

17.5.2. TRP → SRPS: NEW RESEARCH IN PHYSICS AND COSMOLOGY

As stated above in Guideline 8, steps 1–3, our fundamental problematic includes a second set of approaches: In what ways would a revised eschatology, in which the existing universe as the creation is to be transformed into the new creation, lead in turn to a revised philosophy of nature, to criteria of theory choice among current theories in science, or to the construction of new scientific research programs? These questions are the primary focus of this essay and will occupy its remaining portions.

Clearly we have a procedural problem. Current eschatology has not yet been reconstructed. What then can we do for now? We might start with a more limited approach: begin with our existing eschatology and the ele-

41. Polkinghorne has made very similar suggestions to these, see for example [77].

42. See [84].

ments of continuity, discontinuity, and their preconditions listed above and see what SRPs they might suggest. For the purposes of this essay, I will pick one that seems particularly promising for physics: *temporality.* We will therefore start with a theological understanding of "time and eternity," the *locus classicus* by which theology considers time.[43] We will then explore what implications or suggestions this understanding might have for current physics and cosmology. In the process we will look for those aspects of temporality that are already present in the world and which constitute elements of continuity and of discontinuity in respect to the eschatological transformation of the world. This, in turn, should lead to interesting questions about the way physics works with these aspects of temporality. We will also consider whether there are other aspects of temporality in the world that physics may have overlooked but which, from the perspective of eschatology, might be expected to exist and whether it would be fruitful if physics were to consider including them. In the process our goal will be to generate concrete suggestions for possible directions toward physics research programs.[44]

17.5.2.1. Eternity and Temporality in the New Creation

A crucial argument shared widely among contemporary theologians[45] is that eternity is a richer concept of temporality than timelessness or unending time. In essence, eternity is the source of time as we know it, and of time as we will know it in the new creation. Eternity is the fully temporal source and goal of time. Barth calls it "supratemporal," Moltmann calls it "the future of the future," and Peters refers to the future as coming to us *(adventus)* and not merely that which tomorrow brings *(futurum).* Pannenberg claims that God acts proleptically from eternity: God reaches back into time to redeem the world, particularly in the life, ministry, death, and Resurrection of Jesus. In this approach the relation between "time and eternity" is modeled

43. The focus on "time and eternity" is particularly appropriate for our task here since most contemporary theologians root their eschatology within the framework of *creatio ex nihilo.* Thus much of what they claim about "time and eternity" applies to the universe as it is (i.e., as creation), and not just as it will be (i.e., as "new creation").

44. Another way of saying this is to ask what an expanded scientific conception of nature would be like if we were to inherit it from the eschatological idea of "new creation" instead of entirely from a theology of creation as it currently exists, and what ramifications this might have for current science.

45. For helpful overviews of trinitarian theologians on the problem of "time and eternity," see Peters [74] and Russell [81].

on the relation of the finite to the infinite. Here the infinite is not the negation of the finite (as in the view of eternity as timelessness); instead the infinite includes but ceaselessly transcends the finite.

This view of eternity includes at least five distinct themes:

- co-presence of all events: distinct events in time are nevertheless present to one another without destroying or subsuming their distinctiveness;
- "flowing time": each event has a "past/present/future" (ppf) structure often referred to as the "arrow of time";
- duration: each event has temporal thickness in nature as well as in experience; events are not pointlike present moments with no intrinsic past or future;
- prolepsis: the future is already present and active in the present while remaining future; and
- global future: there is a single future for all of creation.

17.5.2.2. Continuity, Discontinuity, and Their Preconditions Regarding Temporality

The temporality of the present creation is characterized by two of these five themes:

- flowing time; and
- duration.

Thus flowing time and duration will be elements of continuity in the transformation of the universe. Guidelines 5 and 6 suggest that the universe includes the "transcendental" conditions for the possibility of the other three themes to be found in the new creation, namely:

- co-presence;
- prolepsis; and
- global future.

I will first focus on the problem of "flowing time," where the transformation will also involve discontinuity in a subtle way. Later I will return to co-presence, prolepsis, and global future.

The combination of flowing time and co-presence in the eternity of the new creation is meant to indicate that its mode of temporality can include what is of real value about time: that each event has an identity in itself, characterized in part by its own unique set of past and future events. The event as present is "actual"; events that are past in relation to the present event are "determinate but no longer actual," and events that are future in relation to the present event are "potential and indeterminate." I will refer to this as the "inhomogeneous temporal ontology" of flowing time, and designate it as the "ppf" structure of time often referred to as "the arrow of time." But temporality in the present world also includes what is tragic, painful, exilic, and existentialist about time: the lack of "co-presence," that is, the isolation of each present event and the inaccessibility of other events in their actuality to the present event in its actuality. Whereas we routinely experience the world as spatially simultaneous (*pace* special relativity),[46] meaning that separate events "in space" share a common present, events in the past and future are never "simultaneous." In this sense, the flow of time that we now experience is a "broken" or "distorted" form of the genuine flow of time that awaits us in eternity, where the characteristic of "co-presence" will allow us to experience all events together while retaining their separate temporal character (their unique set of "ppfs"). We could say that in the new creation, "flow" and "co-presence" function together in offering genuine temporality, what Boethius called "the simultaneous presence of unlimited life." Thus in the new creation, "flow" keeps "co-presence" from reducing to "*nunc*" while "co-presence" keeps "flow" from reducing to a stream of isolated moments.

In short, *the theme of "flowing time" is an irreducible part of the theological concept of creation.* At the same time, it is a theme that must allow for its "fulfillment" in its transformation into the "flowing time" of new creation's time, particularly with the addition of "co-presence." This will lead us directly to our conversation with science and its view of time in nature: Does it support an objective "arrow of time" in physics?

The second issue is with "duration." Many theologians claim that genuine temporality involves temporal duration: the "present" does not have a "pointlike" structure but a "thickness" in time. This concept is closely related to, but distinct from, the "ppf" structure of the present moment discussed above; it has to do, primarily, with the moment itself. As we will see in dis-

46. Again, all of this must be reconstructed in a "relativistically correct way" in future work.

cussing Pannenberg's arguments, where the theme is most clearly spelled out, the claim being made is that objective time in nature, and not just our subjective experience of time, is one of duration as well as flow. This again leads us to our conversation with science and its view of time in nature: Can one construct a theoretical/mathematical approach to temporal duration that would be of value to physics? I will return to the theme of duration after treating "flowing time" below, where I will also give a more expanded account of the theological argument.

17.5.2.3. Flowing Time in Physics: Special Relativity, General Relativity, Quantum Mechanics, and Thermodynamics

A cursory overview of four areas in contemporary physics—special relativity (SR), general relativity (GR), quantum mechanics (QM), and thermodynamics—suggests that flowing time is an illusion. The equations of SR, GR, and QM are time reversible; those of thermodynamics are time irreversible, but thermodynamics is reducible to dynamics through statistical mechanics. The situation, however, is more complicated and requires further discussion before specific kinds of TRP → SRPs can be considered. I will focus on SR and QM here and extend the discussion to GR and thermodynamics in future work.

As is well known, SR can be given two competing interpretations, as a recent debate between Chris Isham and John Polkinghorne vividly illustrates [47]:

Block universe (philosophy of being): All events in space-time have equal ontological status/are equally "real" and "actual," and flowing time is an illusion of subjectivity; a crucial argument is the invariant space-time interval.[47] This view is sometimes referred to as the "spatialization of time."

Flowing time (philosophy of becoming): Events have the usual "ppf" structure, though the notion of "present" or "the future"/"the past" has to be nuanced in light of the "elsewhen" problem of SR and the fact that time passes at different rates for clocks in relative motion. A crucial argument is the invariance of causality within the lightcone.[48] This view is sometimes referred to as the "dynamization of space."

47. A staunch defense of the "philosophy of being" is given by Costa de Beauregard [15].
48. A classic work is Capek [11].

There are two kinds of SRPs that could be generated by the theological preference for "flowing time" in the context of SR:

SRP 1: Follows step 1 (path 6) and step 2 (path 7) in that it finds ways to show predictive preference for the flowing time interpretation over the "block universe" interpretation (i.e., find arguments for an objective "arrow of time" based on SR) based on wider considerations than SR alone. Alternatively, it distinguishes between what Ellis calls "the world of possibilities" and "the world of actualities" and shows how this bears on the comparative evaluation of the two interpretations.[49]

SRP 2: Follows step 3 (path 8) to find ways to revise SR to support flowing time over "block universe." Isham essentially laid the gauntlet at the feet of the "flowing time" advocates in his comment here: "Do the opponents of the block universe seek merely to reinterpret the existing theories of physics, or do they make the much stronger claim that their metaphysical views can be sustained only by changing the theories?"[50]

In my opinion SR is a kinematic theory whose status is so secure by both a wealth of data from differing domains in physics and from a variety of theoretical approaches (as John Lucas and Peter Hodson thoroughly argue)[51] that it would be fruitless to try SRP 2. SRP 1, however, might be worth considering. One could explore whether relativistic quantum mechanics and quantum field theory, or perhaps GR, offer a basis for a more compelling case for a "flowing time" interpretation over a "block universe" interpretation.

Quantum mechanical formalism presupposes "flowing time," but the interpretation of this formulation is highly debated. According to the "standard" Copenhagen interpretation, quantum mechanics consists of two parts: 1) the reversible time development of the wave function ψ of a quantum mechanical system as governed by the Schrödinger equation, whose "square" corresponds to the probability that an observable will have a particular value upon measurement, and 2) the irreversible process of "measurement," which is not governed by the Schrödinger equation but which is essential if empirical values for dynamical variables like position and mo-

49. Ellis [24], see also [22] and article in [85].

50. Isham and Polkinghorne [47], p. 141. Although joint authorship is ascribed to the article, it is clear who wrote this comment.

51. Lucas and Hodgson [59], see especially fig. 5.1.1, p. 152.

mentum are to be obtained. It is the irreversibility of measurement that introduces a fundamental "quantum mechanical arrow of time" into our view of nature.[52] The problem, of course, is that this way of formulating the issue of the irreversibility of time presupposes the Copenhagen interpretation, with its "double ontology" of quantum system and classical measuring apparatus.

If we wish to pursue the argument for an "objective arrow of time" in nature in light of QM, what are our options?

SRP 3: Follows step 1 (path 6) to find a way to interpret QM such that it does, finally, provide an objective basis for an arrow of time that is empirically fruitful.

SRP 4: Follows step 2 (path 7) to modify QM in one of a number of ways such that it will provide an objective basis for an arrow of time. There are at least two options that have been pursued: 1) nonlinear versions of the Schrödinger equation, and 2) stochastic versions of Schrödinger equation.

SRP 5: Follows step 3 (path 8) to replace QM by a new theory that equals (and possibly surpasses) its predictive power and provides an objective basis for an arrow of time. Such a theory will have to take into account the empirical violations of Bell's theorems, which rule out "local, realist" theories ("hidden variables" theories).

17.5.2.4. Duration in Physics: The Hidden Aspect of Time in Nature

Duration is the second major topic that characterizes the temporality of the present creation from a theological perspective. In some versions of Christian eschatology duration is found not only in our subjective experience but also in nature: the physical "present" does not have a "pointlike" structure but a "thickness" in time. This concept is closely related to, but distinct from, the "ppf" structure of the present moment discussed above; it has to do, primarily, with a proposed structure for the "present" moment itself. In future work it will be interesting to explore a variety of ways in which duration as a theological as well as a philosophical concept might serve heuristically to suggest interesting questions and possible research programs in the natural sciences, particularly as duration is developed in the thought

52. For recent discussions of the relation between quantum physics and time's arrow, see [55, 93].

of theologian Wolfhart Pannenberg and philosopher Alfred North Whitehead.

17.5.2.5. Preconditions in Physics for Co-presence, Prolepsis, and Global future

We turn now to the preconditions for the possibility of continuity: co-presence, prolepsis, and global future.

"Co-presence" can be defined as the temporal presence to one another of distinct events in time, a presence that does not destroy or subsume their distinctiveness (e.g., their unique "ppf" structure).[53]

It is challenging to think of positive preconditions on the nature of flowing time that make its transformation into co-presence possible. We might, however, be able to formulate a "negative" precondition: there should be nothing about flowing time that makes its transformation into co-present flowing time intrinsically impossible or the concept of such a transformation entirely unintelligible.

The theological meaning of prolepsis is that God acts from the eternal future in the present. The primary instance is God's act in raising Jesus from the dead and, in turn, the ongoing transformation of creation into new creation. Clearly the theological meaning of "future" is distinct from the scientific meaning; yet the attempt here is to find ways to bring the two meanings into a more fruitful relationship. Thus there may be features in physics and cosmology that might function as physical preconditions for the possibility of proleptic action of God. The causal structure of the universe might be more subtle than the "arrow of time" discussion allows; it might not be entirely inconsistent with the idea that the future transformation of the universe by God effects the present. For brevity, I will simply list examples of how these features have been discussed in physics here:

- backward causality: Feynman/Wheeler advanced potentials; tachyons;
- violations of local causality: backward propagation of sound exceeding the speed of light in de Sitter space-time either with a non-zero L or with space filled with an incompressible fluid [62, 25];
- violations of global causality: Gödel's universe; essential singularities

53. A possible mathematical model for the concept of co-presence is a non-Hausdorff manifold in which distinct events are not separable topologically.

> (holes) in ordinary (FLRW) space-time related to time/space orientability; chronology condition; causality condition; future/past distinguishing condition; strong causality condition; stable causality condition; the existence/nonexistence of a Cauchy surface (or a partial Cauchy surface); and so on.[54]

Please note that while these have all been discussed as theoretical possibilities, they have not in any case been shown to occur in the physics universe. Furthermore, these topics become much more complex as we move to quantum gravity/quantum cosmology [9, 43, 45].

A common theme in Christian eschatology is the unity of the future: the creation will be transformed into a single global domain so that all creatures in the new creation can be in community. As above, in discussing the possible preconditions for such a global future I want to underscore the difference between its scientific and theological meanings. Nevertheless, it is intriguing to ask about the future as understood in physics and cosmology. Is there a unique global causal future that could serve as a precondition for the transformation of the future into the future of the new creation?

According to SR, there is no unique global causal future. The causal futures of any two events P and Q will share some events in common, but there will always be other events that lie either in the causal future of P but not of Q, or of Q but not of P. GR, however, provides a possible "precondition" for the theme of the eschatological global future: it shows that the topology of the universe is contingent on the distribution of matter, and it allows for a variety of topologies for the future, including ones in which geodesics do not separate to arbitrary distances.[55]

Regarding the points made above concerning co-presence, prolepsis, and the global future, we may offer the following SRP:

SRP 6: Follows guideline 8 to ask, to what extent, if any, do these and other features in physics and cosmology point to the complex character of time and time's arrow as indicative of the preconditions for co-presence, prolepsis, and the global future?

54. For definitions see Hawking and Ellis [37].

55. To Tipler's credit one can also see how the topology of the future could, to a certain extent, be affected by life in the universe, giving the future what he called the "Omega point."

17.6. Conclusions and Future Directions

This essay has attempted to show that a sustained discussion of resurrection, eschatology, and cosmology are of crucial importance not only to "theology and science" but to several central areas in Christian theology today, that progress might be possible if we expand the methodology from its previous format to make the discussion genuinely interactive, and that possible areas for research in both theology and in science might result, which would be of value to both fields. We have taken only very preliminary steps, but hopefully they have shown that such progress is at least potentially possible, and that the directions for future work are becoming clear.

REFERENCES

1. Alighieri, D., "The Paradiso," in *The Divine Comedy,* trans. John Ciardi (W. W. Norton, New York, 1970), canto XX, vs. 88–90.

2. Ayala, F. J., "Darwin's Devolution: Design without Designer," in *Evolutionary and Molecular Biology: Scientific Perspectives on Divine Action,* eds. R. J. Russell, W. R. Stoeger, S.J., and F. J. Ayala (Vatican Observatory, Vatican City State/Center for Theology and the Natural Sciences, Berkeley, Calif., 1998).

3. Barbour, I. G., *Religion in an Age of Science,* The Gifford Lectures, 1989–90 (San Francisco: Harper and Row, 1990).

4. Barbour, I. G., "Five Models of God and Evolution," in *Evolutionary and Molecular Biology: Scientific Perspectives on Divine Action,* eds. R. J. Russell, W. R. Stoeger, S.J., and F. J. Ayala (Vatican Observatory Publications, Vatican City State/Center for Theology and the Natural Sciences, Berkeley, Calif.,1998).

5. Barrow, J. D., *Impossibility: The Limits of Science and the Science of Limits* (Oxford University Press, Oxford, 1998), 155–89, esp. 181.

6. Barrow, J. D., "Cosmic Questions," paper from the AAAS Program of Dialogue on Science, Ethics, and Religion, Washington, D.C., April 14–19, 1999. Barrow's article, now titled "Cosmology, Life, and the Anthropic Principle" has since been published in *Annals of the New York Academy of Sciences,* vol. 950: *Cosmic Questions,* ed. J. B. Miller (New York Academy of Science, New York, 2001).

7. Barrow, J. D., and Tipler, F. J., *The Anthropic Cosmological Principle* (Clarendon Press, Oxford, 1986).

8. Birch, C., "Neo-Darwinism, Self-Organization, and Divine Action in Evolution," in *Evolutionary and Molecular Biology: Scientific Perspectives on Divine Action,* eds. R. J. Russell, W. R. Stoeger, S.J., and F. J. Ayala (Vatican Observatory Publications, Vatican City State/Center for Theology and the Natural Sciences, Berkeley, Calif., 1998).

9. Birrell, N. D., and Davies, P. C. W., *Quantum Fields in Curved Space* (Cambridge University Press, Cambridge, 1982).

10. Bynum, C. W., *The Resurrection of the Body in Western Christianity, 200–1336* (Columbia University Press, New York, 1995).

11. Capek, M. "Time in Relativity Theory: Arguments for a Philosophy of Becoming," in *The Voices of Time: A Cooperative Survey of Man's Views of Time as Expressed by the Sciences and by the Humanities,* ed. J. T. Fraser (University of Massachusetts Press, Amherst, 1966, 1981), 434–54.

12. Clayton, P. *Explanation from Physics to Theology: An Essay in Rationality and Religion* (Yale University Press, New Haven, Conn., 1989).

13. Cohn, J. D. "Living with Lambda," *Astroph/9807/128 V2 Preprint* (1998).

14. Davis, S. Kendall D., S.J., and O'Collins, G., eds., *The Resurrection* (Oxford University Press, Oxford, 1997).

15. de Beauregard, O. C., "Time in Relativity Theory: Arguments for a Philosophy of Being," in *The Voices of Time: A Cooperative Survey of Man's Views of Time as Expressed by the Sciences and by the Humanities,* ed. J. T. Fraser (University of Massachusetts Press, Amherst, 1966, 1981), 417–33.

16. Deason, G. B., "Protestant Theology and the Rise of Modern Science: Criticism and Review of the Strong Thesis," *CTNS Bulletin, 6.4,* 1–8 (Autumn 1986).

17. Drees, W. B., *Beyond the Big Bang: Quantum Cosmologies and God* (Open Court, La Salle, Ill., 1990).

18. Dyson, F., "Time without End: Physics and Biology in an Open Universe," *Rev. Mod. Phys., 51,* 447–60 (1979).

19. Dyson, F., *Disturbing the Universe* (Harper and Row, New York, 1979).

20. Dyson, F., *Infinite in All Directions* (Harper and Row, New York, 1988).

21. Ellis, G. F. R., *Before the Beginning: Cosmology Explained* (Boyars/Bowerdean, New York, 1993).

22. Ellis, G. F. R., "Ordinary and Extraordinary Divine Action: The Nexus of Interaction," in *Chaos and Complexity: Scientific Perspectives on Divine Action,* eds. R. J. Russell, N. C. Murphy, and A. R. Peacocke, Scientific Perspectives on Divine Action Series (Vatican Observatory Publications, Vatican City State/Center for Theology and the Natural Sciences, Berkeley, Calif., 1995), 384.

23. Ellis, G. F. R., "The Thinking Underlying the New 'Scientific' World-Views," in *Evolutionary and Molecular Biology: Scientific Perspectives on Divine Action,* eds. R. J. Russell, W. R. Stoeger, S.J., and F. J. Ayala (Vatican Observatory Publications, Vatican City State/Center for Theology and the Natural Sciences, Berkeley, Calif., 1998), 251–80.

24. Ellis, G. F. R., "Natures of Existence (Temporal and Eternal)," ch. 18 in this volume.

25. Ellis, G. F. R., Hwang, J. C., and Bruni, M., *Phys. Rev. D, 40,* 1819–26 (1989).

26. Ellis, G. F. R., and Stoeger, W. R. (S.J)., "Introduction to General Relativity and Cosmology," in *Quantum Cosmology and the Laws of Nature: Scientific Perspectives on Divine Action,* eds. R. J. Russell, N. C. Murphy, and C. J. Isham, Scientific Perspectives on Divine Action Series (Vatican Observatory Publications, Vatican City State/Center for Theology and the Natural Sciences, Berkeley, Calif., 1993).

27. Folse, H. J., "Complementarity, Bell's Theorem, and the Framework of Process Metaphysics," *Proc. Stud., 11, no. 4,* 259–73 (Winter 1981).

28. Foster, M., "The Christian Doctrine of Creation and the Rise of Modern Science," in *Creation: The Impact of an Idea,* eds. D. O'Connor and F. Oakley (Charles Scribner's Sons, New York, 1969).

29. Geroch, R. P., "Singularities in Closed Universes," *Phys. Rev. Lett., 17,* 445–47 (1966).

30. Goldsmith, D., *Einstein's Greatest Blunder? The Cosmological Constant and Other Fudge Factors in the Physics of the Universe* (Harvard University Press, Cambridge, Mass., 1995).

31. Hartle, J. B., and Hawking, S. W., "Wave Function of the Universe," *Phys. Rev. D., 28,* 2960–75 (1983).

32. Hartshorne, C., "Bell's Theorem and Stapp's Revised View of Space-Time," *Proc. Stud., 7, no. 4,* 183–91 (Winter 1977).

33. Haught, J. F., *The Promise of Nature: Ecology and Cosmic Purpose* (Paulist Press, New York, 1993), 124–25, 130.

34. Haught, J. F., *Science and Religion: From Conflict to Conversion* (Paulist Press, New York, 1995), 174–75.

35. Geroch, R. P., "Singularities in Closed Universes," *Phys. Rev. Lett. 17,* 445–47 (1966).

36. Hawking, S. W., "The Occurrence of Singularities in Cosmology. III. Causality and Singularities," *Proc. Roy. Soc. Lond., A 300,* 187–201 (1967).

37. Hawking, S. W., and Ellis, G. F. R., *The Large Scale Structure of Space-Time,* Cambridge Monographs on Mathematical Physics Series (Cambridge University Press, Cambridge, 1973).

38. Hawking, S. W., and Penrose, R., "The Singularities of Gravitational Collapse and Cosmology," *Proc. Roy. Soc. London, A 314,* 529–48 (1970).

39. Hoyle, F., Burbidge, G., and Narlikar, J. V., *A Different Approach to Cosmology: From a Static Universe Through the Big Bang Towards Reality* (Cambridge University Press, Cambridge, 2000).

40. Hoyle, F., and Narlikar, J. V., *Action at a Distance in Physics and Cosmology* (W. H. Freeman and Company, San Francisco, 1974).

41. Hoyle, F., and Narlikar, J. V., *The Physics-Astronomy Frontier* (W. H. Freeman and Company, San Francisco, 1980).

42. Huchingson, J. E., *Religion and the Natural Sciences: The Range of Engagement* (Harcourt Brace Jovanovich College Publishers, Fort Worth, 1993).

43. Isham, C. J., "Creation of the Universe as a Quantum Process," in *Physics, Philosophy, and Theology: A Common Quest for Understanding,* eds. R. J. Russell, W. R. Stoeger, S.J. and G. V. Coyne, S.J. (Vatican Observatory Publications, Vatican City State, 1988), 375–408.

44. Isham, C. J., "Quantum Theories of the Creation of the Universe," in *Quantum Cosmology and the Laws of Nature: Scientific Perspectives on Divine Action,* eds. R. J. Russell, N. C. Murphy, and C. J. Isham, Scientific Perspectives on Divine Action Series (Vatican Observatory Publications, Vatican City State/Center for Theology and the Natural Sciences, Berkeley, Calif., 1993), 49–90.

45. Isham, C. J., *Lectures on Quantum Theory: Mathematical and Structural Foundations* (Imperial College Press, London, 1995).

46. Isham, C. J., and Butterfield, J., "Topos Perspective on the Kochen-Specker Theorem. I. Quantum States as Generalized Valuations," *Int. J. Theor. Phys., 37, no. 11,* 2669–733 (1998).

47. Isham, C. J., and Polkinghorne, J. C., "The Debate Over the Block Universe," in *Quantum Cosmology and the Laws of Nature: Scientific Perspectives on Divine Action,* eds. R. J. Russell, N. C. Murphy, and C. J. Isham, Scientific Perspectives on Divine Action Series (Vatican Observatory Publications, Vatican City State/Center for Theology and the Natural Sciences, Berkeley, Calif., 1993), 134–44.

48. Isham, C. J., and Savvidou, K. N., "Time and Modern Physics" (A Blackett Laboratory Preprint, London, 2000), 1–17.

49. Jones, W. B., "Bell's Theorem, H. P. Stapp, and Process Theism," *Proc. Stud., 8, no. 1,* 250–61 (Spring 1978).

50. Kaiser, C. B., *Creation and the History of Science,* The History of Christian Theology Series, no. 3 (Eerdmans, Grand Rapids, Mich., 1991).

51. Klaaren, E. M., *Religious Origins of Modern Science: Belief in Creation in Seventeenth-Century Thought* (William B. Eerdmans, Grand Rapids, Mich., 1977).

52. Kolb, E. W., and Turner, M. S., *The Early Universe* (Addison-Wesley Publishing Company, Reading, Mass., 1994).

53. Kragh, H., *Cosmology and Controversy: The Historical Development of Two Theories of the Universe* (Princeton University Press, Princeton, N.J., 1996).

54. Krauss, L. M., "The End of the Age Problem, and the Case for a Cosmological Constant Revisited," *CWRU-P6-97 / CERN-Th-97/122 / Astro-Ph/9706227 Preprint* (1997).

55. Leggett, A., "Time's Arrow and the Quantum Measurement Problem," in *Time's Arrows Today: Recent Physical and Philosophical Work on the Direction of Time,* ed. S. F. Savitt (Cambridge University Press, Cambridge, 1995), 191–216.

56. Lindberg, D. C., and Numbers, R. L., eds., *God and Nature: Historical Essays on the Encounter Between Christianity and Science* (University of California Press, Berkeley, 1986).

57. Linde, A. D., "Particle Physics and Inflationary Cosmology," *Physics Today, 40, no. 9,* 61–68 (1987). See also A Guth: astro-ph/0101507 at http://xxx.lanl.gov.

58. Lorenzen, T., *Resurrection and Discipleship: Interpretive Models, Biblical Reflections, Theological Consequences* (Orbis Books, Maryknoll, N.Y., 1995).

59. Lucas, J. R., and Hodgson, P. E., *Spacetime and Electromagnetism* (Clarendon Press, Oxford, 1990).

60. Macquarrie, J., *Principles of Christian Theology,* 2nd ed. (Charles Scribner's Sons, New York, 1977, 1966), ch. 15, esp. 351–62.

61. Misner, C. W., Thorne, K. S., and Wheeler, J. A., *Gravitation* (W. H. Freeman and Co., San Francisco, 1973).

62. Murphy, G. L., "Positivity and Causality: Classical Theory (Preprint)" (1978).

63. Murphy, N., *Theology in the Age of Scientific Reasoning* (Cornell University Press, Ithaca, N.Y., 1990).

64. Murphy, N., *Anglo-American Postmodernity: Philosophical Perspectives on Science, Religion, and Ethics* (Westview Press, Boulder, Colo., 1997).

65. Murphy, N., and Ellis, G. F. R., *On the Moral Nature of the Universe: Theology,*

Cosmology, and Ethics, Theology and the Sciences Series (Fortress Press, Minneapolis, Minn., 1996).

66. North, J. D., *The Measure of the Universe: A History of Modern Cosmology* (Dover Publications, Inc., New York, 1965, 1990).

67. O'Collins, G., S.J., *Jesus Risen: An Historical, Fundamental, and Systematic Examination of Christ's Resurrection* (Paulist Press, New York, 1987).

68. Pannenberg, W., "Theological Questions to Scientists," in *The Sciences and Theology in the Twentieth Century,* ed. A. R. Peacocke (University of Notre Dame Press, Notre Dame, Ind., 1981).

69. Pannenberg, W., *Metaphysics and the Idea of God* (Eerdmans, Grand Rapids, Mich., 1990).

70. Pannenberg, W., *Systematic Theology,* trans. G. W. Bromiley (Eerdmans, Grand Rapids, Mich., 1991), 1.

71. Pannenberg, W., *Toward a Theology of Nature: Essays on Science and Faith,* ed. T. Peters (Westminster/John Knox Press, Louisville, Ky., 1993).

72. Peacocke, A., *Theology for a Scientific Age: Being and Becoming—Natural, Divine, and Human,* enlarged ed. (Fortress Press, Minneapolis, Minn., 1993).

73. Penrose, R., "Gravitational Collapse and Space-Time Singularities," *Phys. Rev. Lett.,* 14, 57–59 (1965).

74. Peters, T., *God as Trinity: Relationality and Temporality in the Divine Life* (Westminster/John Knox Press, Louisville, Ky., 1993), 175–76.

75. Peters, T., ed., *Science and Theology: The New Consonance* (Westview Press, Boulder, Colo., 1998).

76. Polkinghorne, J. C., *The Faith of a Physicist: Reflections of a Bottom-up Thinker,* Theology and the Sciences Series (Fortress Press, Minneapolis, Minn., 1994).

77. Polkinghorne, J. C., "Eschatology: Some Questions and Some Insights from Science," in *The End of the World and the Ends of God: Science and Theology on Eschatology,* eds. J. Polkinghorne and M. Welker (Trinity Press International, Harrisburg, Pa., 2000).

78. Prigogine, I., *From Being to Becoming: Time and Complexity in the Physical Sciences* (W. H. Freeman and Co., San Francisco, 1980).

79. Richardson, W. M., and Wildman, W. J., eds., *Religion and Science: History, Method, Dialogue* (Routledge, New York, 1996).

80. Russell, R. J., "The God Who Transcends Infinity: Insights from Cosmology and Mathematics into the Greatness of God," in *How Large Is God?* eds. J. M. Templeton and R. L. Herrmann (Philadelphia: Templeton Foundation Press, 1997).

81. Russell, R. J., "Time in Eternity: Special Relativity and Eschatology," *Dialog, 39, no. 1,* 46–55 (March 2000).

82. Russell, R. J., "Did God Create Our Universe? Theological Reflections on the Big Bang, Inflation, and Quantum Cosmologies," in *Annals of the New York Academy of Sciences,* vol. 950: *Cosmic Questions,* ed. J. B. Miller (New York Academy of Sciences, New York, 2001).

83. Russell R. J., "Theology and Science: Current Issues and Future Directions," on the CTNS website, www.ctns.org (http://www.ctns.org/Publications/ Russell_Paper/ russell_paper.html).

84. Russell, R. J., "Bodily Resurrection, Eschatology, and the Challenge of Physical Cosmology," in *Resurrection: Theological and Scientific Assessments,* eds. T. Peters, R. J. Russell, and M. Welker (Eerdmans, Grand Rapids, Mich.: forthcoming 2002).

85. Russell, R. J., Clayton, P., et al., eds., *Quantum Mechanics: Scientific Perspectives on Divine Action,* Scientific Perspectives on Divine Action Series (Vatican Observatory Publications, Vatican City State/Center for Theology and the Natural Sciences, Berkeley, Calif., 2001).

86. Russell, R. J., Murphy, N., et al., eds., *Neuroscience and the Person: Scientific Perspectives on Divine Action,* Scientific Perspectives on Divine Action Series (Vatican Observatory Publications, Vatican City State/Center for Theology and the Natural Sciences, Berkeley, Calif., 1999).

87. Russell, R. J., Murphy, N. C., and Isham, C. J., eds., *Quantum Cosmology and the Laws of Nature: Scientific Perspectives on Divine Action,* Scientific Perspectives on Divine Action Series (Vatican Observatory Publications, Vatican City State/Center for Theology and the Natural Sciences, Berkeley, Calif., 1993).

88. Russell, R. J., Murphy, N. C., and Peacocke, A. R., eds., *Chaos and Complexity: Scientific Perspectives on Divine Action,* Scientific Perspectives on Divine Action Series (Vatican Observatory Publications, Vatican City State/Center for Theology and the Natural Sciences, Berkeley, Calif., 1995).

89. Russell, R. J., Stoeger, W. R. (S.J.), and Ayala, F. J., eds., *Evolutionary and Molecular Biology: Scientific Perspectives on Divine Action,* Scientific Perspectives on Divine Action Series (Vatican Observatory Publications, Vatican City State/Center for Theology and the Natural Sciences, Berkeley, Calif., 1998).

90. Schneiders, S., "The Resurrection of Jesus and Christian Spirituality," in *Christian Resources of Hope* (Columba, Dublin, 1995), 81–114.

91. Shimony, A., "Quantum Physics and the Philosophy of Whitehead," in *Search for a Naturalistic World View,* vol. 2: *Natural Science and Metaphysics* (Cambridge University Press, Cambridge, 1965, 1993).

92. Sklar, L., "The Elusive Object of Desire: In Pursuit of the Kinetic Equations and the Second Law," in *Time's Arrows Today: Recent Physical and Philosophical Work on the Direction of Time,* ed. S. F. Savitt (Cambridge University Press, Cambridge, 1995), esp. 212–15.

93. Southgate, C., Deane-Drummond, C., et al., eds., *God, Humanity and the Cosmos: A Textbook in Science and Religion* (Trinity Press International, Harrisburg, Pa., 1999).

94. Stamp, P., "Time, Decoherence, and 'Reversible' Measurements," in *Time's Arrows Today: Recent Physical and Philosophical Work on the Direction of Time,* ed. S. F. Savitt (Cambridge University Press, Cambridge, 1995), 191–216.

95. Stapp, H. P., "Quantum Mechanics, Local Causality, and Process Philosophy," *Proc. Stud., 7, no. 4,* 173–82 (Winter 1977), ed. William B. Jones.

96. Stoeger, W. R., S.J., "Contemporary Physics and the Ontological Status of the Laws of Nature," in *Quantum Cosmology and the Laws of Nature: Scientific Perspectives on Divine Action,* eds. R. J. Russell, N. C. Murphy, and C. J. Isham, Scientific Perspectives on Divine Action Series (Vatican Observatory Publications, Vatican City State/Center for Theology and the Natural Sciences, Berkeley, Calif., 1993), 209–34.

97. Stoeger, W. R., S.J., "Key Developments in Physics Challenging Philosophy and Theology," in *Religion and Science: History, Method, Dialogue,* ed. W. M. Richardson and W. J. Wildman (Routledge, New York, 1996), 183–200.

98. Stoeger, W. R., S.J., "Scientific Accounts of Ultimate Catastrophes in Our Life-Bearing Universe," in *The End of the World and the Ends of God: Science and Theology on Eschatology,* ed. J. Polkinghorne and M. Welker (Trinity Press International, Harrisburg, Pa., 2000).

99. Stoeger, W. R. (S.J)., and Ellis, G. F. R., "A Response to Tipler's Omega-Point Theory," *Science and Christian Belief, 7, no. 2,* 163–72 (1995).

100. Tipler, F. J., *The Physics of Immortality: Modern Cosmology, God, and the Resurrection of the Dead* (Doubleday, New York, 1994).

101. Trefil, J. S., *The Moment of Creation: Big Bang Physics from Before the First Millisecond to the Present Universe* (Macmillan, New York, 1983).

102. Trefil, J., and Hazen, R. M., *The Sciences: An Integrated Approach,* 2nd ed./updated ed. (John Wiley and Sons, New York, 2000), 347–70.

103. Weinberg, S., *Gravitation and Cosmology: Principles and Applications of the General Theory of Relativity* (John Wiley and Sons, New York, 1972).

104. Whitehead, A. N., *The Principle of Relativity with Applications to Physical Science* (Cambridge University Press, Cambridge, 1922).

105. Whitehead, A. N., *Science and the Modern World* (Free Press, New York, 1925).

106. Whitehead, A. N., *Process and Reality* (Macmillan, New York, 1929).

107. Whitehead, A. N., *Process and Reality,* corrected ed., ed. D. R. Griffin and D. W. Sherburne (Free Press, New York, 1978), 94–95, 238–39, 254.

108. Wildman, W. J., "Evaluating the Teleological Argument for Divine Action," in *Evolutionary and Molecular Biology: Scientific Perspectives on Divine Action,* eds. R. J. Russell, W. R. Stoeger, and F. J. Ayala, Scientific Perspectives on Divine Action Series (Vatican Observatory Publications, Vatican City State/Center for Theology and the Natural Sciences, Berkeley, Calif., 1998).

109. Will, C. M., *Theory and Experiment in Gravitational Physics* (Cambridge University Press, Cambridge, 1981).

110. Wolterstorff, N., *Reason Within the Bounds of Religion* (Eerdmans, Grand Rapids, Mich., 1976).

[18]

NATURES OF EXISTENCE (TEMPORAL AND ETERNAL)

George F. R. Ellis

18.1. Introduction

Ultimately, one's view of the nature of the far future, both as regards the universe and as regards humanity, depends on one's view on the nature of existence, that is, of the kinds of ontology possible. Some kinds of entity may be expected to endure forever, others not, regardless of the fate of the physical universe. Closely related is the issue of what kind of entities may preexist the universe and in some sense make it come into existence, for whatever preexists may perhaps be expected to postexist as well. Here one runs into the usual paradoxes in terms of referring to what occurs and exists: "before the beginning" and "occurs" do not make sense when time does not yet exist, nevertheless one feels compelled to consider some kind of temporal existence independent of the existence of the universe and its time rela-

Note: I thank Phil Clayton, Bill Stoeger, and Nancey Murphy for helpful comments on an earlier draft.

tions.[1] Language cannot easily handle this situation, and we are feeling for a mode of explanation in a domain where ordinary ideas of time and existence may not apply. Accepting this irremovable problem, we can try to see what understanding we may gain.

Preliminary to doing so, we need to look at issues about what should be considered "real" not only in the physical world, but in the world of human interactions in which we live and move and have our being. We consider this first, before turning to the issues of ontology and the far-future universe.

18.2. Hierarchical Structure and Existential Levels

Matter is hierarchically structured, with living beings composed of cells that are made of molecules composed of atoms that comprise nuclei and electrons; the nuclei being made of protons and neutrons that are made of quarks. The successive levels of order entail chemistry being based on physics, material science on both physics and chemistry, geology on material science, and so on.[2] Given the number of lower-level constituents involved (1 gram of hydrogen contains 6×10^{23} molecules, for example), a detailed description of the lower levels is impracticable in most situations, and in conditions close to equilibrium is replaced by a statistical description, with the laws of statistical physics relating the lower- to the higher-level behaviors.

The profound discovery of molecular biology is that through the extraordinary nature of biological molecules, a fully mechanistic description applies at the microlevel of living organisms also, including human beings and the human brain.[3] However, biological systems are open systems that are far from thermal equilibrium, so statistical physics does not apply. Rather what happens is governed by detailed structural relations and molecular interactions.

At the classical levels of structure (i.e., levels where we can use phenomenological relations that ignore quantum theory), the relation of micro- to macro- structure is in principle deterministic, given a complete description of the microstructure and the lower-level causal relations. Thus a fundamental feature of the structural hierarchy in the physical world is *bottom-up*

1. Which could conceivably be associated either with higher dimensions, or with other universes in an ensemble.

2. See for example Peacocke [36] and Ellis [15] for a more detailed discussion.

3. See for example Campbell [11].

action: what happens at each higher level is based on causal functioning at the level below, hence what happens at the highest level is based on what happens at the bottom-most level. This is the profound basis for reductionist world views. On this basis, there is a tendency for people to speak as if the higher-level structures are not real ("you are nothing but a collection of atoms"): only the microstructure truly exists, because it is seen as the basis of the causality at all levels.

The first point in response is that we don't even know what the lowest levels of structure are (maybe they are quarks, but maybe not), so if taken seriously this view reduces us to a state of impotence—we don't know what is really real! But additionally, this absolutist view simply does not take seriously the existential status of the higher levels. In contrast, the view taken in this essay is that quarks and electrons are real, as are protons and neutrons, tables and chairs, planets and stars, ants and people. The fact that each is comprised of lower-level entities does not undermine its status as existing in its own right. At a certain level this statement is obvious: I am presently sitting on a chair. To deny this simple fact is an exercise in futility that ignores the evidence. Daily life has its own obvious existential status; those who deny this are playing philosophical games that are uninteresting because they fail to account for features of life that are of major significance to us.

The higher-level features are based in the *ordered relations* between the component parts; these relationships are not "nothing," on the contrary, they are the essential essence that enables the higher-order structure to exist and have its own levels of emergent order and behavior. These higher levels of order are not implied by the lower-level microphysics alone; indeed the language applicable at the microlevels does not even have the ability to describe the higher-level structure and behavior, for the applicable concepts and associated order are not microlevel concepts. They cannot be comprehended at that level.[4] Furthermore, the lower levels place remarkably little constraint on what happens at the higher levels, where independent levels of causality and phenomenology come into play. Thus, for example, I can type any text I like on this computer; the deterministic action of the forces between the particles comprising the machine at the microlevel allow this. The

4. For example, a spark plug is not a concept that can be described in the language of quantum field theory.

phenomenological causal relations valid at each level in the hierarchy can be studied in their own right; for example, the computer with which I write this essay acts as a word processor with well-established behavioral rules, but at a lower level functions in terms of a machine code with its own precise language. Indeed *we can attain and confirm high representational accuracy and predictive ability for quantities and relations at each level, independent of any knowledge of interactions at lower levels,* giving well-validated and reliable descriptions at higher levels accurately describing the various levels of emergent nonreducible properties and meanings.

The physical reasons allowing this independence of higher-level properties from the nature of lower-level constituents have been discussed by Anderson [3], Schweber [46], and Kadanoff [25]. This is a fundamentally important property of physics, underlying our everyday lives and their reality. Its source is the nature of quantum field theory applied to the microproperties of matter as summarized in the standard model of particle physics; a significant role is played by the scaling behavior of matter characterized by the renormalization group [46]. The overall result is that effective field theories arise that describe the behavior of matter at each level [23], or more generally, phenomenological relations that give a reliable description of higher-level behavior. The higher-level behaviors are largely independent of the detailed nature of their component parts at lower levels. An example (already mentioned) is digital computers, with their hierarchical, logical structure expressed in a hierarchy of computer languages that underlie the top-level user programs [50]. The user does not need to know machine code, and indeed the top-level behavior is independent of which particular hardware and software underlie it at the machine level. The computer has a reality of existence at each level that enables one to deal with it meaningfully as an entity at that level (a "virtual machine"), independent of the details of its lower-level structure. Another example is that a motor mechanic does not have to study particle physics in order to ply her trade.

The key point underlying the apparent effectiveness of higher-level order in causal terms is the complementary feature to bottom-up action, namely *top-down action,* which is ubiquitous in biology and in practical applications of physics [10, 18, 27, 32, 33, 36, 37]. By top-down action, higher-level entities coordinate vast numbers of actions at the lower level so as to produce the desired effect at the higher levels; thus bottom-up action takes place in a way that is controlled in a top-down manner. This is how the higher-level entities are able to act in a way that is meaningful at the higher level, so that there is an ontological reality of higher levels of the hierarchy, which can be considered as comprising entities existing in their own right and with their own effective laws of behavior. An example is that cited earlier of the functioning of a computer: electrons flow in the computer circuits according to the instructions entered via keystrokes or mouse movements; they do so in such a way as to carry out those instructions reliably. They would flow differently were different keystrokes to be entered. This feature completely changes the nature of causal interactions in a hierarchical system (see fig. 18.1), enabling the higher-causal levels to be effective causally in the context of a system where bottom-up action underlies what happens at the higher levels.

Top-down (holistic) action is central both in physics [28] and in molecular biology, even though the main emphasis in most texts is on the bottom-up (mechanistic) aspects. The first central theme of evolutionary biology is the development of particular DNA codings (the particular sequence of bases in the DNA) through an evolutionary process that results in adaptation of an organism to its ecological niche [11]. This is a classical case of top-down action from the environment to detailed biological microstructure—through the process of adaptation, the environment (along with other causal factors) fixes the specific DNA coding [10]. There is no way you could ever predict this coding on the basis of biochemistry or microphysics alone. The second central theme of molecular biology is the reading of DNA by an organism in the developmental process [53, 12]. This is not a mechanistic process, but is context dependent, with what happens before having everything to do with what happens next. The central process of developmental biology, whereby positional information determines which genes get switched on and which do not in each cell, so determining their developmental fate, is based on the existence of gradients of positional indicators in the body [53]. Without this feature organism development in a structured way would not be possible. Thus the functioning of the crucial cellular mechanisms determining the type of each cell is controlled in an explicitly top-down way. At a more macrolevel, recent research on genes and various hereditary diseases shows that existence of the gene for such diseases in the organism is not a sufficient cause for the disease to in fact occur: outcomes depend on the nature of the gene, on the rest of the genome, and on the environment [52].

FIGURE 18.1. *Bottom-up and Top-Down Action*

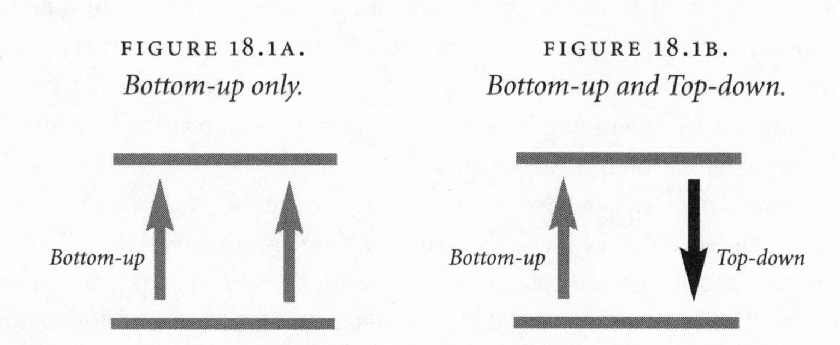

The fundamental importance of top-down action is that it changes the overall causal relation between upper and lower levels in the hierarchy of structure and organization, cf. the difference between fig. 1a and fig. 1b. In particular it enables existence of multi-level feedback loops.

The macrosituations determine what happens, not microfeatures by themselves, which do work mechanistically but in a context of higher-level meaning that largely determines the outcome. And note particularly that the macro-environment includes the result of conscious decisions (the patient will or will not seek medical treatment for a hereditary condition, for example), so these too are a significant causal factor.

18.3. Human Volition

The last point emphasizes that consciousness brings a whole new series of effects into the causal network. When a human being has a plan in mind (say a proposal for a bridge to be built) and this is implemented, then enormous numbers of microparticles (comprising the protons, neutrons, and electrons in the sand, concrete, bricks, and so on that become the bridge) are moved around as a consequence of this plan and in conformity with it. Thus in the real world, the detailed microconfigurations of many objects (which electrons and protons go where) is in fact to a major degree determined by the macroplans that humans have for what will happen, and the way they implement them. Some specific socially important examples of top-down action involving goal-choice are:

The internet: This embodies local action in response to information requests, causing electrons to flow in meaningful patterns in a computer's silicon chips and memory and activating memory locations thousands of miles away, when one reads web pages. This is a structured influence at a distance owing to channeled causal propagation and resulting local physical action.

Hiroshima: The dropping of the nuclear bomb at Hiroshima in 1945 was a dramatic macro-event realized through numerous micro-events (fissions of uranium nuclei) occurring because of a human-based process of planning and implementation of those plans.

Global warming: The effect of human actions on the earth's atmosphere, through the combined effect of human causation, moves very large numbers of microparticles (specifically, CFCs) around, thereby affecting the global climate. The macroprocesses at the planetary level cannot be understood without explicitly accounting for such human activity [45].

The foundation for all of this is the top-down action in the human body, where the brain controls the functioning of the parts of the body through a hierarchically structured control system with multiple levels of feedback loops, which uses decentralized control principles to spread the computational and communication load and increase local response capacity [6]. It is a highly specific system in that dedicated communication links convey information from specific areas of the brain to specific areas of the body, enabling brain impulses to activate particular muscles (by coordinated control of electrons in myosin filaments in the bundles of myofibrils that constitute skeletal muscles) in order to carry out consciously formulated intentions. This is the feature that underlies the discussion of the right-hand branch of the causal hierarchy proposed in Murphy and Ellis [34], where the various aspects of human consciousness with a nonreducible nature[5] act down on the physical environment in an effective way, and are themselves hierarchically structured in a causal fashion in terms of goals and aims, with ethics as the topmost causal level.

Furthermore, through this process there is top-down action by the mind on the body, and indeed on the mind itself, both in the short term (immediate causation through the structural relations embodied in the brain and body) and in the long term (structural determination through imposition of repetitive patterns). An example of the latter is how repeated stimulation of the same muscles or neurons encourages growth of those muscles and neu-

5. As in other cases discussed above, the relevant concepts (intention, fear, love, hate, etc.) cannot even be expressed in terms of the lower-level concepts (electrons, atoms, neurons, etc.)

rons. This is the underlying basis of both athletic training and of learning by rote, but it is much more, for it applies to all habitual behavior; indeed, the nature of our chosen goals and even our morality is in the long term hard-wired into our neural circuits, and so realized in the physical microstructure of our brains [34]. A related area of importance that is only now beginning to be investigated by Western medicine is the effect of the mind on health [30, 49].

The implication is that human ideas and emotions are causally effective in the real world, in terms of determining the motions of atoms and molecules, and so must be allowed an existence in their own right.[6] They are not reducible to the world of atoms and molecules, being essentially different not only from the material stuff of the world but also from the structural relationships between such components (which enable them to exist but are not identical to them). They come into being through a psychosocial process [8] controlled by the nature of our brains, in turn determined through our evolutionary history. Thus, for example, a virtue such as humaneness is a capacity generated by genetic inheritance [21]. But this capacity only develops in the context of particular environments and experience, which shape the way genetic inheritance evolves into patterns of understanding and behavior through a neural developmental process in each individual with a broadly Darwinian nature [13]. This kind of developmental determination through interaction with the environment, rather than some more direct genetic specification, is required both by the lack of sufficient genetic information to be able to determine neuronal connections in any detail[7] and by the imperative of constructing a good neural fit to the specific social milieu and environment in which the organism lives. The consequence is that *environmental and social forces play a major role in shaping the mind* through this process. Thus there is major top-down action by the environment—physical and social—on the human mind, as well as by the mind on itself.

6. There is nothing strange about this idea; it is obvious from everyday experience. What *is* strange is that in the face of scientific reductionism one should have to explicitly state this everyday fact.

7. The roughly 10^4 genes available to construct the human brain are hugely less than the 10^{11} neurons in the brain and the vastly greater number of possible connections between them—there is about one gene per 10^7 neurons!

18.4. A Holistic View of Causality

Given the multiple levels of causation in complex systems, it is important that we realize the multiple kinds of explanation of any event. First, in looking at complex systems, we must include top-down (systemic/holistic) effects and explanation as well as bottom-up (analytic/reductionist) effects and explanation. For example, Russell Ackoff [2] comments in an illuminating way on systemic explanation: you can explain why a jumbo jet flies by analyzing the forces on the wing that keep it in the air (a bottom-up, analytic explanation) or by remarking that it flies because it has been designed to fly (a top-down, systemic explanation). Both are true explanations.

Second, we must remember that every human endeavor has an explanation in terms of many factors:

(a) *microstructure:* physics, chemistry, biochemistry, microbiology;
(b) *macrostructure:* psychology, sociology, economics, politics;
(c) *historical context:* evolutionary origins, cosmological basis;
(d) *phenomenology:* its intrinsic merit in its own terms.

For example, we can to some degree "explain" the thoughts of a developmental biologist, engineer, or economist as being

(a) due to the forces between electrons in her neurosystem, according to Maxwell's equations; due to detailed molecular interactions realizing the principles of physical chemistry; due to the interaction of various electric and chemical potentials, summarized in the Hodgkin-Huxley equations; and so on (bottom-up explanations);

(b) due to her psychological makeup, which channels her thoughts in various directions, which are shaped by her education and culture, and perhaps to a degree by her socioeconomic class and political disposition (top-down systemic explanation);

(c) due to her evolutionary origins as a member of a hominid species, where specific sociobiological processes led to the basic structure of her brain and perhaps specific brain modules (evolutionary explanation);

(d) due to the internal logic of developmental biology, engineering, or economics, and the relevant data she discovers through her studies, together gradually cohering to form a theory that both respects the data and provides an overall explanatory pattern for the subject (the internal logic of the subject considered in its own right).

These all represent causal forces at work, and so are all correctly identified as explanatory factors that can help us understand what is happening; consequently, *none of them by itself provides the total explanation, which is given by the confluence of all these factors* (one can at any moment focus on any particular one of them and ignore all the others; that is then the "explanation" in that context, where all the others are taken for granted, but it is only effective because all the other features are active in the background). And the same overall complex pattern of causes will be at work in any subject: physics, mathematics, religion, and evolutionary biology, for example.

We can focus on any of these causal factors (evolution, for example), while taking the others for granted, and study its particular effects; this gives a valid partial explanation of what occurs. *The problem comes when we claim this is the total explanation,* that is, that we need take none of the other factors into account. Indeed, *fundamentalism,* properly considered, is just *the insistence that some single aspect is the whole*—it is the claim that some partial explanation is all there is to consider, denying the need to take into account the whole set of causes in operation. We need to resist the strong tendency to adopt this position. A full causal explanation must take all causes into account. Nevertheless, in understanding the subject in its own right, it is (d), its own internal logic and structure in relation to the relevant data set, that counts most in each case (the other factors are only relevant here if they can be shown to have distorted the internal analysis from what it ought to have been). And this applies just as much to ethics or religion as it does to physics or evolutionary biology. Each must be looked at in its own internal terms as well as in terms of evolutionary psychology or sociology. In no case can either of the latter provide a complete explanation.[8]

8. Thus, for example, the attempt to give a *total* explanation of religion or ethics in terms of evolutionary biology is just as misplaced as a similar attempt to explain evolutionary biology itself in such terms.

18.5. A Holistic View of Ontology

Closely related to issues of explanation are issues of ontology, delimiting what is real, that is, what actually exists, as compared with what is seen as insubstantial because it is taken as essentially comprised only of lower-level objects, and so is without serious existential status in its own right. The overall question, then, is *to what entities should we assign an ontological reality?* The discussion above of human intentions and emotions suggests that there are not only different levels of reality in a hierarchically structured system, all present at the same time in a cohesive way, but that there may be different kinds of reality—different worlds, as suggested *inter alia* by Plato, Popper, Eccles, and Penrose—that must each be taken seriously.

To approach this question, I *take as given the reality of the everyday world*—tables and chairs, and the people who perceive them—and then assign a reality additionally to each kind of entity that can have a demonstrable causal effect on that everyday reality. Thus, I suggest *an entity must be assigned an ontological reality if it has a presence that results in detectable causal effects in the real everyday world.* In this case it represents a clearly existing causal factor, able to influence what happens. The key point then is that given this definition, if any of these kinds of entities are omitted from the overall scheme of understanding, then causality is necessarily incomplete. Hence there is a strong imperative to recognize all of them and take them seriously.

The problem now is to characterize the various kinds of independent reality that may exist in this sense. Taking into account the causal efficacy of all the entities discussed above, I propose as a possible completion of the proposals by Popper and Eccles [43] and Penrose [38, 39] that the following four worlds are ontologically real. These are not different causal levels within the same kind of existence, rather they are quite different kinds of existence, but related to each other through causal links [39]. The challenge is to show firstly that each is indeed ontologically real, and secondly that each is sufficiently and clearly different from the others that it should be considered as separate from them.

First, and foundationally, we have

World 1: The physical world of energy and matter, hierarchically structured to form lower- and higher-causal levels whose entities are all ontologically real. This is the basic world of matter and interactions between matter, based at the microlevel on elementary particles and fundamental forces, and providing the ground of physical existence.[9] The macroworld attains its specific properties in virtue of the specific structural relationships existing between the microcomponents [2, 27]. It comprises three major parts:

World 1a: Inanimate objects (both naturally occurring and manufactured);

World 1b: Living things, apart from humans (amoeba, plants, insects, animals, etc);

World 1c: Intelligent beings, with the unique property of being self-conscious. All these objects are made of the same physical stuff, but the structure and behavior of inanimate and living things (enabled by their constitutional relations and described respectively by physics and inorganic chemistry, and by biochemistry and biology) are so different that they require separate recognition, particularly when self-consciousness and purposive activity (described by psychology and sociology) occur. The different levels of the hierarchy are each real, with their own phenomenology and associated descriptive language; top-down and bottom-up action takes place between these levels in the hierarchy, as discussed above.

Second, I assign an ontological reality to human thoughts and to intentions, plans, and emotions. As discussed in the previous section, these demonstrably have causative effects on the course of macro-, and hence micro-, events. Thus I include them as truly existing macrolevels of causation, because our understanding of events dictates much that happens in the causality chain, as do our plans and emotions, for example, the existence of jumbo jet aircraft—and the nuclear explosion at Hiroshima. Thus we recognize

9. The nature of the ontology at the quantum level is disputable because of quantum uncertainty, wave-particle duality, and the phenomenon of entanglement; also the ontological nature of the physical laws describing the physical properties of matter is unknown (we do not know if the laws are prescriptive or descriptive). Nevertheless, there is clearly a micro-physical existence underlying the existence of macro-physical features. We do not have to understand the microphysical ontology of either particles or forces in order to comprehend the effective macro-physical laws.

World 2: The world of individual and communal consciousness, which consists of ideas, emotions, and social constructions. This again is ontologically real (it is clear that these all exist), and are causally effective.[10]

This world of human consciousness can be regarded as comprising three major parts:

World 2a: Human information, thoughts, theories, and ideas;
World 2b: Human goals, intentions, sensations, and emotions;
World 2c: Explicit social constructions.

These worlds are different from the world of material things, and are realized through the human mind and society. They are not identical to each other: world 2a is the world of rationality, world 2b is the world of intention and emotion, and so comprehends nonpropositional knowing, while world 2c is the world of consciously constructed social legislation and convention. Although each of these aspects is individually and socially constructed in a complex interaction between culture and learning [8], as discussed above they are indeed each capable of causally changing what happens in the physical world, and each has an effect on each of the others.

In more detail, the rationality of world 2a[11] is hierarchically structured, with many different components. It includes words, sentences, paragraphs, analogies, metaphors, hypotheses, theories, and indeed the entire bodies of science and literature, and refers both to abstract entities and to specific objects and events. It is necessarily socially constructed on the basis of varying degrees of experimental and observational interaction with world 1, which it then represents with varying degrees of success. World 2a is represented by symbols, particularly language and mathematics, which are arbitrarily assigned (e.g., in terms of various alphabets) and which can themselves be represented in various ways (sound, on paper, on computer screens, in digital coding, and so on). Thus each concept can be expressed in many different ways, and is an entity in its own right independent of which particular way it is coded or expressed.

10. The causal hierarchy that Nancey Murphy and I have developed, described in [34], has a dividing hierarchy of the sciences, separating out the natural sciences and the social/ human sciences at the higher levels, precisely in order to give expression to the causal reality of these features.

11. This is Popper and Eccles's world 2 and Penrose's mental world. Insofar as that concept is well defined, this is also the world of *memes.*

These concepts sometimes give a good correspondence to entities in the other worlds, but *the claim of ontological reality of entities existing in world 2a (specific concepts or theories) makes no claim that the objects they refer to are real.* Thus this world equally contains concepts of rabbits and fairies, galaxies and UFOs, science and magic, electrons and ether, unicorns and apples; the point being that all of these certainly exist *as concepts.* That statement is neutral about whether these concepts correspond to objects or entities that exist in the real universe (specifically, whether there is or is not some corresponding entity in world 1) or whether the theories in this world are correct (that is, whether they give a good representation of world 1or not). All the ideas and theories in this world are ontologically real in that they are able to cause events and patterns of structures in the physical world (e.g., they may appear as entries in a dictionary or encyclopedia or other book and hence are encoded in print; indeed each one of the above ideas occurs as a reference in this paragraph!). The fact that they have these multiple representations shows that they cannot be reduced to the physical concept of brain states, as is sometimes suggested; rather, brain states provide one of a variety of classes of physical representations of these abstract concepts.

World 2b[12] is a world of motivation and senses that is also ontologically real, for it is clear that they do indeed exist in themselves. In world 2b, we find the goals and intentions that cause the intellectual ideas of world 2a to have physical effect in the real world.

World 2c is the world of language, customs, roles, laws, and so on, which shape and enables human social interaction.[13] It is developed by society historically and through conscious legislative and governmental processes. It gives the background for ordinary life, enabling worlds 2a and 2b to function, particularly by determining the means of social communication.

Next, we have

World 3: The world of Aristotelian possibilities characterizes the set of all physical possibilities, from which the specific instances of what actually happens in world 1 are drawn. This world of possibilities is ontologically real because of its rigorous prescription of the boundaries of what is possible—it

12. This is Popper and Eccles's world 3 .

13. See Aberle [1] for a discussion of such features, and Berger and Luckmann [8] for how they are effected in society.

provides the framework within which world 1 exists and operates, and in that sense is causally effective. It can be considered to comprise two major parts:

World 3a: The world of physical possibilities, delineating possible physical behavior;

World 3b: The world of biological possibilities, delineating possible biological organization.

These worlds are different from the world of specific existing material things for they provide the background within which that world exists. In a sense they are more real than that world, because of the rigidity of the structure they impose on world 1. There is no element of chance or contingency in them, and they certainly are not socially constructed (although our understanding of them is so constructed). They rigidly constrain what can happen in the physical world, and are different from each other because of the great difference between what is possible for life and for inanimate objects.

In more detail, world 3a delineates possible physical behavior (it is a description of all possible motions and physical histories of objects). Thus it describes what can actually occur in a way compatible with the nature of matter and its interactions (e.g., motions that conserve energy and momentum); only some of these configurations are realized through the historical evolutionary process in the expanding universe. This world delineates all physically possible actions (different ways particles, planets, footballs, automobiles, aircraft can move, for example); from these possibilities, what actually happens is determined by initial conditions in the universe, in the case of interactions between inanimate objects, and by the conscious choices made, when living beings exercise volition.

These possibilities are characterized by the laws of behavior of matter, and indeed most physicists would think of this world as simply being determined or represented by the laws of physics, which rigidly determine the behavior of matter; hence an alternative would be to speak of the world of laws of physics. However, the ontological nature of those laws is disputed, whereas the existence of these limitations (consequent on the nature of matter) is indisputable. That is why I prefer to introduce here the space of possibilities, rather than the set of physical laws of behavior that may determine those possibilities (but I am not excluding the latter; see below). The issue of where these limitations exist and what the precise nature of their existence is, is unclear: there are various views. However, they are inescapably built into the nature of matter, whatever description one uses. If one believes the laws of physics exist in a Platonic sense, spaces of possibilities are deeply embedded in their descriptions: examples are the "block universe" of special relativity, the phase spaces of dynamical systems theory, and the Hilbert spaces of quantum theory. If one believes that the behavior of matter is somehow dictated to be as if the laws of physics exist, even though they do not, then these limits (and so the associated spaces of possibilities) are incorporated in a clear sense in the existence of the specific behavior of matter. To delimit the ontological nature of these worlds of possibilities in this case, one would have to solve the intractable problem of why, given this view, all matter has the same specific, mathematically describable behavior everywhere in the universe, and that resolution to this problem will indicate the locus of existence of these limitations.

World 3b delineates possible biological organization (and so is a representation of all possible organisms). This defines the set of potentialities in biology by giving rigid boundaries to what is achievable in biological processes. Thus it constrains the set of possibilities from which the actual evolutionary process can choose—it rigorously delineates the set of organisms that can arise from any evolutionary history whatever (one cannot, for example, have an ant the size of an elephant, or an animal that functions without energy intake). This "possibility landscape" for living beings, determined to a considerable degree by the underlying space of physical possibilities, underlies evolutionary theory, for any mutation that attempts to embody a structure that lies outside its boundaries will necessarily fail. Thus even though it is an abstract space in the sense of not being embodied in specific physical form, it strictly determines the boundaries of all possible evolutionary histories. In this sense it is highly effective causally.

Only some of the organisms that can potentially exist are realized in world 1 through the historical evolutionary process; thus only part of this possibility space is explored by evolution in any particular world. When this happens, the information is coded in the hierarchical structure of matter in world 1, and particularly in the genetic coding embodied in DNA, and so is stored via ordered relationships in matter; it then gets transformed into various other forms (folded proteins, for example) until it is realized in the structure of an animal or plant. In doing so it encodes both a historical evolutionary sequence, and structural and functional relationships that emerge in the phenotype and enable its functioning, once the genotype is read. This is the way that numerous directed feedback systems (such as occur in each living cell and in animal physiology) and the idea of purpose (as in animal behavior) can enter the biological world, and so distinguish the animate from the inanimate. The structures occurring in the nonbiological world can be complex (a mountain chain, for example), but they do not incorporate purpose or order in the same sense.[14] Just as world 3a can be thought of as encoded in the laws of physics, world 3b can be thought of as encoded in terms of biological information [26, 41], which is a core concept in biology, distinguishing the world of biology from the world of nonliving entities, which incorporates no corresponding concept.[15]

Finally, we have

World 4: A Platonic world of (abstract) realities that are independent of human existence, but not embodied in physical form. They can have causal effects in the physical world. This world (Platonic realities discovered by human investigation) is perhaps the most contentious. The prime contenders to inhabit such a world are mathematical forms:

World 4a: Mathematical forms. The existence of a Platonic world of mathematical objects is strongly argued by Penrose [38, 39], the point being that major parts of mathematics are discovered rather than invented (rational numbers, zero, irrational numbers, and the Mandelbrot set being classic examples). They have an abstract rather than embodied character, this abstract nature being manifested in that the same abstract quantity can be represented and embodied in many different symbolic and physical forms; the same underlying structure and pattern is present in all these representations. They are not determined by physical experiment, but are independent of the existence and culture of human beings, for the same features will be discovered by intelligent beings in the Andromeda galaxy as here, once their mathematical understanding is advanced enough (which is why they are advocated as the basis for inter-stellar communication); for example, all intelli-

14. Unless designed by intelligent beings precisely in order to store and interpret complex information.

15. And is of course irreducible: it cannot be expressed in terms of lower-level concepts.

gences will discover the same values for the numbers π, e, and √2, even if they code these values in different forms.

This world is partially discovered by humans (not all of it is yet known!), and so is represented by our mathematical theories in world 2; that representation is a cultural construct, but the underlying mathematical features they represent are not—indeed, like physical laws, they are often unwillingly discovered, as for example in the cases of zero [48] and irrational numbers. This world is causally efficacious in terms of the process of discovery and description (one can for example print out the values of irrational numbers such as p, or graphic versions of the Mandelbrot set, in a book).

A key question is what if any part of logic, probability theory, and physics should be included here. In some as yet unexplained sense, the world of mathematics underlies the world of physics [39]. Many physicists at least implicitly (and sometimes explicitly) assume the existence of a Platonic world underlying physics: namely,

World 4b: Physical laws, which underlie the nature of physical possibilities (world 3a).

Further proposals are

World 4c: A Platonic world of concepts and ideas, which provide a foundation for worlds 2a and 2c; and

World 4d: Platonic aesthetic forms, which provide a foundation for our sense of beauty.

These possibilities will not be pursued further at this point, although the idea of worlds 4b and 4c will be picked up later on. It is sufficient for my present purpose to note that the existence of a world of mathematical forms, which (in agreement with Penrose) I strongly support, establishes that this category of world indeed exists and has causal influence.

Now with this background, I can make the main claim:

The overall family of worlds: All these worlds exist—worlds 1 to 4 are ontologically real and are distinct from each other. The reason for this claim is that (1) they are all causally effective; omitting any of them will result in an incomplete causal account of actions and events in the physical world, as argued above; and (2) they are clearly distinct from each other in nature. Indeed, they are each of a different nature from the others—possibilities do not exist in the same sense as realized possibilities, for example—so the sense of "existence" in each case is different; nevertheless the claim is that this existence is required in each case by any complete causal scheme for happenings in the physical world. It is not clear at this point whether the world of possibilities (world 3) exists independently of world 1, which realizes those possibilities; I will argue later that this is indeed the case. However, the causally effective existence of world 3 can be argued independent of the resolution of that issue. It is also not clear "where" worlds 3 and 4 exist. I will return to that question in the context of the discussion of theology below.

The claim made here clearly denies simple-minded materialism, for it assigns an essential existence to a variety of entities that do not live in world 1 (this world is all that would be recognized by such a viewpoint). It also rejects both strong relativism and any strong reductionism by recognizing the reality of the separate levels of the hierarchy of material structure and by assigning an existential reality to the ordering relationships (and in the case of living entities, to the biological information) that give these levels their essential nature.[16]

What then of epistemology? Given the existence of the various worlds mentioned above, the proposal here is that *epistemology is the study of the relation between world 2 and worlds 1, 3, and 4.* It attempts to obtain accurate correspondences to quantities in all the worlds by means of entities in world 2a.

16. In Ruse's terms [44]: *Ontological reductionism* ("one tries to show everything comes from or consists of one or just a few basic substances") is accepted here in that the stuff of all matter, including living things, is taken to be just the particles of elementary particle physics (which are *not* just a few!), but is rejected in that the ordering relations between them are also seen as having an existential status over and above that of their constituent particles, these relations being crucial in the nature and functioning of all complex systems, including life. *Methodological reductionism* ("one urges explanation in terms of the smallest possible entities") is accepted as an initial principle of analysis, but it is understood that it is bound to fail in the end—unless one accepts macrosystems as "smallest possible entities"—because it does not take account of top-down action. *Theoretical reduction* (where one tries to explain one theory in terms of another) is again recognized as an idea one should pursue as far as possible, but it is recognized that it is bound to fail in practice even in the arena of physics [28], where the effective theories replacing fundamental theories are only partially explained by those fundamental theories (see [23] and the example of the arrow of time), let alone in biology, the functioning of the mind (e.g., see the case of the Hodgkin-Huxley equation [47]), and social systems.

This implicitly or explicitly divides world 2a theories and statements into (1) true/accurate representations, (2) partially true/misleading representations, (3) false/poor/misleading representations, and (4) ones where we don't know the situation. These assessments range from statements such as "It is true her hair is red" and "There is no cow in the room" to "Electrons exist," "Newtonian theory is a very good description of medium-scale physical systems at low speeds and in weak gravitational fields" and "The evidence for UFOs is poor."[17]

This raises interesting issues about the relation between reality and appearance. It is of course known that everyday life gives a quite different appearance to reality than implied by our knowledge of microscopic physics—as Eddington pointed out [14], a table is actually mostly empty space between the atoms that make up its material substance, but in our experience it is a solid hard object. Daily life experience is determined by the nature of our sensory systems and is shaped by the time and spatial averaging of physical features that inevitably occurs in their operation; we cannot easily discern features of the universe on either a very small scale, or a very large scale by our own sensory systems alone—we need expensive scientific instruments to do so. One can learn much about object-appearance relations through solid state physics and physical chemistry (dealing with the appearance of everyday physical objects), astrophysics (dealing with detection probabilities and appearances of the same astronomical object seen by different detectors, giving very different views of the same object at different wavelengths), and relativity theory (giving rigorous rules for the changes in appearances of objects when one changes the reference frame). Cosmology also gives useful analogies and arguments about the relation between evidence and theory, particularly because there exist regions beyond the cosmological horizon that we cannot see or examine by any astronomical observation whatever [17, 19].

Similar issues arise in epistemology in much broader contexts, demon-

17. This approach thus explicitly entails a *correspondence theory of truth*—our ideas map a meta-physically independent world of things (world 1) into a space of concepts (world 2a) in a way that successfully represents relations in that world in terms of a complex of concepts and relations between them, rather than a *coherence theory of truth* ("the true is that which fits in with the rest, and the false or illusory that which does not") [44]. It is quite unclear to me what coherent relation the latter somewhat ill-defined statement has to ontology.

strating the folly of equating existence with observational proof of existence. Indeed, there is a widespread tendency to equate epistemology and ontology (as in both logical positivism and in extreme relativism). This is a logical error, and a variety of examples can be given where it seriously misleads. This is related to a confusion between world 2 and the other worlds discussed here, which seems to underlie much of what has happened in the so-called "science wars" and the notorious Sokal affair.[18] The proposal here strongly asserts the existence of independent domains of reality (worlds 1, 3, and 4) that are not socially constructed, and implies that we do not know all about them, and indeed cannot expect ever to understand them fully. That ignorance does not undermine their claim to exist in a way that is independent of human knowledge and understanding. Any explicit or implicit claim that we can equate epistemology and ontology is in the end simply an example of human hubris.

18.6. The Universe: Origins and the Far Future

With this understanding of ontology, we can return to the issue of what kinds of entities might preexist and/or postexist the universe. We discuss worlds 1, 3, and 4 here, and turn to the much more controversial issue of world 2 in the next section.

Clearly world 1 (particles and forces) comes into being with the creation of the universe, and comes to an end if the universe ends; that world remains in existence forever if the universe does not end, even if the particles themselves decay away. However in some sense, world 3 (Aristotelian possibilities) and world 4 (Platonic entities) transcend physical existence, relating to its eternal underlying structure rather than its transient and contingent physical existence. Thus on many views they may be expected both to precede the existence of the universe and to last after its end, in the case of a universe that comes to an end, or to continue eternally unchanged and watchful in an ever-expanding universe, long after the last life has ended and all matter has decayed away. In an ensemble of universes of the kind discussed by Martin Rees (this volume), both kinds of future fate would happen indefinitely often in the many universes that would occur.

18. See http://www.physics.nyu.edu/faculty/sokal/index.html and http://zakuro.math.tohoku.ac.jp/~kuroki/Sokal/

Much of physics is underpinned by quantum field theory, which, in its full form applied to the standard model of particle physics [40] (and I would argue that one cannot talk meaningfully about quantum theory in the abstract: it only gains its full meaning in the context of the theory to which it is applied), is immensely complex. It conceptually involves *inter alia*

- Hilbert spaces, operators, commutators, symmetry groups;
- particles/waves/wave packets, various kinds of spinors, quantum states/wave functions;
- parallel transport/connections/metrics in spaces of various dimensions;
- the Dirac equation and interaction potentials;
- broken symmetries and associated particles; and
- variational principles (Lagrangians and Hamiltonians) that seem to be logically and/or causally prior to all the rest.

Derived (effective) theories, including classical (nonquantum) theories of physics, equally have complex abstract structures underlying their use: force laws, interaction potentials, metrics, and so on.

The issue here is, *What is the ontology/nature of existence of all this quantum apparatus?* We seem to have two options [31]:

(1) These are simply our own mathematical and physical constructs that happen to characterize reasonably accurately the physical nature of physical quantities.

(2) They represent a more fundamental reality as Platonic quantities that have the power to control the behavior of physical quantities (and can be represented accurately by our descriptions of them).

On the first supposition, the "unreasonable power of mathematics" to describe the nature of the particles is a major problem—if matter is endowed with its properties in some way we are unable to specify, but not determined specifically in mathematical terms, and its behavior happens to be described accurately by equations of the kind encountered in present-day mathematical physics,[19] then that is truly weird! Why should it then be possible that *any*

19. For example, a Yang-Mills type theory associated with a group structure with SO(3)×SU(2)×U(1) as a subgroup, or a string theory of all forces associated with particular Calabi-Yau spaces.

mathematical construct whatever gives an accurate description of this reality, let alone ones of such complexity as in the standard theory of particle physics? Additionally, it is not clear on this basis why all matter has the same properties—why are electrons here identical to those at the other side of the universe? Why do they all obey the same force law? What is it that maintains *all* the matter in the universe under the sway of that specific behavior, which can be accurately described by this arcane mathematics?

On the second supposition, this is no longer a mystery—the world is indeed constructed on a mathematical basis, and all matter everywhere is identical in its properties. But then the question is how does that come about? How are these mathematical laws imposed on physical matter? And which of the various alternative forms (Schrödinger, Heisenberg, Feynman, Hamiltonian, Lagrangian) is the "ultimate" one? What is the reason for variational principles of any kind?

Either way, we face the fundamental metaphysical issues [15, 17, 19, 34]: *What chooses this set of laws/physical behaviors, and holds it in existence/in operation?* These issues arise specifically in terms of current scientific speculations on the origin of the universe by numerous workers (Tryon, Hartle, Hawking, Gott, Linde, Turok, Gasperini, et al.).[20] Most of these proposals either envisage a creation process starting from some very different previous state, which state then itself needs explanation, and is based in the validity of at least some of the present laws of physics (which are therefore invariant to some major change in the status of the universe, such as a change of spacetime signature from Euclidean to hyperbolic), or else represent what is called "creation from nothing," but in fact envisages some kind of process based on most or all of the apparatus of quantum field theory mentioned above—which is far from nothing! In both cases *the laws of physics in some sense preexist the origin of (the present phase of) the universe, for they are presumed to preside over its creation.*

In the case of ever-existing universes, the same essential issue arises: Why do these specific laws (and their associated possibility spaces) preside over their existence and evolution? Thus whatever ontology the laws of physics may have, on these views both world 3 and world 4 preexist the universe ontologically (the latter in some mysterious way either underlying the mathe-

20. A summary of the numerous present proposals is given in [19].

matical nature of world 3, or else deeply tuned to its nature by an amazing coincidence), and their ontological status is not changed by its coming into existence. It is natural to assume they would postexist it too: whether the universe survives forever or not, in the far future these eternal tendencies would remain unchanged and undiminished.[21]

On the other hand, if these laws come into being together with the existence of the universe, they cannot preside over its coming into existence: something else must do so, originating both the universe and these laws themselves and giving the latter their nature and power over the physical world. Thus if one is to be able in any sense to "explain" this origin, a mysterious metaworld 3 and metaworld 4 would be precursors for the existence of worlds 3 and 4, giving them their shape and structure. (If we do not assume something of this kind, there is no reason why there should be any laws of physics at all, or any ordered behavior of any kind in the universe, and there would be no order at all in the kind of multiverse envisaged by Martin Rees.) These metaworlds then must in some sense preexist the universe—existing in a Platonic space of some kind, in other universes in a multiverse, in higher dimensions, or in some other way with a kind of transcendent existence that has superior status to the transient and contingent status of our own universe, which had a beginning of some kind even if it does not have an end.[22]

Thus attempts at providing a physical explanation for the beginning of the universe end up suggesting some kinds of entities that preexist the universe and would continue to exist after all matter has decayed away, and perhaps even after the universe itself had come to an end. Some kind of transcendent existence is hinted at through the coming into being of a transient kind of existence.

18.7. Consciousness and the Far Future

At a first glance, world 2 (the world of consciousness) must come into existence with conscious life and die out with the ending of that life; its existence will be of a very transient kind.[23] Thoughts and emotions did not exist

21. Some theories consider laws of physics that change. But they then are assumed to change according to some higher-order physical law or theory that is unchanging. That is then what remains unmoved as all else decays.

22. The popular eternal inflationary universe models still require a start (see [22, 9]).

23. Other essays in this volume consider the possibility of intelligent life, or at least complex

before physically based intelligent beings came into existence, arising from the ashes of supernova explosions that for the first time ever in the history of the universe established habitable planets where complex life forms could come into being. Such beings certainly did not occur in the hot big bang phase of the early universe, when heavy elements had not yet been created and neither atoms nor molecules could survive the bombardment of high-energy photons, so nothing like a nervous system could then exist. Fully self-conscious existence came into being on Earth only a few million years ago; animal intentionality (clearly also physically effective) came into being much earlier, but nevertheless a very long time after the start of the expansion of the universe. The same will be true on all other planets anywhere else in the universe. The existence of world 3 is possible only from the time the agents of intelligent thought have come into being, and until the time of their extinction.

Deeper reflection suggests the issue is not so simple. The profound initial underlying question is the following: Can ideas/concepts come into being truly *ex nihilo* or do they require some kind of pre-image, or preexisting potentiality, in order to be realized? (I turn to the issues of emotion and motivation in the following section.)

Physical objects can only come into being in accord with the Aristotelian possibility space discussed above, which therefore in some sense contains a pre-image of all possible objects, from which active agents select specific ones to be actualized. The process of biological evolution selects specific possibilities for actualization from all those that could potentially exist, as defined by the preexisting space of biological possibilities. Physical and biological existence and action only take place in this context. A corresponding question here is, if there is no precursor of any kind, can the truly new in the domain of concepts possibly come into being? Do they not also have to actualize some preexisting potentiality, rather than truly represent the creation and existence of something completely new and without any kind of precursor?

---

information processing, continuing forever on the basis of the energy requirements for this to happen. However, that is just a necessary condition; it is far from sufficient. Unless one establishes that physical objects with the required complexity could also exist forever, this potentiality for continuing thought processes will not be realized. Furthermore, I do not find credible the idea that human life can in some meaningful sense "migrate" to computer systems, or be recreated in them, as some have suggested (see the article by Boden in this volume).

The capacity to have the new idea has to be there, clearly, and in that sense it is already preexistent in the structure of the brain; but does perhaps the germ of the idea also have to be there in some possibility space in order for it to be realized? Support for this view comes from the strong AI position.[24] The point is that any realizable digital computer must have a finite size and be made of a finite number of components, and so has a finite number of states. Select any maximum size you like for such a computer (the size of the earth, the size of the solar system, the size of the galaxy—it does not matter). As long as there is such a maximum size, the number of possible connections and states is finite, and so the number of possible concepts or thoughts that it can express is finite.[25] Taking all possible such states and connections also defines a finite dimensional space; this in turn defines the space of all possible thoughts and concepts, if the strong AI position is true. That space, in some way related to the Platonic world 4c mentioned above, exists in a way that is quite independent of any specific attempts to explore or realize it and does not depend for its existence on the physical universe (indeed, it is closely related to Penrose's Platonic world 4a of mathematics). The specific thoughts that are actually realized in the real universe are in effect chosen for actualization from this larger space of possible thoughts. One cannot have thoughts from outside this domain, which preexists every human brain. Brains are not its originators, they are its explorers

Even if one does not support the strong AI position, a similar argument may still be applicable to the human brain, or indeed the brain of any conceivable physically embodied intelligent being, because it too must have a finite maximum size and finite number of possible connections between component cells. Furthermore, although they may operate on analogue principles, they will be composed of units based on quantum physics principles, and so will allow only a finite number of distinct brain states when we consider the ensembles of all possible physically achievable human brains. It seems likely that the argument above may hold more or less unchanged.

Thus it can be argued that the preexistence of the world of possible ideas is implicit in the fact that brains are physically embodied, taken together

24. I do not support this position, but the argument it leads to is significant.

25. The Turing machine concept of an infinite tape is unrealizable. Thus what can be achieved in reality is less than what can be achieved in principle by Turing machines.

with the preexistence of the world 3 of physical possibilities (or a Platonic world 4b of physical laws) as argued above. I personally would go further and suggest one could argue the case on purely philosophical grounds anyway, independent of any issues to do with physical embodiment; that is, one can suggest it is not even coherent to suggest that world 2 ideas can truly arise out of nothing; rather, that ideas of a kind never seen before cannot be genuinely new human inventions arising out of nothing. As in the particular case of mathematics, while they appear novel, they must in fact necessarily be human discoveries of entities with a preexisting nature. Thus one can argue on this basis for the existence of a Platonic world 4c, as mentioned above.

Clearly this purely philosophical stance is controversial, but the argument based on physical embodiment is coherent. It suggests that there is a transcendent space of ideas that preexists the universe, and so will also survive in the far future after the last intelligent being is gone. If this is true, ideas exist forever, not just in the sense that once they have come into being they are written forever into the history of the universe, but rather that they exist as abstract entities in some kind of eternal Platonic world similar to Penrose's Platonic world of mathematics (world 4a, discussed above). However, this proposal is not essential to my main argument. I put it forward both for its intrinsic interest, and because if true, it provides some additional support for the overall view (the reality of multiple kinds of existences) put forward in this essay.

18.8. Ethics and Theology

The previous section concerned the intellectual side of human life, but not issues of emotion, will, value choices, and ethics. These are crucial both for the conduct of our daily lives, and for the long-term future of humanity. Indeed, one can make a case that the long-term survival of humanity (say, longer than another thousand years) depends on the majority of humankind making an ethical transition that will replace self-seeking with a much more generous world view, embodying a potentiality for self-sacrificial ("kenotic") behavior [34, 42] and including the ability to apportion blame to oneself and a readiness for forgiveness and reconciliation [24]. If this does not take place, someone will use the technological power at our disposal to destroy us all. There certainly exist people with that tendency, as is shown both by the examples of the attacks of September 11, 2001, on New

York and Washington, D.C., and the existence of so many people who create computer viruses.

The key question, then, is *what is the origin and ontological status of a viable ethics to shape the values that guide the way we live?*, so completing the set of causal relations underlying effective human action.[26] Just like physics, astronomy, mathematics, evolutionary biology, and any other branch of human knowledge or experience, it will have the form it does partly for psychological reasons, partly for sociological reasons, partly for reasons to do with evolutionary biology, and partly to do with its own intrinsic nature (cf. section 18.4 above). Each of these causal effects will give a partial explanation, and consequently none will represent the total cause. But if it is to have the normative value required of a genuine moral vision that can transform lives and truly alter behavior, it must have an indubitable intrinsic value that transcends the purely social and psychological: it must have a quality that is deemed to be right and proper and indeed worth dying for because it is intrinsically good. A deep moral vision will have a truly transformative character.[27]

It can be argued [34] that to do justice to this true nature of morality, encompassing its deepest aspects rather than just the modicum of norms required for society to function moderately well, moral visions must exist in a universal, nonrelative way—indeed they must have a transcendent character. Our values do, of course, also have effective causes in terms of both social pressures and evolutionary history. In considering the way this leads to our value systems, it is important that *the emotional reward for virtue is the major motivator for cultivating virtue:* that is, there is an evolutionary and developmental transition to motivation by internal (mental) rewards, instead of just external (physical) rewards [21]. This is fully in accord with some current views on the development of the brain [35] suggesting that emotion serves as the reward system (and so acts as the fitness test) in the developmental process of each mind. But then the issue becomes what decides which actions generate a positive emotional reward and which a negative one? And that ties one into the cultural environment in which one is

26. See the branching causal hierarchy of Murphy and Ellis [34].

27. As embodied in Peter Berger's concept [7] of an "ecstasy" that changes socially accepted world views on the nature of reality.

brought up and which one is immersed in, including the religious culture of the community (one gets different emotional responses in a criminal culture and mainstream American culture, different answers if one belongs to the Taliban or to mainstream Muslim culture). The dominant culture in which one is immersed determines what acts tend to generate positive emotional responses from those around one; and that then largely determines one's own emotional responses and hence one's ethical orientation. Thus we have here a further major example of top-down action: the *effect of a society's culture, including its religious disposition, on the world view and thought patterns of individuals who live in that society* [8].

The main issue, however, is untouched by these features. The core point is that, irrespective of where our moral tendencies come from (what fraction is of evolutionary origin, what fraction of socio-environmental origin, and what fraction is shaped by one's own individual choices in response to these various tendencies), *neither customs of society at large nor our internal reactions determine what is in fact right or wrong*—they determine customs and habits, but not normative values. One's internal emotional reward system may be operating appropriately (e.g., Martin Luther King) or it may not (e.g., Hitler, Osama Bin Laden), and we can judge to what degree societal and internal value systems correspond to what is right and wrong by bringing in moral values that are separate from each of the above factors that help shape our moral tendencies. One can argue logically about values to some degree (e.g., certain actions will tend to create peace and others to create war) *but we can never avoid the need to bring in a value system from outside that judges whether these outcomes are good or not,* bringing in the crucial ethical dimension (logic per se cannot decide whether peace or war are "good" or even "desirable," because "is" cannot determine "ought"; Monod's hope that science itself can provide the needed values is a fallacy). A process that is solely logical, evolutionary, or socially determined does not actually determine what is right and wrong; that comes from somewhere else, and we discern it by *a process of mindfulness applied in the moral sphere.* That process will be fed into by the evolutionary and cultural influences, but is separate from them.

The proposal here is that this genuine and independent moral dimension is recognized, and encompassed by inclusion in the overall scheme of things of a world 5:

World 5: The world of underlying purpose, the set of values and meanings expressing the purpose of existence. This is ontologically real, is effective through revelation and discovery, and provides the basis of ethics and morality. The key assumption is that *this world exists in an ontologically real way* [34], which is of course a highly contentious claim. It is proposed to be of an abstract character similar to world 4 (the Platonic world of logical and mathematical ideas) but is different in two very important respects.

First, it adds to the various transcendent dimensions of existence examined so far, a recognition of the dimensions of purpose and meaning, implying that ethical choices are of real significance and value. Thus just as recognition of the other transcendental worlds (worlds 3 and 4) lays a proper foundation for understanding causality in the physical world, recognition of world 5 lays a proper foundation for understanding ethics in the world of human action by recognizing the reality of highest level goals and meaning.

One can use arguments like those of the previous section to propose that these must be recognized as having a real existential nature (for indeed these are included in the world of ideas); but additionally their normative nature makes them quite different from other concepts, for they represent desirable ideals relating to action choices. We recognize them based on our experiences of morality in the world. These experiences, for example the recognition of good and evil in human actions, have as solid a claim to reality as any other of our experiences. Without something like what is proposed here, one is simply unable to characterize any action whatever—for example, the September 11th terrorist attacks on the United States of America—as either "good" or "evil"; one can simply characterize it as something we like or we don't like, but essentially can say no more than that. Recognition of world 5 is effectively the statement that we can indeed, with sufficient discernment and reflection, characterize some acts as good and some as evil in a way that has real moral meaning.[28]

28. Here I explicitly reject the *ontological reductionist* claim that morality can be completely explained in terms of sociobiology [44]. Those who take this stance need to explain how morality can continue to have a truly normative claim once its origin is viewed in this way. In particular, (1) Have they abandoned the right to call any act whatever "evil," in some sense that still has cross-cultural normative meaning? (2) Do they believe there is some basis where proponents of the eugenics movement, and other such evolutionary based proposals for a coercive new moral order, can be queried?

Second, and crucial in order that the previous claim be believable, is the nature envisaged here for this moral order: namely [16, 34, 42], *it is persuasive rather than coercive in nature.*[29] The suggestion is that while the values embodied in world 5 embrace justice as well as love, they centrally have a kenotic rather than coercive character—that is, a generously giving nature, engaged by spiritual inquiry and practice [51] rather than by logic and intellectual analysis. Indeed, at its core is a paradoxical nature associated with kenosis and self-sacrifice ("he who would save his life must lose it"), in a profound sense contradicting the world of logical argumentation. However, understanding of the transformatory nature of this world is available through experience (as illustrated, for example, in the musical *Les Miserables* based on Victor Hugo's novel of the same name).

The claim is that it is this kind of vision that distinguishes true morality from many other versions of morality that have been supported by social power structures through the ages, built on coercion and fear. In the view proposed here, those are rejected as false visions of morality. Furthermore, in this view, any explanation that does not include the self-sacrificial nature of kenosis, and so omits highest level moral action—the transformational nature of true morality—is simply inadequate. It omits what true morality is all about.

Accepting its existence, this world influences behavior in the physical world through people's adoption of this preferred set of values and meanings. Thus just as we discern the nature of physical structure and mathematics (worlds 1, 3, and 4) by a process of investigation and discovery (rather than by a process of purely social construction) and then incorporate it in our thinking and actions, the same is true for the meanings and associated ethical values in world 5, which are ultimately discerned through a combination of personal experience and social interaction. They are then causally efficacious because ethics is the topmost level of the hierarchy of goals [34] and so shapes all the underlying goals in important ways, indeed playing a crucial role in the nature of choices made in world 2b, which then determine the effect of human actions on world 1.

Given this understanding, we are then in a position to define the proper

29. Worlds 3 and 4 are coercive in that their claims cannot be denied or questioned: they have a rigid and inviolable grip on causality.

nature of ethics:[30] *Ethics is a study of relations between world 5 and world 2. It attempts to relate accurately the preferred values of world 5 to goals and choices in world 2b and world 2c.* Thus it explicitly or implicitly characterizes goals and choices in world 2b and world 2c as (1) ethical, (2) unethical, (3) neutral, or (4) undecidable, according to their compatibility with the values and meanings embodied in world 5. Note that this characterization cannot be made without having an ontologically existent world 5, that is, a set of values and meanings that are preferred in the sense of having a true normative character, as opposed to sets of values that are derived *only* by some combination of social and psychological construction and sociobiological evolution. These can all be argued to fail to have a true moral character, (see e.g., Murphy [32]).

The essential claim being made is that underlying existence there is more than the blind forces and logic encapsulated in the transcendent worlds 3 and 4, but rather something corresponding to the full depths of human nature, underlying the highest aspects that we are capable of experiencing and understanding—and that indeed these highest level experiences could not meaningfully come into being and be recognized were there not some such underlying aspect of existence; they could not come into existence of themselves without some such precursor in the underlying structure.

Without an assumption of this kind, our ontological scheme will be unable to comprehend the true depths of human nature, whose existence is undoubted, for it is recorded in all the great plays, novels, and art of humanity, as well as in our history. Acknowledging its existence is a rejection of all those thin scientific world views that deny this marvelous depth of understanding and experience has real meaning, instead trying to claim that the only features that are truly real are those that can be encompassed by our partial and faulty mathematico-physical models of the world.[31] Those views in effect claim that only intellectual understanding has real meaning, and emotion and values are secondary and derivative. The view proposed here strongly denies that position.

30. In obvious parallel to the characterization of epistemology given above, see section 18.4.

31. See, for example, Atkins [4]. The models are partial because of the idealizations they embody, and faulty because of the divergences that occur in them, as well as foundational problems related to such issues as the nature of quantum measurement and the existence of the arrow of time.

When we assert the positive position that all the worlds 1–5 exist, we are rejecting a reductionist viewpoint in favor of an integrative view that can be beautifully expressed as follows:

> I have never liked the phrase that says we're just made of dust and return to dust. We are energy, which is interchangeable with light. We are fire and water and earth. We are air and atoms and quarks. Moreover, we are dreams, hopes and fears, held together by wisdom and driven apart by folly. So much more than dust. The biblical verse should say, "Miracle thou art and to Mystery returneth."[32]

But what then underlies the possible existence of this layer of meaning and purpose? It is not derivable from any of the other transcendent worlds considered so far. Its most plausible basis [34] is some kind of transcendent entity that incorporates such values in its very being, which is much the same as what is usually encompassed in the concept of God, the prime mover or cause, and so originator of meaning and values as well as existence.

So there is a final kind of world to consider:[33]

Metaworld 0: The underlying fundamental reality: the world of God, partially described by theology and entailing the fundamental meanings of the universe and of life.

The key assumption of the theist position is that this world exists. This is highly debatable precisely because it cannot be proved or disproved beyond doubt by scientific, logical, or philosophical argument. The essence of the problem is the transcendent yet hidden nature of this metaworld: it has a character of a totally different kind to the rest, indeed is not properly describable in terms of the entities they comprehend, but is nevertheless able to interact with them and influence them in important ways—in particular being the ground of their existence.

World 0 is quite different in quality than all the others; it provides the ul-

32. From Fulghum [20]. I thank Gillian Hawkes for this reference.

33. Any attempt at complete explanation must include the deepest layer of causality, that is, the issue of ultimate causation in relation to cosmology and the laws of physics. The metaphysical issue of purpose or chance arises here [15, 16]. I do not wish to argue that issue here, but rather to state that it cannot be avoided if one wants to look at deep issues of causation. Many biologists writing in the area take the existence and nature of the universe and the laws of physics for granted—but those are precisely the issues where the deep metaphysical questions come in.

timate context in which they are set. Worlds 1 to 5 are parts or aspects of the "one world" that is the creation of God, depending on world 0 for their existence. Thus in this view, God is *transcendent* and *acts foundationally* through creating worlds 4, 3, and 1, thus enabling world 2; and is also *immanent* and *causally efficacious* in world 2 through revelation, communicating the nature of world 5, and hence through consequent human action influences world 1 (as, for example, in the construction of churches and temples, and the spending of resources on missionary and charitable activity).

The reason for belief in world 0 is usually deeply rooted in an individual's life experience of a self-authenticating nature. However, it can also to a degree be supported by consideration of a vast range of data on claimed religious or spiritual experiences [51], together with a wider variety of *intimations of transcendence* [18], ranging from aesthetic and artistic experiences through experiences of community and love. None of these is by itself logically conclusive, but together they form an overall interpretative pattern of explanation that has considerable weight in terms of explaining a broad band of evidence from human life and experience. It provides the only easily intelligible explanation of "where" the ideal worlds 3 and 4 exist: namely, they exist "in the mind of God." Additionally, the age-old issue of design remains with us in a modern form, namely in the guise of the anthropic issue: *Why does the specific chosen set of laws and initial conditions that has come into being allow intelligent life to exist?* [5, 16]. The implication of the present view is that the laws and conditions are there by design, in the good old theistic sense: God is the final cause, with the laws of physics (and consequent laws of biochemistry) the efficient cause. The proposal to explain this nature of existence scientifically by means of some kind of ensemble of universes, or "multiverse," remains a possible logical alternative, but it also remains unproven. While it carries a certain weight, it is not testable in the usual scientific sense, so its scientific status is unclear. And in the end it only postpones the problem; for then the issue is why this kind of multiverse, and not another? Why a multiverse that allows *any* universe that permits life, rather than one that does not? And finally, is the multiverse itself possibly a creation of God?

Overall, there is no logical, philosophical, or scientific proof either that this anthropic view is true or that it is false; and that this should be so is consistent with the noncoercive nature of the vision put here [16, 34, 42]. One either assents to it or not, on the basis of one's life experience and overall world view. From an intellectual viewpoint, this is a choice one can make freely—it is an option that is logically consistent and supported by a substantial amount of evidence [18]. One can claim that world 5 and metaworld 0 exist without contradicting irrefutable evidence[34] or committing a major logical fallacy. One is similarly free to deny their existence.

18.9. The Far Future and Hope

So where does this leave us as regards the far-future universe? Well, what preexists will presumably remain in the future. I have argued above that there are good grounds for believing that the transcendent worlds of possibilities and Platonic realities (worlds 3 and 4) preexist and so will survive forever,[35] even as everything else decays, and that they would in some sense continue in existence even if the universe came to an end. These kinds of existence are implicit in a scientific understanding of the world around us, even if this is not often recognized.

But these are rather austere kinds of existence, even if they have the seeds of full humanity contained within. I have further argued that the moral force and meaning that humans discern is also embodied in an existence (world 5) that preexists the universe and hence will also continue to exist even when humanity and other intelligent life comes to an end, so that the rich depth of struggle and understanding embodied in human life also has a hope of continuing in a meaningful sense. These kinds of existence are implicit in any view that believes that moral struggles are real meaningful engagements, rather than being just human minds playing games that have a surface appearance of morality, but in a deep sense deny it any reality.

Finally, I have argued that while one cannot attain any certainty, one is entitled—if one finds it a compelling vision supported by the evidence contained in one's life experience—to believe that the underlying basis of meaning and morality is indeed a Creator who embodies the beauty of a kenotic

34. The major possible counterevidence comes from the problem of evil, and particularly that committed in the name of religion; for arguments in this respect and related references, see [34] (and note also that those who reject the arguments given above about morality have no ground for denoting any action as evil—which is the basis for deploying this argument). The corresponding problem for those rejecting this view is the mirror-image problem of good (why does it exist in abundance?).

35. Perhaps more accurately, their effective presence in the universe will survive.

nature in his/her being. While this belief is supported by various lines of evidence, they are not conclusive; belief in such a kenotic Creator is based on faith. It can provide the seeds for hope that despite the surface appearances, which may lead to a despair in the face of ultimate desolation and decay, the far-future evolution of the universe is in fact filled with light and joy. The paradoxical nature of a kenotic ethic and practice, which can transform defeat into victory, allows even the possibility that in some sense we do not understand our individual spirits may remain after death, as believed by many world religions. Science provides no reason to believe this is true, and indeed makes it seem rather unlikely, but cannot conclusively prove it false either.[36]

Whether this is true or not, in some broader sense the existential entities that presided over the start of the universe, ensuring it was able to provide hospitable habitats for life and that physical conditions existed that ensured intelligent life would come into being and lead to the rise of the highest levels of courage and ability in science, art, and ethics, will still preside over its far future and ultimate fate. It is not unreasonable to believe this entity has the character of a Spirit of some kind, as envisaged in Genesis 1 and John 1, rather than being just a blind set of impersonal laws. In this situation, rationality and hope can both maintain their integrity as we contemplate the far future and leave us with the belief that all may after all be well even then, as stated eloquently by George Fox, the founder of the Quakers, in his dying words:

> I am glad I was here. Now I am clear, I am fully clear . . . All is well; the Seed of God reigns over all and over death itself. And though I am weak in body, yet the power of God is over all, and the Seed reigns over all disorderly spirits.[37]

Indeed, something like this must be the case if a moving spirit with a kenotic nature underlies the existence of the universe. There is sufficient evidence in human experience that this is indeed the case, that one can dare to hope—and have faith—that it is true.[38]

36. Despite some highly exaggerated claims, science has made no serious progress in resolving the central problem of consciousness. Even how to approach it is strongly contested.

37. George Fox, *Journal* (1691). Item #11 in [29].

38. It is with good reason that St. Paul emphasized that along with the primary quality of love, faith and hope are cardinal virtues (1 Cor. 13:13).

REFERENCES

1. Aberle, D. F., et al., "The Functional Pre-requisites of a Society," *Ethics, 60* (1950).

2. Ackoff, R. L., *Ackoff's Best: His Classic Writings on Management* (John Wiley and Sons, New York, 1999).

3. Anderson, P., "More Is Different," *Science, 177,* 377 (1972). Reprinted in Anderson, P. W., *A Career in Theoretical Physics* (World Scientific, Singapore, 1994).

4. Atkins, P. W., "The Limitless Power of Science," in *Nature's Imagination: The Frontiers of Scientific Vision,* ed. J. Cornwell (Oxford University Press, Oxford, 1995), 122–32.

5. Barrow, J. D., and Tipler, F. J., *The Anthropic Cosmological Principle* (Oxford University Press, Oxford, 1986).

6. Beer, S., *Brain of the Firm* (John Wiley and Sons, New York, 1981).

7. Berger, P., *Invitation to Sociology* (Doubleday, New York, 1963).

8. Berger, P., and Luckmann, T., *The Social Construction of Reality* (Penguin, London, 1979).

9. Borde, A., Guth, A. H., and Vilenkin, A., "Inflation Is Not Past Eternal," gr-qc/0110012 (2001).

10. Campbell, D. T., "'Downward Causation' in Hierarchically Organised Biological Systems," in *Studies in the Philosophy of Biology: Reduction and Related Problems,* eds. F. J. Ayala and T. Dobzhansky (Macmillan, London, 1974), 179–86.

11. Campbell, N. A., *Biology* (Benjamin Cummings, Redwood City, Calif., 1990).

12. Coen, E., *The Art of Genes* (Oxford University Press, Oxford, 1999).

13. Edelman, G., *Neural Darwinianism* (Oxford University Press, Oxford, 1989).

14. Eddington, A. S., *The Nature of the Physical World* (Cambridge University Press, Cambridge, 1928).

15. Ellis, G. F. R., *Before the Beginning: Cosmology Explained* (Bowerdean/Boyers, London, 1993).

16. Ellis, G. F. R., "The Theology of the Anthropic Principle," in *Quantum Cosmology and the Laws of Nature,* ed. R. J. Russell (University of Notre Dame Press, Notre Dame, Ind., 1993).

17. Ellis, G. F. R., "The Different Nature of Cosmology," *Astron. Geophys., 40,* 4.20–4.23 (1999).

18. Ellis, G. F. R., "Intimations of Transcendence: Relations of the Mind and God," in *Neuroscience and the Person,* eds. R. J. Russell, et al. (University of Notre Dame Press, Notre Dame, Ind., 1999), 449.

19. Ellis, G. F. R., "Before the Beginning: Emerging Questions and Uncertainties," in *Toward a New Millenium in Galaxy Morphology,* eds. D. Block, I. Puerari, A. Stockton, and D. Ferreira (Kluwer, Dordrecht, 2000).

20. Fulghum, R., *From Beginning to End: The Rituals of our Lives* (Ivy Books, New York, 1996).

21. Goodenough, U., talk at SSQ Meeting, 2001.

22. Guth, A., "Eternal Inflation," talk given at AAAS Program of Dialogue on Science, Ethics, and Religion, Washington, D.C., April 14–16, 1999. Guth's title has since

been published in *Annals of the New York Academy of Sciences,* vol. 950: *Cosmic Questions* (New York Academy of Sciences Press, New York, 2001), 66–83; astro-ph/0101507.

23. Hartmann, S., "Effective Field Theories, Reductionism, and Scientific Explanation," *Stud. Hist. Phil. Mod. Phys., 32B,* 267 (2001).

24. Helmick, R. G., and Petersen, R. L., *Forgiveness and Reconciliation: Religion, Public Policy, and Conflict Transformation* (Templeton Foundation Press, Philadelphia, 2001).

25. Kadanoff, L., *From Order to Chaos. Essays: Critical, Chaotic, and Otherwise* (World Scientific, Singapore, 1993).

26. Kuppers, B. O., *Information and the Origin of Life* (MIT Press, Cambridge, Mass., 1990).

27. Kuppers, B. O., "Understanding Complexity," in *Chaos and Complexity,* eds. R. J. Russell, et al. (University of Notre Dame Press, Notre Dame, Ind., 1995), 93.

28. Laughlin, R. B., "Fractional Quantisation," Rev. Mod. Phys., 71, 863–74 (1999).

29. London Yearly Meeting of the Religious Society of Friends, *Christian Faith and Practice in the Experience of the Society of Friends* (Headley Brothers, London, 1966).

30. Moyers, B., *Healing and the Mind* (Doubleday, New York, 1995).

31. Murphy, N., "Divine Action in the Natural Order: Buridan's Ass and Schroedinger's Cat," in *Chaos and Complexity,* eds. R. J. Russell, et al. (University of Notre Dame Press, Notre Dame, Ind., 1995), 325.

32. Murphy, N., "Supervenience and the Non-reducibility of Ethics to Biology," in *Evolutionary and Molecular Biology,* eds. R. J. Russell, et al. (University of Notre Dame Press, Notre Dame, Ind., 1998), 466.

33. Murphy, N., "Supervenience and the Downward Efficacy of the Mental: A Non-reductive Physicalist Account of Human Action," in *Neuroscience and the Person,* eds. R. J. Russell, et al. (University of Notre Dame Press, Notre Dame, Ind., 1999), 147.

34. Murphy, N., and Ellis, G. F. R., *On the Moral Nature of the Universe* (Fortress Press, Minneapolis, Minn., 1995).

35. Panksepp, J., and Panksepp, J., "The Seven Sins of Evolutionary Psychology," *Evol. Cog., 6,* 108 (2001).

36. Peacocke, A. R., *An Introduction to the Physics and Chemistry of Biological Organisation* (Oxford University Press, Oxford, 1989).

37. Peacocke, A. R., *Theology for a Scientific Age* (Fortress Press, Minneapolis, Minn., 1993).

38. Penrose, R., *Shadows of the Mind* (Oxford University Press, Oxford, 1994).

39. Penrose, R., *The Large, the Small, and the Human Mind* (Cambridge University Press, Cambridge, 1997).

40. Pesken, M. E., and Schroeder, D.V., *An Introduction to Quantum Field Theory* (Perseus Books, Reading, Mass., 1999).

41. Pickover, C. A., *Visualizing Biological Information* (World Scientific, Singapore, 1995).

42. Polkinghorne, J., ed., *The Work of Love: Creation as Kenosis* (SPCK, London, 2001).

43. Popper, K., and Eccles, J. C., *The Self and Its Brain: An Argument for Interactionism* (Springer, Berlin, 1977).

44. Ruse, M., *Res. News Opp. Sci. Theol., 2,* 26 (September 2001).

45. Schellnhuber, H. J., "Earth System Analysis and the Second Copernican Revolution," *Nature, 402, no. 6761,* C19 (2 December 1999).

46. Schweber, S., "Physics, Community, and the Crisis in Physical Theory," *Physics Today, 46,* 34 (1993).

47: Scott, A., *Stairway to the Mind* (Copernicus-Springer, Berlin, 1995), 50–54, 207–10.

48. Seife, J., *Zero: The Biography of a Dangerous Idea* (Penguin, London, 2000).

49. Sternberg, E., *The Balance Within: The Science Connecting Health and Emotions* (W. H. Freeman, San Francisco, 2000).

50. Tannebaum, A. S., *Structured Computer Organization* (Prentice Hall, Englewood Cliffs, N. J., 1990).

51. Wakefield, G., ed., *A Dictionary of Christian Spirituality* (SCM Press, London, 1983).

52. Weiss, K. M., "Is There a Paradigm Shift in Genetics? Lessons from the Studies of Human Diseases," *Mol. Phylo. Evol., 5,* 259 (1996).

53. Wolpert, L., et al., *Principles of Developmental Biology* (Oxford University Press, Oxford, 1998).

APPENDIX

Olaf Stapledon (1886–1950)

Stephen R. L. Clark

Anyone intrigued by fantasies or scientific speculations about the very far future owes a debt to Olaf Stapledon, whose *Last and First Men* [21] and *Starmaker* [24] consolidated that long-term perspective before the Second World War. Robert Crossley, one of Stapledon's biographers (see [5]), has remarked that he has often found that people think that Stapledon was Scandinavian, or else American. J. B. S. Haldane, on reading *Last and First Men*, grew indignant that he had not heard before of someone he assumed must be a research scientist. In fact, Stapledon was a philosopher, educated at Balliol College before the First World War, and in the ambulance corps during it; he was born and bred in the Wirral (the peninsula between the Dee and the Mersey in northwest England), and was for many years a notable figure in the literary and philosophical culture of Merseyside. His doctorate, for work in philosophical psychology, was earned at Liverpool. He taught for the university's extramural department, and for the Workers' Educational Association. His archives, including many of the detailed notes he made for his classes, are housed in Liverpool University Library. Brian Aldiss's suggestion that "his contacts with the outside world were few; he preferred to

Note: A slightly expanded version of "Olaf Stapledon (1886–1950)" [7], here reprinted with the permission of the editor of *Interdisciplinary Science Reviews;* see also my *A Parliament of Souls* [6].

dream and cultivate his garden"[1] is peculiarly wrong-headed, as his letters and lectures make quite clear.

Stapledon was a seminal writer of what—against his own preference—has come to be called science fiction, and a representative (but also deeply critical) member of a literary and philosophical movement that has influenced more than fiction: namely, optimistic technohumanism. Loyalty to the (Human) Spirit—easily equated with a preference for "science" over "superstition"—led writers like Wells, Chardin, Bernal, and Haldane, as well as Stapledon, to imagine a future in which we had remade our natures and regularly remade worlds. Where Stapledon differed was in his doubt, educated by a wide reading in past philosophy, that we either could or should control everything. He also recognized the possibility that there were other creatures, far away from here and of a different biological kind, who yet served the Spirit. Loyalty to the Spirit also required an ironically worshipful submission to a reality that always transcends our grasp, and that may—in the end—be better known by other means than the intellectual. "Although, when clerics expound their faith," he said in a review of Wells's *Star-Begotten,*[2] "I fly to line up behind Mr. Wells, I am an erring disciple. For, when he in turn explains, I feel a restless expectation of a something more which is never forthcoming. He is too ready to assume that an idealization of the positivistic, scientific mood, which is mainly a product of the nineteenth century, really can adequately suggest the essence of the truly human mentality."

He was a philosopher inspired by scientific speculation and older theory alike. Fashions in philosophy have shifted several times since the 1930s: "logical positivists," who claimed (self-refutingly) that only observational and analytic sentences were even meaningful (a claim that was, by its own lights, therefore meaningless), paid no attention to Stapledon's worries about the reality of time, the relationship of mind and brain, and the nature of spiritual maturity. Present-day philosophers have returned to those ancient problems that, he said, would still trouble thinkers at the far end of hu-

1. Aldiss [2]. Aldiss rightly criticizes many of Fiedler's [8] mistakes, but is hardly better informed about Stapledon's life and reputation.

2. Stapledon [25]. Wells's book has a theme not unlike Stapledon's own *A Man Divided* [34] and *Last Men in London* [22], but the two writers had quite different notions of what a real awakening would be.

mankind's existence ([21], p. 283), and can still learn from Stapledon. They might learn, for example, the folly of those such as the Fourth Men, the Great Brains, who "casually solved, to [their] own satisfaction at least, the ancient problems of good and evil, of mind and its object, of the one and the many, of truth and error" ([21], p. 213), without any genuine spirituality. The danger is that they may also neglect the writings by which Stapledon was instructed, namely the great Greek philosophers, especially Plato, the Stoics, the Skeptics, and Plotinus. For Plotinus, the last great classical philosopher, so Armstrong has written, "Philosophical discussion and reflection are not simply means for solving intellectual problems (though they are and must be that). They are also charms for the deliverance of the soul" [3]. Many passing remarks in Stapledon's fiction earn longer discussions elsewhere—the reality of time, the possibility of a world composed solely of sounds, the nature of the mind and its engagement with material reality. They never become merely technical discussions.

> Theories, theories, myriads upon myriads of them, streamed over me like wind-borne leaves, like the contents of some titanic paper-factory flung aloft by the storm, like dust-clouds in the hurricane advance of the mind. Gasping in this vast whirling aridity, I almost forgot that in every mote of it lay some few spores of the organic truth, most often parched and dead but sometimes living, pregnant, significant. ([21], p. 379)

But though Stapledon's academic publications are not as ephemeral as some have thought,[3] his best-known work is still *Last and First Men* [21], a "history" of the next several million years whose literal inaccuracy about the immediate future is no real flaw. Britain may not have fought a destructive war with France, and the Chinese don't wear pigtails any longer. But maybe we are indeed now living through the gradual, global victory of a blend of vulgar piety and technical brilliance: the sort of religiosity that sees religious merit in wealth and technocratic progress. And even if this too turns out false, the main message of that saga is still worth remembering: not that of Wells in either his hopeful or his despairing mood, but rather that world views change with changes in biological and historical setting, that the Truth lies beyond us all, and yet may be the object of our desire.

It does sometimes seem that Stapledon wishes to persuade us that all sys-

3. See especially *Philosophy and Living* [26].

tems of law and metaphysics rest upon the material necessities of the time. The Holy Empire of Music, for example, which prevails among the Third Men for a thousand years, purports to be funded on "the fiction that every human being was a melody, demanding completion within a greater musical theme of society," but actually derives its power from the biological realities of that species ([21], p. 199). He does not conclude, however, to the self-refuting thesis that there is no escape from relativities. "It was necessary that man should be forced for a while to stand on his own feet, arrogantly but totteringly, so as to rediscover in blood and tears his own weakness and folly, and his inability to save himself without the passion born of a vision of something which in some important sense is beyond himself" [33]. That we cannot lay hold firmly on Truth Absolute is itself a truth that shows we are not bound within the circles of our time and kind (for if we were we could not know we had not grasped the truth).

In his later life Stapledon was identified as a fellow traveler, though on no better grounds than that he attended (as perhaps he should not have done) the Soviet-backed Cultural and Scientific Conference for Peace, and retained his (broadly) socialist ideals and recognition of the material context even of sound thought. He was not alone. A great many thinkers of the 1930s after all believed it necessary to say what suffering, what indignity the old regime created, and saw marxist revolutionaries as the most active and least self-regarding heralds of a better order. A great many thinkers also thought that we should soon have a complete and unified theory of life, the universe and everything. Once that was achieved we need never be surprised by anything again, nor need we fear that our power be less than our desire. All understanding increases power: perfect understanding would allow us unlimited power, either to change the world or (at the least) to change our selves, or our subjects' selves, so that they no longer nursed impossible desires. The common good would be well served when we had bred individualism and superstition out of the race. "They were preparing to take charge of mankind, to make the planet into a single well-planned estate, and to re-orientate human nature. But what kind of world would they desire to make, whose knowledge was only of numbers?" (Stapledon [21], p. 571). Even decent and intelligent writers allowed themselves such thoughts as must now make the blood run cold:

The task confronting biology, physiology and medicine is not only to master scientifically the maladies and phenomena of counter-evolution (sterility and physical weakening) which undermine the growth of the noosphere—but to produce by various means (selection, control of the sexes, action of hormones, hygiene, etc) a superior human type. . . . What attitude should the advancing sector of humanity adopt towards static and decidedly unprogressive ethnic groups? (Teilhard de Chardin, writing in 1937)[4]

Or consider Haldane's wishful fantasy of future evolution, when such self-regarding sentiments as pride, "a personal preference concerning mating," and pity ("an unpleasant feeling aroused by the suffering of other individuals") have been bred out of the species [9]. Stapledon, to similar effect, incorporated in some of his forward-looking fantasies doctrines of a superior humanity disdainful of our small-minded morals that were soon to be dreadfully embodied in the here and now. His utopian researchers regularly experiment on children, eliminate "the unfit" and see themselves as heralds of a greater dawn licensed to do just what they please to those they judge inferior. Readers sometimes fail to notice this. One of Brian Aldiss's most extraordinary critical judgments, for example, is his description of *Odd John* [23] as a "pleasant superman tale," with a "light and cheerful mood."[5] John is a multiple murderer who treats human beings as vermin, experimental animals, or pets, and is critical of totalitarian barbarism only because it does not allow for the untrammeled existence of Homo Superior. He and his companions end in mass suicide rather than collaborate with ordinary people, whom they despise as an "inferior species." They are "aristocrats" only in the sense condemned by Stapledon even when he admits the value of an intellectual aristocracy ([27], p. 98).

For Stapledon did not really believe in "scientific socialism." He attended peace conferences not "to cut some kind of figure in a world whose attention he had failed to capture,"[6] but because he thought it desperately important (as of course it was) that people try to understand each other. "To avoid a savage religious war it is desperately urgent that each side make a serious at-

4. Cited by Speaight [20].

5. Aldiss [1], and in Aldiss [2]: "*Odd John* is a worthwhile contribution in the poor Little Superman line."

6. As Aldiss suggests in [2], p. 1008. It is a matter of record that he *had* captured as much attention as Aldous and Julian Huxley, Joad or Lawrence: see, for example, Coates [4].

tempt to understand and respect the most cherished values of the other and to see itself through the other's eyes"[7]—and so to learn humility. "Marxists of the harsher kind . . . bring revolution into disrepute. They give some excuse for the fear that if they were in power they would prove as insensitive and ruthless as their opponents."[8] The future offered us annihilation, totalitarian rule, or a new kind of human world whose lineaments we could detect in personal love and humane endeavor. Eugenicists of the kind all too familiar in the 1930s imagined that they could breed a "better" sort of humanity, but all the experiments that Stapledon imagines end in calamity, partly because the eugenicists themselves have no clear or sound ideas of what "better" means, and partly, as Plato knew, because the world disrupts our careful plans. Shall we breed for manual dexterity, or musical sensitivity, or ecological wholeness, or Great Brains ("huge bumps of curiosity equipped with cunning hands"),[9] or ecstatic flight (all of which ideals—and others—are explored in *Last and First Men*)? Alternate futures are represented, for example, in *Darkness and the Light.*[10] But even in the better future his superior beings, or self-styled supermen, invariably confront a reality that surpasses them, that cannot be wholly understood or controlled. When they reach out to the Maker and Sustainer of Worlds, the Starmaker, near the very end of cosmic evolution, they find themselves as far away from It as ever. According to Leibniz, another philosopher with a gift—largely undeveloped—for speculative fiction, "there is no Spirit, however exalted, who does not have an infinite number of others superior to him. However, although we are much inferior to so many intelligent beings, we have the privilege of not being visibly over-mastered on this planet, on which we hold unchallenged supremacy" [12]. Stapledon chose to represent beings that overmastered us, only to find themselves as distant from the One as ever.

7. Addressing the New York conference in 1949: cited by Fiedler [8], p. 24. Fiedler's attempt to psychoanalyze Stapledon is, to me, wholly unconvincing: proof, indeed, that Stapledon was right to think that our grasp of truth is conditioned by our own spiritual condition.

8. Stapledon [26], p. 38. That they had *already* proved this was apparently not something that Stapledon quite realized.

9. Stapledon [21], p. 212. These beings Stapledon depicts as devoid of any instinctive responses save curiosity and constructiveness, operating "a very accurate behavioristic psychology," but with no inner understanding of their subjects (civil and experimental).

10. Stapledon [28]. The better future is brought about by Tibetan missionaries, the worse by a romantic sadomasochism.

The Infinite is not a Very Big Number, and Odd John, despite the sycophantic praise of that novel's narrator, is no nearer Truth than we. When Stapledon's Last Men peer into a pool on Neptune, admiring their distant cousin, Homunculus, they relive Stapledon's own visit to the pools of Anglesey.[11] And both, I suspect, repeat the image used by Plato in his *Phaedo,* that we live ourselves in the dells and rock pools of a richer world. "What a world this pond is! Like the world you are to plunge into so soon" (Stapledon [21], p. 344). And the Last Men themselves are to discover that their world is set within immensity.

But though our worlds are small ones, lost in time and space, it still matters how we live in them. There is still a distinction to be made between different kinds of community, different kinds of whole, different pictures of the Spirit. For Stapledon did not conclude that because the Truth was always greater than our thought of it, there was therefore *nothing* to be said for one side against another. "When scientific detachment supports a simple materialistic metaphysic, and denies right and wrong and all the higher reaches of human experience, it takes the first step toward social disaster" (Stapledon [26], p. 85). And a few years later: "If Churchill's mentality, with all its faults, is not objectively more developed and less perverted than Hitler's, not only is this war [1939–45] not worth fighting, but also the struggle in every human mind in all ages of history between the somnolent and the lucid ways of behaving is based on an illusion" [30]. He often suggested that the future would be a struggle between communities that did and communities that did not deny the value and particular powers of individuals. The "right" kind of community was one in which human persons remained individuals even though they felt themselves as members of a larger whole. The spirit that wakes successively in mortal individual, bomber crew, nation, humankind through the story of *Death into Life* [31] is one that arises out of genuine communication, real friendships. "Genuine sociality tends now to be rejected in favour of the more primitive kind of sociality based on sheer animal gregariousness, in which the dominant motive is not mutual respect but the will to conform to the behaviour of the group, and to enforce con-

11. Compare *Last Men in London* [22], p. 343, and McCarthy [18], p. 32: a much better book than Fiedler's. See also Stapledon [24], p. 204, where the greatest of galactic intelligences watch the rise and fall of civilizations as "might we ourselves look down into some rock-pool where lowly creatures repeat with naive zest dramas learned by their ancestors aeons ago."

formity" (Stapledon [26], p. 88). The merely aggregative, domineering unities that enlightened humans strive against in the shape of Martian invaders, or mad empires (in *Last and First Men* or *Starmaker*), do not allow for personal love, because, as Haldane hoped, there are no individual persons there. Stapledon always reverted to "the little glowing atom of community,"[12] his marriage to Agnes, and would accept no system that denied its value. Readers drew the wrong lesson if they thought that the here and now no longer mattered. Doris Lessing's *Canopus in Argos* sequence is a truer representation of the Stapledonian perspective.[13]

But there is still a problem about Stapledon's vision that can be explored through a look at one of his less well-known fables, *The Flames* [32], which encapsulates many of his concerns and characteristic ironies. Patrick McCarthy correctly observes that "as an example of controlled and sustained irony *The Flames* is without parallel among Stapledon's works" ([18], p. 115). Thos, the critical minnow-watcher, presents and comments on the narrative of Cass, the speculative generalist. Cass, after years of seeking to see and understand things "from the inside," by telepathic or mystical means, seems to himself to have been addressed by a living flame, hidden in a pebble plucked from a cold, snow-shrouded landscape. It turns out that there are such creatures, salamanders, born in the sun's troposphere and condemned to live out a cold and intermittent existence on solid earth since the planets were formed. The late world war, and its manifold fires, have brought them out of hibernation in the dust of the air,[14] and they sense the possibility of forming a symbiotic alliance with us: we to provide the environment within which they can live, they to provide the mental stability and community awareness we lack. This sort of symbiotic pattern is many times repeated in Stapledon's work. If we cannot agree, the flames' other option is to instigate nuclear spasm: "then at last, with the whole planet turned into a single atomic bomb, and all the incandescent continents hurtling into space, we should

12. Stapledon [24], p. 333; see also Crossley [5].

13. See Doris Lessing, *Archives Re: Colonized Planet 5: Shikasta* [13], the third novel of the sequence. *The Sirian Experiments* [14] describes one of the "Mad Empires"; the fourth, *The Making of the Representative for Planet 8* [15], well captures Stapledon's image of the spirit's triumph at world's end.

14. An idea perhaps echoed from Leibniz, who suggested that every animate being had existed forever in seminal form, only roused to moral life occasionally.

have for a short while conditions almost as good as those of our golden age in the sun" ([32], p. 61). Cass is on the point of agreeing to act as the flames' ambassador, when he learns that his own marriage had been deliberately destroyed by the flames (and his wife incidentally driven to suicide) so that he might be a suitably single-minded instrument of their purposes (the Neptunian hero of *Last Men in London* had done the same, less violently, to his victim, unrebuked).[15] Cass concludes that this proves the flames' real ill will, destroys the flame with a glass of cold water (the flames turn to revivable dust if slowly extinguished, but perish forever if suddenly doused), and sets himself to warn humankind of the deadly peril they stand in. He is eventually incarcerated in an asylum, where the flames again convince him that their intentions are good. Meanwhile, however, their more lucid companions, in the sun, have (as usual in Stapledon) discovered that Reality "was wholly alien to the spirit, and wholly indifferent to the most sacred values of the awakened minds of the cosmos,"[16] and are undergoing a desperate religious war in which the flames' original pious agnosticism is lost. Cass, now converted to that typically Stapledonian position, is threatened by flames converted in their turn to a militant theism. Cass dies in a fire, victim of homicide or, as Thos supposes, his own deranged endeavor.

The swirling confusion of Stapledonian history is here compressed into one man's life, and the struggle to live lucidly, without self-deception, is depicted without sentimental gloss. Who is deceived? Who is sane? Allegorically, of course, the flames are simply those technological powers whose use may lead to utopia or to disaster. Or else they are a shifting image of the individual-in-community, less inclined than we to imagine that they are abstract individuals, rather than elements within the global, or the stellar, community, and by the same token all too ready to ignore the needs and passions of each such element, and fall into the little death of the hive-mind. Or else again, they are images of the division that concerned Stapledon so often, between sleep and awakened life. The story even allows him what he does not

15. He offers a typically pompous justification of his "mental vivisection" ([21], p. 391), that it was a necessary part of a very lofty task to which Paul would have agreed "in his best moments." Stapledon's Neptunian, and the eventual cosmic spirit, eventually realize that they themselves are being treated to an identical torment.

16. Stapledon [32], p. 79. The same "discovery" is imagined in *Darkness and the Light* [28], that we are no more than snowflakes by the feet of battling titans.

attempt elsewhere, the thought that present individuals are fallen creatures, forever reaching toward a perfection they have lost: "Each new experience came to us with a haunting sense of familiarity and a suspicion that the new version was but a crude and partial substitute for the old" ([32], p. 37).

The ultimate unreality of time's passage, Stapledon's other constant theme, is not represented in *The Flames,* though the mere fact that millennia of high endeavor are compressed and mirrored in the last few months of one man's life is a little reminder that time does not advance, and the collapse even of the high solar civilization of the flames (contacted at long last by the terrestrial exiles) a warning that there is no security within time.

The flames entertain the project of initiating nuclear spasm, "through loyalty to the spirit in us," if they should decide that the human species was doomed to self-destruction sooner or later ([32], p. 60), rather as the Fifth Men of Stapledon's other future history destroy the inhabitants of Venus, on the plea that they are less developed and failing creatures: this slaughter, incidentally, produces in its agents, on the one hand, an "unreasoning disgust with humanity," and on the other, a "grave elation" expressing itself in the thought that "the murder of Venerian life was terrible but right." Odd John assures his biographer bluntly, "If we could wipe out your whole species, we would" ([23], p. 216), but, in the end, he and his fellows refrain.

The unresolved ambiguities of the story—are the flames trustworthy or not; is Cass insane or not; is the god of humane devotion certain to be victorious or not; is "agnostic piety, an inarticulate worship of I know not what for being I know not what, except that it is worshipful" ([30], p. 42) coherent or humane; do personal ties count for more or less than an imagined general good; are the demands of the heart to be accepted alongside the judgments of the mind or not—are what makes it art rather than academic philosophy—though this does not make it less *philosophical* in a broader sense. The flames' philosophizing too "was more imaginative and less conceptual than [ours], more of the nature of art, of myth-construction, which [they] knew to be merely symbolical, not literally true."[17] They recognize a lack in themselves, of precise analysis and practical intelligence, which they hope will be compensated within a future symbiosis. This too is a common theme in Stapledon: the arachnoids and ichthyoids who are the core of the eventual

17. Stapledon [32], p. 34: Plato thought much the same about his speculations.

galactic and cosmic spirit embody the active and contemplative virtues, and are lost without each other. The plant-men who inhabit certain small, hot worlds, similarly, have day-time and night-time phases: "During the busy night-time they went about their affairs as insulated individuals," and during the day they are united in contemplative ardor with the cosmos. "In the day-time mode [a plant-man] passed no moral judgement on himself or others. He mentally reviewed every kind of human conduct with detached contemplative joy, as a factor in the universe. But when night came again, bringing the active nocturnal mood, the calm, day-time insight into himself and others was lit with a fire of moral praise and censure" ([24], p. 109f). The plant-men fail, because after a prolonged attempt to live without contemplation, detached (literally!) from their roots, they swing to the opposite extreme. "Little by little they gave less and less energy and time to animal pursuits, until at last their nights as well as their days were spent wholly as trees, and the active, exploring, manipulating, animal intelligence died in them forever" ([24], p. 111). A similar fate awaits the contemplative lemurs at the hands of our own more "practical" ancestors ([24], p. 456), while our "practical" species goes down in ruin for its failure to remember its place in the world.

The tension is not one that Stapledon ever rationally resolved: on the one hand, the active intelligence wills to produce as fine a future as possible; on the other, the contemplative intelligence, on Stapledon's account, recognizes that this already is the finest universe possible, however ill it suits our animal passions. "The Man Who Became a Tree," one of his short stories, ends with the tree's acceptance even of the woodsman's axe. In *Odd John,* similarly, John's murderous activity is contrasted with the contemplative Islam of an old boatman, as "superior" as John but wholly disinclined to try and make any difference to the world. "Allah wills of his creatures two kinds of service. One is that they should toil to fulfil his active purpose in the world. The other is that they should observe with understanding and praise with discriminating delight the excellent form of his handiwork" ([23], p. 194). This distinction is elaborated in *Saints and Revolutionaries* [27], together with many warnings of the revolutionary's, and the saint's, perversions. Both are necessary, but each seems to exclude the other.

The question is: Can we retain a real devotion to the Spirit while conducting ourselves like Odd John, the Fifth Men, or the flames? Can we commit genocide "through loyalty to the Spirit," as Haldane, Teilhard de

Chardin, and Himmler all imagined, or knowingly destroy the sanity of our children or the lives of "brutes" for the sake of an ideal? "Were the masters of Buchenwald my ministers?" asks Stapledon's imagined God with heavy irony in his last work [35]. Surely not: any advantage won by deceit or violence is "outweighed by a greater hurt in the future, namely damage to the tradition of kindliness and reasonableness."[18] Conversely, can we retain a real devotion to humane endeavor if we teach ourselves that "in the universe as a whole all suffering and all evil contribute to the development of mind, or the awakening of spirit"? ([26], p. 46).

Stapledon recognized in himself an opposition between the little, frightened animal and the realistic intelligence, between everyday concerns and a memory of our cosmic place, even between cooperative action and contemplative joy, between "the saint" and "the revolutionary": all oppositions that Fiedler sought to psychoanalyze away, and that Stapledon sought to resolve through critical intelligence and faith in the light. "Little by little [Paul] came to think of this Neptunian factor in his mind as truly 'himself', and the normal Terrestrial as 'other, something like an unruly horse which his true self must somehow break in and ride" ([21], p. 393)—as Plato had said before him. It is understandable that some people should resent that intelligence: "If we were all on board ship and there was trouble among the stewards I can just conceive their chief spokesmen looking with disfavour on anyone who stole away from the fierce debates in the saloon or pantry to take a breather on deck. For up there, he would taste the salt, he would see the vastness of the weather, he would remember that the ship had a whither and a whence. He would remember things like fogs, storms, and what had seemed in the hot, lighted rooms down below to be merely the scene for a political crisis would appear once more as a thin egg-shell moving rapidly through an immense darkness over an element in which men cannot live."[19] Lewis, though he has been falsely accused of misrepresenting Stapledon, here speaks with a Stapledonian voice. Like Stapledon's bird-men, Lewis's Ransom (in *Out of the Silent Planet*) can accept even his own death if it takes

18. Stapledon [26], p. 101: though this reliance on utilitarian reasoning is not in fact a satisfactory answer to the violent revolutionary.

19. Lewis, *Of Other Worlds* [17], p. 59f. Lewis makes clear (on p. 77) that his villainous scientist, Weston (in *Out of the Silent Planet* [16]), was inspired by Haldane's writings, not Stapledon's, despite Aldiss's wholly inaccurate observation that "this pillorying represents about the peak of Stapledon's fame" ([2], p. 1008).

place in the heavens: his Earth-bound, muddy self feels differently. Or consider Machiavelli's praise of ardent scholarship, a vocation now derided by political "realists," social climbers, disgruntled students, and other philistines:

> On the coming of evening I return to my house and enter my study; and at the door I take off the day's clothing covered with mud and dust . . . and put on garments regal and courtly; and reclothed appropriately, I enter ancient courts of ancient men where, received by them with affection, I feed on that food which only is mine and which I was born for, where I am not ashamed to speak with them and ask them reasons for their actions; and they in their kindness answer me; and for four hours of time I do not feel boredom, I forget every trouble, I do not dread poverty, I am not so frightened by death; I give myself entirely over to them.[20]

But it was not scholarship alone that Stapledon and his Neptunian sought to practice: what they sought was insight into truth, a way of waking up from ordinary self-concerns. Plotinus's image of that great awakening was of the living unity-in-diversity of Intellect, described in terms of the interpenetration of a community of living minds. It is an image found in other traditions also: God, so Tibetan lamas told Francisco Orazio, "is the community of all the holy ones" [11]. This Intellect has its being and its purpose in contemplation of the rationally incomprehensible One, that from which the universe, itself alive in every part, takes its beginning. The individual soul is "carried out by the surge of the wave of Intellect itself and lifted on high by a kind of swell, and sees suddenly, not seeing how" [19]. What it sees is not to be identified with any human love, but no one can ever see it who does not practice virtue and humane benevolence. Love is only spiritual when "there is a waking to discover and value something more than the particular beloved individual and the particular common 'we'" ([33], p. 79), but it can never quite forget its particular beginning if it is to avoid a long decline into the hive-mind.

> Two lights for guidance. The first our little glowing atom of community, with all that it signifies. The second the cold light of the stars, symbol of the hypercosmical reality, with its crystal ecstasy.[21]

20. Machiavelli to Vettori, 10 December 1513: cited by Hillman [10].

21. Stapledon, *Starmaker* [24], p. 333. It is perhaps also significant that the Neptunian narrator of humankind's story is left at last beside his lover, an astronomer: "we have ranged in our work

Conversely, a Neptunian or higher spirit wishing to visit us, to "tune into" our inner being, must select

> that mode of the primitive which is . . . characterized by repressed sexuality, excessive self-regard and an intelligence which is both rudimentary and in bondage to unruly cravings. . . . He must also reconstruct in himself the unconscious obsession with matter, or rather with the control of matter by machinery and chemical manipulation. He must conceive also the mind's unwitting obeisance and self-distrust before its robot offspring. ([24], p. 382)

So also Plotinus.

Plotinus, in short, is the source and inspiration of much that is of lasting value in Stapledon. Those who think of him merely as a propagandist for a great science-inspired mythology neglect the serious philosophical concerns and scholarship that moved him. He did not only look forward, and like the Last Men saw nothing odd in exploring the past imaginings and theories of humanity to find some clue to what was yet to be, and what is always, outside time. "For you," the flames tell humankind, "the golden age is in the future (or so you often like to believe); for us, in the past" ([32], p. 28). For Stapledon time itself was not all that important, and awakened intellects can greet each other over millennia: the brief golden age of the plant-men, whom I described before, is remembered in the awakening world-spirit, is eternally present to the seeing eye.

In one way time is unimportant; in another it contains all we know. In their last despair the Last Men speak of the

> many million, million selves; ephemeridae, each to itself, the universe's one quick point, the crux of all cosmical endeavour. And all defeated! It is forgotten. It leaves only a darkness, deepened by blind recollection of past light. Soon, a greater darkness! Man, a moth sucked into a furnace, vanishes; and then the furnace also, since it is but a spark islanded in the wide, the everlasting darkness. If there is a meaning, it is no human meaning. Yet one thing in all this welter stands apart, unassailable, fair, the blind recollection of past light. ([21] p. 605)

Or as Stapledon put it at the close of the superdog Sirius's tragic history, when Sirius's human lover sings his requiem: "The music's darkness was lit

---

very far apart . . . But now we will remain together till the end. There is nothing more for us to do but to remember, to tolerate, to find strength together, to keep the spirit clear so long as may be . . .": [24], p. 604.

up by a brilliance which Sirius had called 'colour', the glory that he himself, he said, had never seen. But this, surely, was the glory that no spirits, canine or human, had ever clearly seen, the light that never was on land or sea, and yet is glimpsed by the quickened mind everywhere."[22] Stapledon professed to have no *theory* of that glory, and recognized the need to retain a properly skeptical attitude to that and all other idealisms. That it was both real and vital he insisted to the end. It is the part of speculative philosophy, he wrote in his notes for a summer school in Ormskirk, to see things whole, and of critical philosophy to see things clearly. The goal of both must be to wake up into lucidity from our usual self-preoccupation, and in waking to realize how sleepy we still are.

22. *Sirius* [30], pp. 187f. Sirius is an intelligent, language-using dog: far more intelligent, and open-minded, than most of the scientists, priests, and politicians who pontificate about him.

REFERENCES

1. Aldiss, B., *Billion Year Spree* (Weidenfeld and Nicholson, London, 1973), 235f.

2. Aldiss, B., "Review of L. A. Fiedler's *Olaf Stapledon: A Man Divided*," *Times Literary Supplement* (23 September 1982), 1007f.

3. Armstrong, A. H., "Plotinus," in *Cambridge History of Later Greek and Early Medeaieval Philosophy*, ed. A. H. Armstrong (Cambridge University Press, Cambridge, 1970), 195–271, esp. 260, after Plotinus's *Ennead* V.3.17.

4. Coates, J. B., *Ten Modern Prophets* (Muller, London, 1944).

5. Crossley, R., ed., *Talking across the World: The Love Letters of Olaf Stapledon and Agnes Miller, 1913–19* (University Press of New England, Hanover and London, 1987).

6. Clark, S. R. L., *A Parliament of Souls* (Clarendon Press, Oxford, 1990), 39–46.

7. Clark, S. R. L., "Olaf Stapledon (1886–1950)," *Int. Sci. Rev.*, *18*, 112–19 (1993).

8. Fiedler, L. A., *Olaf Stapledon: A Man Divided* (Oxford University Press, Oxford, 1983).

9. Haldane, J. B. S., *Possible Worlds* (Chatto and Windus, London, 1930; 1st published 1927), 303.

10. Hillman, J., *Re-Visioning Psychology* (Harper and Row, New York, 1975), 199.

11. Kant, I., *Kant's Political Writings*, ed. H. Reiss (Cambridge University Press, Cambridge, 1970), 107.

12. Leibniz, G. W., *New Essays on Human Understanding*, eds. J. Bennett and P. Remnant (Cambridge University Press, Cambridge, 1981), 490 (4.17.16).

13. Lessing, D., *Archives Re: Colonized Planet 5: Shikasta* (Cape, London, 1979).

14. Lessing, D., *The Sirian Experiments* (Cape, London, 1981).

15. Lessing, D., *The Making of the Representative for Planet 8* (Cape, London, 1982).

16. Lewis, C. S., *Out of the Silent Planet* (Bles, London, 1937).

17. Lewis, C. S., *Of Other Worlds* (Bles, London, 1966).

18. McCarthy, P., *Olaf Stapledon* (Twayne, Boston, 1982).

19. Plotinus, *Ennead,* VI.7.36, 17–19, trans. A. H. Armstrong (Loeb Classical Library, Heinemann, London, 1988), vol. 7, p. 201.

20. Speaight, R., *Teilhard de Chardin: A Biography* (Collins, London, 1967), 233f.

21. Stapledon, O., *Last and First Men* (Penguin, Harmondsworth, 1972, joint with [22]; 1st published in 1930).

22. Stapledon, O., *Last Men in London* (Penguin, Harmondsworth, 1972, joint with [21]; 1st published in 1932).

23. Stapledon, O., *Odd John* (Methuen, London, 1935).

24. Stapledon, O., *Starmaker* (Methuen, London, 1937).

25. Stapledon, O., in *London Mercury* (1937); reprinted in *Perry Rhodan 86* (Ace, New York, 1976), 33ff.

26. Stapledon, O., *Philosophy and Living* (Penguin, Harmondsworth, 1939).

27. Stapledon, O., *Saints and Revolutionaries* (Heinemann, London, 1939).

28. Stapledon, O., *Darkness and Light* (Methuen, London, 1942).

29. Stapledon, O., *Sirius* (Penguin, Harmondsworth, 1964; 1st published 1944).

30. Stapledon, O., "Morality, Scepticism and Theism," in *Proceedings of the Aristotelian Society* 39 (1944).

31. Stapledon, O., *Death into Life* (Methuen, London, 1946).

32. Stapledon, O., *The Flames* (Secker and Warburg, London, 1947).

33. Stapledon, O., "The Meaning of Spirit," in *Here and Now,* eds. P. Albery and S. Read (Falcon Press, London, 1949), 72ff.

34. Stapledon, O., *A Man Divided* (Methuen, London, 1950).

35. Stapledon, A., ed., *The Opening of the Eyes* (Methuen, London, 1954), 8. (This is an unfinished work of Olaf Stapledon, edited by his widow.)

CONTRIBUTORS

John D. Barrow is a Fellow of Clare Hall and professor of mathematical sciences at Cambridge University, where he leads the Millennium Mathematics Project, a new initiative to improve public understanding and appreciation of mathematics and its applications. He has published more than 320 scientific articles on cosmology and astrophysics, edited three books, and is the author or co-author of fifteen others that have been translated into twenty-seven languages, including *The Anthropic Cosmological Principle, The Artful Universe, Pi in the Sky, Impossibility,* and *The Book of Nothing* that explore the wider historical, philosophical, and cultural ramifications of developments in astronomy. His most recent book is *The Constants of Nature: from alpha to omega.* In 2002 his play *Infinities* was performed at Teatro la Scala in Milan, directed by Luca Ronconi. He has received the Locker Award for Astronomy and was awarded the 1999 Kelvin Medal.

Margaret A. Boden, a professor of philosophy and psychology at the University of Sussex, is the founding dean of Sussex's School of Cognitive and Computing Sciences, a pioneering center for research on intelligence and the mechanism underlying it. Her own research has been especially focused on the phenomena of purpose and creativity, and an early work, *Artificial Intelligence and the Natural Man* (1977 and 1987), has become an academic classic. Holder of a Ph.D. from Harvard and the senior Sc.D. from Cambridge University, Dr. Boden is a fellow of the British Academy and the American Association for Artificial Intelligence.

Steven J. Brams is a professor of politics at New York University and a well-known expert on the theory of games. He has published some two hundred research papers and is the co-editor of two books and the author or co-author of thirteen others. His most recent book (with Alan D. Taylor), *The Win-Win Solution: Guaranteeing Fair Shares to Everybody,* is a blueprint for getting to "yes" in conflict negotiation.

The chemist **A. Graham Cairns-Smith** is an honorary senior research fellow at the University of Glasgow, where he taught for forty years. He is the author of research papers on a variety of scientific topics as well as books, *Genetic Takeover* (1982), and, as coeditor, *Clay Minerals and the Origin of Life* (1986). There are also four books for the more general public: *The Life Puzzle* (1971), *Seven Clues to the Origin of Life* (1985), *Evolving the Mind* (1996), and *Secrets of the Mind* (1999). Another, based on stories from the history of science, is on the way.

Philosopher **Stephen R. L. Clark** has had a long-standing interest in both religion and science fiction. Dr. Clark studied classics at Balliol College, Oxford, and received a Ph.D. in philosophy from Oxford in 1973. He is now the chair of the department of philosophy at the University of Liverpool. He was chief editor of the *Journal of Applied Philosophy* from 1990 to 2000, and has written numerous articles and books, including *Animals and Their Moral Standing* (1997), *How to Think about the Earth: Models of Environmental Theology* (1993), *The Political Animal* (1999), and *God, Religion, and Reality* (1998). His most recent book is *Biology and Christian Ethics*, published in 2000 by Cambridge University Press.

Professor of evolutionary paleobiology at Cambridge University, **Simon Conway Morris** has devoted his research life to the study of the 520-million-year-old Burgess Shale, found between two peaks in the Canadian Rockies, and related fossil-rich formations. Dr. Conway Morris is a fellow of the Royal Society and has held research grants from the society as well as from many other organizations. He has delivered numerous lectures throughout the United Kingdom, Europe, Asia, Canada, and the United States. Dr. Conway Morris is the author of some ninety research papers, the editor of five books, and the author of *The Crucible of Creation* (Oxford University Press, 1998) and *Life's Solution: Inevitable Humans in a Lonely Universe* (Cambridge University Press, 2001).

George V. Coyne, S.J. is the director of the Vatican Observatory and is one of the pioneers in the use of polarimetry, a special technique used in astronomical investigation. A graduate of Fordham University, where he majored in mathematics and earned his licentiate in philosophy, he received his Ph.D. in astronomy from Georgetown University in 1962 and a licentiate in theology from Woodstock College in 1966. Dr. Coyne holds honorary degrees from St. Peter's University (Jersey City) and Loyola University (Chicago) in the United States, the University of Padua, and the Pontifical Theological Academy of Jagellonian University in Cracow. He has published more than one hundred scientific papers and edited a number of books.

A British theoretical physicist based in Australia, **Paul Davies** is currently Professor of Natural Philosophy in the Australian Centre for Astrobiology at Macquarie

University, Sydney. He has contributed to the theory of black holes, early-universe cosmology, and quantum field theory. His current research is on the origin of life. Dr. Davies is the author of more than twenty-five books and one hundred scientific papers. He obtained a doctorate from University College, London, in 1970, and has held academic positions at the universities of Cambridge, London, Newcastle upon Tyne, and Adelaide. Dr. Davies has a strong commitment to bringing science to the wider public, and makes frequent media appearances in various parts of the world. He also writes regularly for newspapers and magazines in several countries. His work in this field was recognized by the award of the 2001 Kelvin Medal by the UK Institute of Physics and the 2002 Michael Faraday Prize by The Royal Society of London. He was also the recipient of the 1995 Templeton Prize for Progress in Religion for his work on the deeper implications of science.

Freeman J. Dyson is widely recognized for his contributions to quantum electrodynamics, the theory of interacting electrons and photons—and perhaps even better known for his creative speculations on subjects ranging from space travel to extraterrestrial civilizations. A graduate of Cambridge University, he taught at Cornell and then at the Institute for Advanced Study in Princeton where he was a professor of physics from 1953 until 1994 when he became professor emeritus. Dr. Dyson is a Fellow of the Royal Society, a member of the U.S. National Academy of Sciences, a foreign associate of the French Academy of Sciences, and an honorary fellow of Trinity College, Cambridge. He has received many honorary degrees and a dozen major science prizes, including the Enrico Fermi Award of the U. S. Department of Energy. In 2000, he won the Templeton Prize for Progress in Religion. His 1979 memoir of his life in science, *Disturbing the Universe,* was nominated for an American Book Award, and he won the National Book Critics Circle Award for Non-Fiction in 1984 for his powerful plea for nuclear disarmament, *Weapons and Hope.* Other books include *Origins of Life* (1986), *Infinite in All Directions* (1988), which won the Phi Beta Kappa Award in Science, *Imagined Worlds* (1997), and *The Sun, the Genome, and the Internet* (1999).

George F. R. Ellis is as widely respected for his anti-apartheid Quaker activism as for his contributions to cosmology. Born in Johannesburg, South Africa, and educated in Natal and Cape Town, he received his Ph.D. in applied mathematics and theoretical physics from Cambridge University in 1964 and has taught in both fields on three continents. For the past decade, he has been a professor of applied mathematics at the University of Cape Town while lecturing throughout the world. Dr. Ellis has served as president of the Royal Society of South Africa and of the International Society of General Relativity and Gravitation. He is a Fellow of the Royal Astronomical Society and the Institute of Mathematics and Its Applications, and has won numerous scientific prizes. Dr. Ellis has written more than two hundred scientific papers and eight major books.

Owen Gingerich is a senior astronomer emeritus at the Smithsonian Astrophysical Observatory and research professor of astronomy and of the history of science at Harvard University. His research interests have ranged from the re-computation of an ancient Babylonian mathematical table to the interpretation of stellar spectra, and he is the co-author of two successive standard models for the solar atmosphere. A graduate of Goshen College and of Harvard, Dr. Gingerich is a leading authority on Johannes Kepler and Nicolaus Copernicus. His 434-page annotated census of extant copies of the latter's *De revolutionibus* was published in 2002.

A professor of philosophy at the Pontifical Academy of Theology in Cracow, Poland, **Michael Heller** is an adjunct member of the Vatican Observatory staff. He also serves as a lecturer in the philosophy of science and logic at the Theological Institute in Tarnow. A Roman Catholic priest, Dr. Heller was ordained in 1959. He was graduated from the University of Lublin, where he earned a master's degree in philosophy in 1965 and a Ph.D. in cosmology in 1966. Dr. Heller is a member of the Pontifical Academy of Sciences. He has published nearly two hundred scientific papers and is also the author of more than twenty books. His most recent volume is *Is Physics an Art?* (1998).

D. Marc Kilgour is a professor of mathematics at Wilfrid Laurier University in Waterloo, Ontario, Canada, director of the Laurier Centre for Military Strategic and Disarmament Studies, and adjunct professor of systems design engineering at the University of Waterloo. A graduate of the University of Toronto where he took an undergraduate degree in engineering and physics and a Ph.D. in mathematics, he has focused his research on decision analysis and has applied game theory and related formal techniques to problems in international security and arms control as well as such other areas as environmental management, negotiation and arbitration, voting, and coalition formation. Dr. Kilgour is the author of four books, including *Perfect Deterrence* (2000) with Frank C. Zagare.

A twenty-year-old prisoner of war interned in England when he began his study of theology and philosophy, **Jürgen Moltmann** has become one of the most respected theologians of our time. For the past thirty years, he has explored the meaning of divine suffering and the unique role of the Cross in disclosing the nature of God. After completing his doctorate in theology at Göttingen University in 1952, he served the Protestant Church in Bremen for five years. After holding several academic positions, he was named professor of systematic theology on the Protestant

Faculty of the University of Tübingen, where he is now professor emeritus. In addition to his monumental study, *The Crucified God* (1974), Dr. Moltmann's most influential works include *Theology of Hope* (1967) and *The Trinity and the Kingdom of God* (1981). His latest book is *The Coming of God* (1997).

One of the world's leading theoretical astrophysicists, **Martin J. Rees,** England's astronomer royal, was for many years the director of Cambridge University's famed Institute of Astronomy. Since 1992 he has been the Royal Society Research Professor at Cambridge and an official fellow at King's College, Cambridge. Dr. Rees has served as president of several scientific societies and is currently a trustee of the British Museum. He has won a dozen major scientific prizes and holds honorary degrees from ten universities. Dr. Rees is the author of some 450 research papers, three technical books, and several books for a lay audience, including *Before the Beginning* (1997), *Just Six Numbers: The Deep Forces That Shape the Universe* (2000), and *Our Cosmic Habitat* (2001). Dr. Rees was knighted by Queen Elizabeth II in 1992.

The founding director of the Center for Theology and the Natural Sciences (CTNS) in Berkeley, California, **Robert John Russell** is a professor of theology and science in residence at the Graduate Theological Union in Berkeley. He is an ordained minister in the United Church of Christ and holds a Ph.D. in physics from the University of California, Santa Cruz. The author of a dozen physics papers and nearly thirty articles on science and religion, Dr. Russell is the co-editor of six books, including *Chaos and Complexity* (1995), which won a Templeton Prize for Outstanding Books in Theology and Science. He is currently preparing a book entitled *Time in Eternity: Eschatology and Cosmology in Mutual Interaction.*

Regius Professor of Divinity at Oxford University, **Keith Ward** is one of Britain's foremost writers on Christian belief and doctrine in the light of modern scientific discoveries and in the context of other faith traditions. He holds a Ph.D. in divinity from Cambridge. Ordained a priest in the Church of England in 1972, he has been canon of Christ Church, Oxford, for the past seven years and is a member of the Council of the Institute of Philosophy and of the Academic Advisory Board of the Oxford Centre for Islamic Studies. The author of numerous works on theology and philosophy, he has just completed a four-volume comparative theology, the final volume of which is *Religion and Community* (2000).

INDEX